NMR IN BIOLOGICAL RESEARCH: PEPTIDES AND PROTEINS

NMR IN BIOLOGICAL RESEARCH: PEPTIDES AND PROTEINS

KURT WÜTHRICH

Institute of Molecular Biology and Biophysics, Swiss Federal Institute of Technology, CH-8049 Zürich, Switzerland

1976

NORTH-HOLLAND PUBLISHING COMPANY
AMSTERDAM · OXFORD

AMERICAN ELSEVIER PUBLISHING COMPANY, INC.
NEW YORK

Library of Congress Catalog Card Number 75-35667

ISBN North-Holland 07204 0389 8
ISBN American Elsevier 0-444-11031-3

Publishers:
NORTH-HOLLAND PUBLISHING COMPANY—AMSTERDAM
NORTH-HOLLAND PUBLISHING COMPANY, Ltd.—OXFORD

Sole distributors for the U.S.A. and Canada:

AMERICAN ELSEVIER PUBLISHING COMPANY, INC.

52 Vanderbilt Avenue
New York, N.Y. 10017

Library of Congress Cataloging in Publication Data

Wüthrich, Kurt.
NMR in biological research.

Includes bibliographies and index.
1. Nuclear magnetic resonance. 2. Peptides. 3. Proteins. I. Title. [DNLM: 1. Nuclear magnetic resonance. 2. Peptides. 3. Proteins. QU55 W959n]
QH324.9.N8W83 1975 574.1'9245'028 75-35667
ISBN 0-444-11031-3 (American Elsevier)

PRINTED IN THE NETHERLANDS

CONTENTS

PREFACE

During the last decade, nuclear magnetic resonance spectroscopy (NMR) has become one of the principal experimental methods for investigations of the molecular structures and intermolecular interactions in biological systems. The biological applications of NMR techniques include studies of the structural and dynamic properties of proteins, nucleic acids, a multitude of low molecular weight compounds of biological, pharmacological and medical interest, membranes and other multimolecular structures, and to some extent intact cells and tissue specimen. This book deals mainly with NMR investigations of peptides and proteins. The emphasis is on the principles and potentialities of the technique, rather than attempting to present a comprehensive review of the results reported so far. Since the physical principles underlying the different biological applications of NMR spectroscopy are the same, and the practical aspects of work with different biological systems are quite similar, much of the material covered applies also for other classes of biological compounds. Thus, as a guide and reference book for biochemists, molecular biologists, and biophysicists working with peptides and proteins, the text should be quite selfcontained in that the confinement to this class of biopolymers made it possible to include advanced topics and recent developments. For work in other areas than peptides and proteins, it should provide a basis for easy access to the specialists reports, e.g. the review articles on different branches of biological NMR work included in the general bibliography (Section I.4). On the other hand, many techniques have been applied in other areas which are also of potential interest for protein research. Some of these have been included in the two last chapters.

Investigations of the relations between covalent structure, molecular conformations and functional properties of proteins play a prominent role in modern biological research. While chemical methods have so far yielded the primary structure of a large number of peptides and proteins, X-ray crystallography has set new standards for the description of the three-dimensional conformations in these molecules. High resolution X-ray data have, however, only been obtained in single crystals, and NMR is at present quite unique in its potential to give structural information on the molecules in solution which is comparable to that obtainable from X-ray crystallography. It is one of the objectives of this book to assess the kind of information on proteins which may presently be obtained from NMR studies, and to outline the prospects for future progress. In this it is also directed to biochemists and molecular biologists who are not directly involved in NMR studies, yet may be faced with a rapidly increasing amount of original literature describing NMR data on systems related to their own research.

High resolution NMR spectroscopy of protons has been a widely used technique for structural studies of small organic molecules since the late 1950's. However, mainly because NMR is an inherently insensitive technique, the extension of its application to macromolecules, and to the observation of nuclei other than protons, have become possible only in the aftermath of recent advances in NMR instrumentation. These are on the one hand the application of superconducting magnets for NMR experiments at very high magnetic fields, and the use of Fourier transform spectroscopy. The present text includes many examples illustrating the ability of modern NMR equipment to give direct and detailed information on the structure and dynamics of biological macromolecules under conditions which may closely resemble the physiological environment of these molecules. Yet it is also one of the principal concerns of the author, to show that the interpretation of the NMR measurements is by no means straightforward, and that without a thorough understanding of the underlying physical processes one can all too easily be mislead. One of the principal criteria for the selection of the material covered in this book was to serve this dual purpose, i.e. on the one hand to report what may be attained by NMR techniques, on the other hand to open the reader's mind for a critical evaluation of current and future achievements.

The main theme of the book is the application of high resolution NMR experiments with the naturally occurring "NMR nuclei" hydrogen-1, carbon-13, and nitrogen-15 for investigations of the molecular conformations of peptides and proteins in solution. The chapters II, III, V, and VII are devoted to this aspect. A phenomenological approach has been chosen which requires little background in physics and mathematics, and care has been taken to include sufficient background information on NMR spectroscopy to enable the biochemists and biologists not previously acquainted with the method, to follow the arguments without reference to additional literature. Numerous experiments are described in much detail, so as to serve as guidelines for those intending to start their own NMR measurements. For a deeper understanding of some of the high resolution NMR phenomena presented in these four chapters, the reader may then turn to the Appendix II, which presents a brief survey of the quantum mechanical background for the interpretation of high resolution NMR spectra in complex spin systems, chapter IV, which introduces some aspects of nuclear spin relaxation, and the Appendix V, where the interpretation of proton and carbon-13 NMR relaxation data is illustrated with numerical examples. To facilitate access to more detailed information on specific topics, each chapter contains a list of review articles and original papers. Besides the "direct" high resolution NMR approach, the use of "NMR probes" plays an important role in biological work. Chapter VI deals with the application of NMR for studies of paramagnetic centers which are often used as NMR probes, and are also an integral component of the active centers in many metalloproteins. Chapter VIII describes the application of various nuclei as NMR probes of macromolecular structures. Here, some experiments with membranes and intact tissues are included to illustrate techniques which may also be of potential interest for studies of proteins. The scope of the book is further extended in chapter IX with a discussion of NMR experiments of proteins in multimolecular systems, such as membranes or complexes formed between proteins and nucleic acids, where the reader will also get aquainted with some NMR spectral features of the other components in these systems. It is to be expected that the future development will be to apply NMR techniques to ever more complex systems, including intact cells and tissues, and quite likely also live organisms.

Many colleagues and friends have indirectly helped to shape the ideas developed in this book. In part their impact is reflected in the reference lists included with each chapter. By the way, rather than attempting to present a complete coverage of the literature, the reference lists are merely intended to provide a selection of representative papers. The author is indebted to all those who granted permission to reproduce figures; details are included in the figure captions. He is particularly obliged to Prof. E. Randall and Dr. M. Llinás for the permission to reproduce some as yet unpublished figures. Again, much as with the references, the selection of the illustrations was rather arbitrary. Many of the figures were prepared specifically for this book, and in several chapters the additional illustrations were taken preferentially from work by the author and his associates. It should at this point also be mentioned, that the authors research programs, which are closely related to the theme of this book, received financial support from the Swiss National Science Foundation.

The author owes much to the cooperation of his postdoctoral associates and graduate students during the time this book was written. The aid by Drs. Larry Brown, Joseph Gilboa, Regula Keller, Antonio De Marco, and by Arno Bundi, Christoph Grathwohl, Robert Hetzel, Jiri Hochmann and Gerhard Wagner in producing experimental data needed for certain illustrations, critically evaluating original papers discussed in the book, and reading individual chapters of the manuscript is gratefully acknowledged. Special thanks go to Prof. R. Schwyzer who contributed much to the authors motivation for work in the field discussed in this book, Dr. Miguel Llinás who read the entire manuscript and contributed a wealth of valuable suggestions, Rudolf Baumann who meticulously prepared all the figures, and Antoinette Schumacher who skilfully and intelligently took on the long task of typing and preparing the manuscript from the handwritten notes. Last but not least I should like to mention the patience exercised by Marianne, Bernhard, and Karin in accepting the curtailments of family life imposed on them during the preparation of the manuscript.

CHAPTER I

INTRODUCTION

This first chapter gives a qualitative introduction to the fundamentals of biological applications of NMR, i.e. on the one hand the occurence of nuclear spin moments in biological material, and on the other hand the physical principles underlying the NMR phenomena. A brief account is then given of the development of biological NMR since the start of the field in the mid-fifties.

1. NUCLEAR SPINS IN BIOLOGICAL MATERIAL

NMR Nuclei

Each of the four most abundant elements in biological material, i.e. hydrogen, carbon, nitrogen and oxygen, has at least one naturally occuring isotope with non-zero nuclear spin, and is thus in principle accessible for NMR observation. Table I-1 gives data on the nuclear magnetism of a number of isotopes. Among the heavier elements, the NMR observation of fluorine, sodium, phosphorous, and chlorine has so far been particularly useful in biological studies.

Even though one has the choice of a sizeable number of isotopes with non-zero spin for biological NMR work, by far the greatest number of experiments have been done with protons. Recently, work with carbon-13, fluorine, and phosphorous has also gained momentum, whereas relatively few practical applications have as yet been described of the many other NMR nuclei in biological materials. This preference for a few isotopes is readily comprehensible if one considers the sensitivity of the NMR method. Quite generally even the NMR signals of the most sensitive nuclei can only be detected in relatively concentrated solutions, and in biological work high resolution NMR with nuclei other than protons and fluorine became practicable only after some very recent advances in NMR instrumentation.

Sensitivity of Nuclei to NMR Observation

The sensitivity for the detection of the nuclear resonances is mainly determined by the line width of the resonances, the natural abundance of the

TABLE I-1

Nuclear Properties of Selected Isotopes

Isotope	Nuclear spin I (in multiples of $h/2\pi$)	Natural abundance (%)	Relative sensitivity for equal number of nuclei at constant field	NMR Frequency at 10 Kgauss (MHz)	Magnetic moment μ (in multiples of the nuclear magneton $eh/4\pi Mc$)	Electric quadrupole moment Q (in multiples of $e \times 10^{-24}$ cm^2)
^{1}H	$\frac{1}{2}$	99.98	1.000	42.57	2.7927	—
^{2}D	1	0.015	9.65×10^{-3}	6.53	0.8574	2.77×10^{-3}
^{11}B	$\frac{3}{2}$	81.17	1.65×10^{-1}	13.66	2.6880	3.55×10^{-2}
^{13}C	$\frac{1}{2}$	1.11	1.59×10^{-2}	10.70	0.7022	—
^{14}N	1	99.63	1.01×10^{-3}	3.08	0.4036	7.1×10^{-2}
^{15}N	$\frac{1}{2}$	0.37	1.04×10^{-3}	4.31	−0.2830	—
^{17}O	$\frac{5}{2}$	0.04	2.91×10^{-2}	5.77	−1.8930	-4×10^{-2}
^{19}F	$\frac{1}{2}$	100	8.33×10^{-1}	40.05	2.6273	—
^{23}Na	$\frac{3}{2}$	100	9.25×10^{-2}	11.26	2.2161	1×10^{-1}
^{25}Mg	$\frac{5}{2}$	10.05	2.68×10^{-3}	2.61	−0.8547	........[a]
^{29}Si	$\frac{1}{2}$	4.70	7.85×10^{-2}	8.46	−0.5548	—
^{31}P	$\frac{1}{2}$	100	6.63×10^{-2}	17.23	1.1305	—
^{33}S	$\frac{3}{2}$	0.74	2.26×10^{-3}	3.27	0.6427	-6.4×10^{-2}
^{35}Cl	$\frac{3}{2}$	75.4	4.70×10^{-3}	4.17	0.8209	-7.9×10^{-2}
^{37}Cl	$\frac{3}{2}$	24.6	2.71×10^{-3}	3.47	0.6833	-6.21×10^{-2}
^{39}K	$\frac{3}{2}$	93.08	5.08×10^{-4}	1.99	0.3909	1×10^{-1}
^{43}Ca	$\frac{7}{2}$	0.13	6.40×10^{-2}	2.86	−1.3153	........[a]
^{79}Br	$\frac{3}{2}$	50.57	7.86×10^{-2}	10.67	2.0990	3.3×10^{-1}
^{81}Br	$\frac{3}{2}$	49.43	9.85×10^{-2}	11.50	2.2626	2.8×10^{-1}
^{111}Cd	$\frac{1}{2}$	12.86	9.54×10^{-3}	9.03	−0.5922	—
^{113}Cd	$\frac{1}{2}$	12.34	1.09×10^{-2}	9.44	−0.6195	—
^{127}I	$\frac{5}{2}$	100	9.34×10^{-2}	8.52	2.7939	2.77×10^{-3}

[a]The quadrupole moments of these nuclei are uncertain.

Note: Nuclei with no spin include ^{12}C and ^{16}O.

isotope, and its "NMR sensitivity" (Table I-1). Quite generally the isotopes with nuclear spin $I > 1/2$ yield rather broad lines, which is due to the electric quadrupole moments in these nuclei. As we shall see later, nuclei with $I > 1/2$ are therefore more suitable for certain spin relaxation studies than for direct NMR observation in macromolecules. Among the nuclei with spin 1/2, ^{1}H, ^{19}F, and ^{31}P occur in high natural abundance, whereas ^{13}C and ^{15}N are present only in very small amounts in natural material. Considering the relative sensitivities in Table I-1, one finds high values for ^{1}H and ^{19}F, whereas the corresponding number is down by a factor of ca. 20 for ^{31}P, ca. 100 for ^{13}C, and ca. 1000 for ^{15}N. The ease with which the NMR signal can be detected in solutions containing fixed molar concentrations of materials with natural isotope composition, is given roughly by the product of natural abundance and NMR sensitivity. These numbers are given in Table I-2 for the isotopes of principal interest for the high resolution NMR studies

TABLE I-2

Approximate relative NMR intensities at constant field and constant molar concentration of material with natural isotope composition for the principal isotopes with $I = 1/2$ (expressed as the product of the natural abundance and the NMR sensitivity).

Nucleus	Relative NMR signal intensity
^{1}H	1
^{13}C	1.7×10^{-4}
^{15}N	3.8×10^{-6}
^{19}F	8.3×10^{-1}
^{31}P	6.6×10^{-2}

discussed in this book. The significance of Table I-2 is perhaps most readily appreciated if one recalls that routine proton NMR studies of organic chemicals are usually done with 0.1-molar or more concentrated solutions. With the use of modern equipment, including in particular facilities for Fourier transform (FT) experiments, ^{1}H- and ^{19}F-NMR studies in less than 10^{-4}-M solutions, and ^{13}C-NMR studies in ca. 10^{-2}-M solutions of natural materials are now practicable. This concentration range can readily be attained with a variety of biopolymers*. Overall, with the presently available

*These considerations on NMR sensitivity have been treated entirely from the aspect of studies in dilute solutions. In view of the high molecular weights encountered in biopolymers this is of principal interest here. Another aspect of NMR sensitivity, which may e.g. be relevant in

instrumentation high resolution NMR spectra of ^{1}H, ^{13}C, and ^{31}P in natural abundance can be employed for structural studies of biological macromolecules. Of the two additional nuclei in Table I-2, the NMR observation of ^{15}N in natural abundance is still very difficult, and ^{19}F is usually only encountered in chemically modified biopolymers.

Isotope Labelling

In many cases where the NMR observation of materials with natural isotope distribution is not informative either because of lacking sensitivity or lack of spectral resolution, studies of compounds enriched or depleted in a given isotope have been successfully employed. Molecules carrying such NMR markers have been obtained by both chemical and biological procedures. From Table I-1 it is apparent that the introduction of a ^{13}C- or ^{15}N-enriched group will in principle be a suitable NMR label. Numerous experiments have been described which used ^{19}F-NMR of partially fluorinated molecules, e.g. peptides including *p*-fluoro-phenylalanine. In other work bacteria were grown on a diet of amino acids of which all but one or two were fully deuterated. In the proteins isolated from these microbes, the proton NMR spectra contained only the resonances of one or two kinds of amino acids, which could thus readily be studied. By various procedures lipid molecules enriched in deuterium or carbon-13, or carrying fluorine atoms in place of some of the protons, have been introduced into membranes, and were subsequently used as NMR reporter groups. It is a big asset of isotope labelling techniques for NMR studies, that the functional properties of the molecules are in most cases essentially not affected. Studies of labelled systems greatly increase the potential applications of NMR in biological studies, and will probably be one of the prominent factors in the future developments of the field.

2. NUCLEAR MAGNETISM AND NUCLEAR MAGNETIC RESONANCE

Spin Angular Momentum

In the preceding section it was pointed out that NMR signals are obtained from those nuclei which possess a non-zero spin, i.e. an angular momentum,

---

metabolic studies, would be the investigation of minute amounts, i.e. of the order of several micrograms, of low molecular weight compounds.

Microcells for a small volume of a fairly concentrated solution of the molecule studied are available for this purpose.

which is characterized by the spin quantum number I. I is an integral or half-integral number (Table I-1), and corresponds to the maximum value of the angular momentum in units of $h/2\pi$, where h is Planck's constant. Quantum theory demands that the allowable nuclear spin states are quantized. Thus the magnetic quantum number m_I can in any given direction, e.g. along the direction of an externally applied magnetic field, only take up one of the $2I+1$ discrete values $+I, +(I-1), \ldots. -(I-1), -I$. Thus if $I = 1/2$, the possible values of m_I are $+1/2$ and $-1/2$, often also referred to as α- and β-spin states, respectively. In the absence of an external magnetic field, the nuclear spin states defined by different values of m_I are energetically degenerate, whereas in a static magnetic field the states with different m_I have different energy. For nuclei with $I = 1/2$, the two possible values of m_I describe states where the nuclear spin is respectively opposed to and aligned with H_0, the latter state being of lower energy (Fig. I-1). The energy

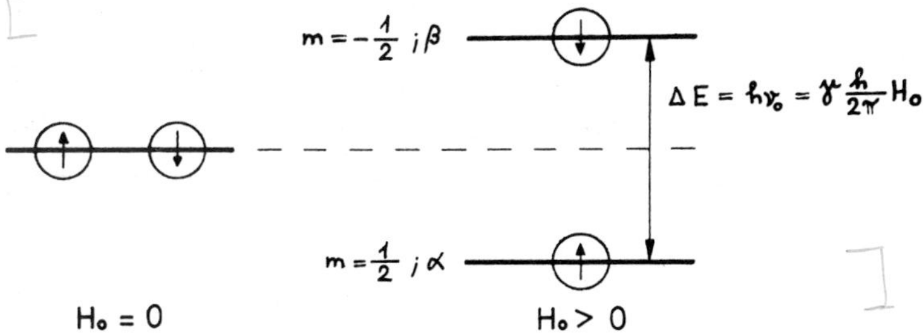

Fig. I-1. Magnetic energy levels for a nucleus with spin $I = \frac{1}{2}$ with and without application of an external magnetic field.

difference ΔE between the upper and lower states is very small, i.e. of the order of a few millicalories per mole, which is an extremely small perturbation in biological systems.

The Resonance Condition

While in the absence of an external magnetic field an equal number of nuclei are in the α- and β-spin states, an excess of nuclei are in the α-state when the sample is exposed to a static external magnetic field. According to the Boltzmann distribution law

$$\frac{n_\beta}{n_\alpha} = \mathrm{e}^{-\Delta E/kT} \tag{I-1}$$

where n_α and n_β are the number of nuclei in the respective levels, and k is the Boltzmann constant, the excess population of the state of lower energy in a typical NMR experiment is of the order of 1×10^{-5}. Because of this population difference the sample becomes magnetized, and a net absorption of energy is observed during irradiation with an electromagnetic field of suitable frequency. This resonance frequency is in the radiofrequency region (Table I-1), and for a magnetic field H_0 is given in hertz (Hz) by

$$\nu_0 = \frac{\Delta E}{h} = \frac{\gamma}{2\pi} H_0. \tag{I-2}$$

Here h is the Planck constant, and γ, the magnetogyric ratio, is a constant for a given nucleus, and is related to the magnetic moment μ and the nuclear spin quantum number I (Table I-1) by

$$\gamma = \frac{2\pi\mu}{Ih}. \tag{I-3}$$

As is readily seen from equation I-2, the nuclear magnetic resonance condition can be attained at fixed radiofrequency ν_0 by sweeping the magnetic field H_0, or at fixed field H_0 by sweeping the frequency ν_0. The resulting NMR signal has a characteristic half width at half height of $\delta\nu_{1/2}$ (Fig. I-2). In the following, three NMR spectral parameters will briefly be discussed, i.e. the chemical shift, the longitudinal spin relaxation time, and the transverse spin relaxation time, which is closely related to $\delta\nu_{1/2}$.

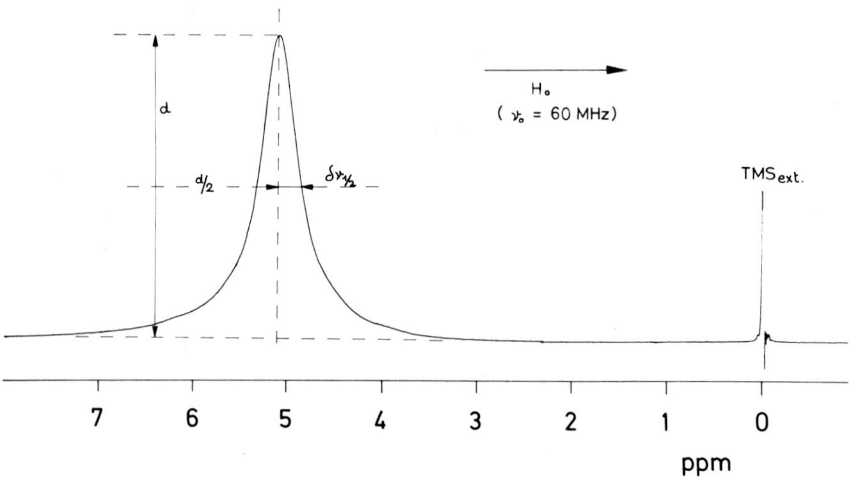

Fig. I-2. Proton NMR signal of H_2O in a ca. 3×10^{-4}-M $MnCl_2$ solution. The horizontal scale is in parts per million from the external reference TMS, where positive numbers indicate shifts to lower field at fixed frequency. It is customary to present NMR spectra so that at fixed frequency H_0 increases from left to right. d is the signal height, and $\delta\nu_{1/2}$ the half width of the line at half height in Hz.

The Chemical Shift

The utility of the NMR method results primarily from the sensitive variation of the resonance frequency with chemical structure and conformation of the molecules and their environment. These so-called chemical shifts arise because a given nucleus is shielded from the externally applied magnetic field to a different extent in different molecules. In practice the observed resonance frequencies ν are measured with respect to a standard reference compound (Fig. I-2), and the chemical shift from this reference signal is described by the shielding constant σ. For liquid samples, and essentially all the work discussed in this book was done with liquid solutions, the magnetic field H_{loc} experienced by the nucleus under investigation is then

$$H_{loc} = H_0(1 - \sigma) \tag{I-4}$$

and the resonance condition is given by

$$\nu = \frac{\gamma}{2\pi} H_0(1 - \sigma). \tag{I-5}$$

Chemical shifts are usually reported in parts per million (ppm) of the resonance frequency of the nucleus (Table I-1). Relative chemical shifts of a given nucleus in different diamagnetic organic molecules are of the order of 10 ppm for ^{1}H and ^{2}D, and several hundred ppm for the other nuclei in Table I-1. In comparison, the resonance frequencies of different nuclei are very widely separated (Table I-1).

Even though much effort has been invested in the theory of NMR chemical shifts, empirical approaches to relate chemical shifts with the molecular structures have up to now been most fruitful. Typical chemical shifts of nuclei in organic molecules will be presented in Chapter II for protons, in Chapter V for carbon-13, and in Chapter VII for nitrogen-15.

Spin Relaxation

Another important feature of the NMR phenomenon, in addition to the resonance frequencies of the observed nuclei, is the nuclear spin relaxation. Whenever the population of the different spin states deviates from the Boltzmann distribution, the system is not in equilibrium, and therefore the populations n_α and n_β change with time until the situation given by Equation I-1 is attained. For example, if we deal with a sample of nuclei with $I = 1/2$ that is initially unmagnetized, and then expose it to a static magnetic field (Fig. I-1), the magnetization process is described by an exponential rise to the equilibrium magnetization

$$n = n_0(1 - e^{-t/T_1}). \tag{I-6}$$

Here, $n = n_\alpha - n_\beta$ is the population difference between the two spin levels, and n_0 is the equilibrium value of this quantity. Equation I-6 shows that the spin lattice relaxation time T_1 characterizes the time needed to magnetize an unmagnetized sample. T_1 is intimately related with the molecular structures and the mobility of the molecules, and has become particularly interesting for investigations of the dynamics of the structures in biological materials.

Since on the average a nucleus remains in a given energy level no longer than for a period of time T_1, one can estimate the minimum line width of the NMR signal (Fig. I-2) from the Heisenberg Uncertainty Principle

$$\mathrm{d}E \cdot \mathrm{d}t \geqslant \frac{h}{2\pi} \tag{I-7}$$

where $\mathrm{d}E$ and $\mathrm{d}t$ are the uncertainties in the measurements of energy and time, respectively. With a line width at half height of $\delta\nu_{1/2}$ Hz, $\mathrm{d}E$ becomes equal to $h\delta\nu_{1/2}$ (Equation I-2), and hence from Equation I-7

$$\delta\nu_{1/2} \geqslant \frac{1}{2\pi T_1}. \tag{I-8}$$

Often NMR lines are broader than indicated by Equation I-8. To account for the additional relaxation processes which are manifested in this line-broadening, the spin–spin relaxation time T_2 is defined so that

$$\delta\nu_{1/2} = \frac{1}{2\pi T_2}. \tag{I-9}$$

T_2 is mainly determined by the rotational and translational Brownian motions of the molecules, and hence is another important source, besides T_1, of information on the dynamic properties of the systems studied.

3. BIOLOGICAL NMR EXPERIMENTS. HISTORICAL BACKGROUND

Principle Advantages of the Method

The first section of this chapter has shown that nuclei with non-zero spin are omnipresent in biological materials. Correspondingly, NMR may in principle quite generally be applied for structural studies of biological macromolecules, and for investigations of intermolecular interactions in solution and in intact tissue specimen. The broad spectrum of applications is one of the reasons which made NMR emerge as one of the most powerful methods for studies of structural and kinetic features in biological systems. Another reason is the high content of information on molecular structures and rate processes in the NMR spectra, which depends on the sensitive manifestation of these phenomena in the NMR spectral parameters introduced in the preceeding section. A third important asset of the method are the non-destructive conditions of the NMR measurements. In many instances the samples can be studied under near-physiological conditions of solvent and temperature, and they are essentially not perturbed by the irradiation at the typical NMR frequencies, where an energy quantum $h\nu$ is of the order of 10^{-5} kT, i.e. $\approx 10^{-6}$ kcal/mole.

Historical Development

It is the purpose of this book to present an up-to-date account of the state of the art. However, the potential advantages of NMR for biological research had long been anticipated, and the first papers on the subject appeared almost twenty years ago. To help the reader in gaining access to the literature which has accumulated in the meantime, it seems therefore appropriate to include also a brief survey of the historical development of the field. In doing so, one has to consider two principal factors which determined the rate of progress. One is the rapid advancement of biochemistry and molecular biology in general, which provided the basis for an increasing supply of biological materials suited for NMR studies. This includes the chemical synthesis of peptides and oligonucleotides, isolation and purification of proteins, nuclei acids, membranes and various tissue specimen, and the preparation of milligram quantities of isotopically labelled systems. When discussing the relations of NMR with the general advancement of the biological sciences, one should also take into account that much of the present spectral interpretations is aided by the availability of sequence data and single crystal atomic coordinates in biopolymers, the majority of which have been accrued over the past fifteen years.

The second and for most purposes decisive factor in the development of biological NMR work was the progress made in NMR methodology and instrumentation. Particularly prominent among the improvements introduced into NMR spectrometers during the past fifteen years are the increased sensitivity and spectral resolution, which both have been and still are limiting many biological applications of the method. This is particularly true for the so-called high resolution NMR experiments, where one observes individual nuclei of the macromolecule of interest.

High Resolution NMR

The first high resolution NMR spectrum of a biopolymer recorded in the literature seems to be that of ribonuclease A reported in 1957 by Saunders, Wishnia and Kirkwood. The sensitivity of the 40 MHz spectrometer used by these authors was obviously sufficient to observe the proton resonances of a one hundreth molar protein solution in D_2O. The spectrum of the 650 protons in ribonuclease A at 40 MHz consisted however of only four broad lines. This poor spectral resolution is not surprising if one considers that the several hundred resonances, which have line widths (Fig. I-2) of 5 to 20 Hz, overlap within the spectral region for protons of ca. 10 ppm, i.e. 400 Hz at a resonance frequency of 40 MHz. On the basis of the proton NMR spectra of amino acids and peptides which had been investigated at about the same time by Takeda and Jardetzky, it could be shown that the observed ribonuclease A spectrum was very similar to that expected from the amino acid composition. On the other hand, with the limited spectral resolution no specific resonance identifications were possible, and no conclusions concerning the biological role of the protein could yet be drawn from the NMR data. Between 1957 and 1965 Kowalsky and Mandel joined in the efforts to unravel the proton NMR spectra of various proteins with molecular weights of ca. 10000–20000. They used spectrometers which operated at 60 and 100 MHz, and were equipped with a computer of average transients to improve the signal-to-noise ratio. These early high resolution NMR investigations yielded much of the basic notions for the interpretation of protein spectra in Chapter III. The big breakthrough in proton NMR studies of proteins, however, came after 1966, when superconducting magnets were used in spectrometers operating at 220 MHz and higher frequencies. With the increased spectral resolution and sensitivity it became now practicable to resolve and identify the resonances of individual protons in a protein, and to derive detailed structural information from these data. As the review articles mentioned in the following section show, most of the studies up until 1970 concentrated on lysozyme, ribonuclease A, myoglobin, hemoglobin, and cytochrome *c*. Since 1970 numerous other proteins have also been investigated. Early biological applications of proton NMR techniques included also high resolution studies of mononucleotides and oligonucleotides, while more recently much emphasis has been on structural studies of transfer RNA's. For other nuclei than protons, the sensitivity criterion (Tables I-1 and I-2) made high resolution NMR spectroscopy of biopolymers impracticable before the introduction of the Fourier transform (FT) method. Commercial FT spectrometers have been available since 1971. At present, studies of other nuclei in native biological materials, in particular ^{13}C and ^{31}P, are an important complementation of the proton NMR data.

As a result mainly of the introduction of the superconducting proton NMR spectrometers, the number of publications on high resolution NMR in biological research increased noticeably after 1968. With the advent of Fourier transform spectroscopy, another quite dramatic rise in the number of investigations occurred since 1971, and is still gaining momentum. Overall, the bulk of the presently available original literature is of very recent origin, mostly more recent than 1969.

Relaxation Enhancement Studies

As an alternative to the high resolution NMR studies, a different NMR approach to the investigation of biological macromolecules can be chosen, which depends to a far lesser degree on optimal sensitivity and resolution of the apparatus. Typical for these experiments are the so-called relaxation enhancement studies. Since an NMR sample contains in addition to a small molar concentration of the macromolecular components one or several low molecular weight species in much higher concentration, e.g. H_2O as the solvent, the resonances of these latter molecules are far easier to observe. Quite often inferences on the structure of the macromolecular components can be made from their influence on the NMR properties of the smaller molecules, in particular the spin relaxation times. As early as 1954, Jacobson, Anderson and Arnold (Nature *173*, 772) observed that DNA markedly altered the proton NMR of the solvent water. In the following years, before NMR instrumentation was really ready for the high resolution NMR approach to macromolecular structures, these techniques had a prominent role in biological research, particularly for studies of paramagnetic materials. At present, experiments of this type are mostly employed for the investigation of interactions between macromolecules and low molecular weight ligands, and for studies of the state of the water or other abundant components in intact tissues.

4. BIBLIOGRAPHY

This section presents a selection of textbooks on NMR spectroscopy, the references to the original articles on early high resolution NMR studies with proteins mentioned in Section I.3, and monographs, proceedings of meetings, and review articles covering the development of biological NMR experiments to the present state.

Textbooks on NMR Spectroscopy

With emphasis on physical principles:

1 Abragam A., The Principles of Nuclear Magnetism. Clarendon Press, Oxford, 1962
2 Slichter C. P., Principles of Magnetic Resonance. Harper and Row, New York, 1963
3 Carrington A. and McLachlan, A. D., "Introduction to Magnetic Resonance with Applications to Chemistry and Chemical Physics". Harper and Row, New York, 1967

With emphasis on applications in organic chemistry:

4 Pople J. A., Schneider W. G. and Bernstein H. J., "High Resolution NMR". McGraw-Hill, New York, 1959
5 Emsley J. W., Feeney J. and Sutcliffe L. H., "High Resolution Nuclear Magnetic Resonance Spectroscopy". Vols. 1 and 2. Pergamon Press, Oxford, 1966
6 Bovey F. A., "Nuclear Magnetic Resonance Spectroscopy". Academic Press, New York, 1969
7 Becker E. D., "High Resolution NMR". Academic Press, New York, 1969
8 Casy A. F., "PMR Spectroscopy in Medicinal and Biological Chemistry". Academic Press, New York, 1971

Original articles on High Resolution NMR of Proteins from the period 1957–1965

9 Saunders M., Wishnia A. and Kirkwood J. G., J. Amer. Chem. Soc. *79*, 3289 (1957): "The Nuclear Magnetic Resonance Spectrum of Ribonuclease"
10 Takeda M. and Jardetzky O., J. Chem. Phys. *26*, 1346 (1957): "Proton Magnetic Resonance of Simple Amino Acids and Dipeptides in Aqueous Solution"
11 Jardetzky O. and Jardetzky C. D., J. Amer. Chem. Soc. *79*, 5322 (1957): "An Interpretation of the Proton Magnetic Resonance Spectrum of Ribonuclease"
12 Jardetzky O. and Jardetzky C. D., J. Biol. Chem. *233*, 383 (1958): "Proton Magnetic Resonance Spectra of Amino Acids"
13 Saunders M. and Wishnia A., Ann. N.Y. Acad. Sci. *70*, 870 (1958): "Nuclear Magnetic Resonance Spectra of Proteins"
14 Wishnia A. and Saunders M., J. Amer. Chem. Soc. *84*, 4235 (1962): "The Nature of the Slowly Exchanging Protons of Ribonuclease"
15 Kowalsky A., J. Biol. Chem. *237*, 1807 (1962): "Nuclear Magnetic Resonance Studies of Proteins"
16 Kowalsky A., J. Biol. Chem. *239*, 3299 (1964): "Studies of the Low Field Proton Magnetic Resonance of Ribonuclease and Simple Compounds"
17 Mandel M., Proc. Nat. Acad. Sci. *52*, 736 (1964): "The Effect of Temperature on the Proton Magnetic Resonance Spectra of Ribonuclease, Oxidized Ribonuclease, and Lysozyme"
18 Mandel M., J. Biol. Chem. *240*, 1586 (1965): "Proton Magnetic Resonance Spectra of Some Proteins"
19 Kowalsky A., Biochemistry *4*, 2382 (1965): "Nuclear Magnetic Resonance Studies of Cytochrome *c*. Possible Electron Delocalization"

Reviews of the early Biological Applications of NMR through 1965

20 Sogo P. B. and Tolbert B. M., Adv. Biol. Med. Physics *5*, 1 (1957): "Nuclear and electron paramagnetic resonance and its application to biology"

21 Jardetzky O. and Jardetzky C. D. In "Methods of Biochemical Analysis" (Interscience, New York 1962) Vol. IX, p. 235: "Introduction to Magnetic Resonance Spectroscopy. Methods and Biochemical Applications"

22 Jardetzky O., Adv. Chem. Phys. *7*, 499 (1964): "The Study of Specific Molecular Interactions by Nuclear Magnetic Relaxation Measurements"

23 Kowalski A. and Cohn M., Ann. Rev. Biochem. *33*, 481 (1964): "Application of Nuclear Magnetic Resonance in Biochemistry"

Proceedings of Meetings on Magnetic Resonance in Biology and review articles 1965–1970

24 Ehrenberg A., Malmström B. G. and Vänngärd T. (eds.): "Magnetic Resonance in Biological Systems", Pergamon Press, New York, 1967

25 Fujiwara S. and Piette L. H. (eds.): "Recent Developments of Magnetic Resonance in Biological Systems", Hirokawa, Tokyo, 1968

26 Ts'o P. O. P., Schweizer M. P. and Hollis D. P., Ann. New York Acad. Sci. *158*, 256 (1969): "Contribution of Nuclear Magnetic Resonance to the Study of the Structure and Electronic Aspects of Nucleic Acids"

27 Inch T. D., Ann. Rev. NMR Spectroscopy *2*, 35 (1969): "Nuclear Magnetic Resonance Spectroscopy in the Study of Carbohydrates and Related Compounds"

28 Roberts G. C. K. and Jardetzky O., Adv. Protein Chem. *24*, 447 (1970): "Nuclear Magnetic Resonance Spectroscopy of Amino Acids, Peptides and Proteins"

29 McDonald C. C. and Phillips W. D. In "Fine Structure of Proteins and Nucleic Acids", G. D. Fasman and S. N. Timasheff (eds.), Dekker, New York, 1970, 1: "Proton Magnetic Resonance Spectroscopy of Proteins"

30 Wüthrich K., Chimia *24*, 409 (1970): "Studien der räumlichen Struktur von Proteinmolekülen mit magnetischer Kernresonanzspektroskopie"

31 Wüthrich K., Structure and Bonding *8*, 53 (1970): "Structural Studies of Hemes and Hemoproteins by Nuclear Magnetic Resonance Spectroscopy"

32 Mildvan A. S. and Cohn M., Adv. Enzymology *33*, 1 (1970): "Aspects of Enzyme Mechanisms studied by Nuclear Spin Relaxation induced by Paramagnetic Probes"

33 Sheard B. and Bradbury E. M., Progr. in Biophysics and Molecular Biology *20*, 187 (1970): "Nuclear Magnetic Resonance in the Study of Biopolymers and their Interaction with Ions and Small Molecules"

34 Cohn M., Quart. Rev. Biophysics *3*, 61 (1970): "Magnetic Resonance Studies of Enzyme-Substrate Complexes with Paramagnetic Probes as illustrated by Creatine Kinase"

35 Burgen A. S. V. and Metcalfe J. C., J. Pharm. Pharmac. *22*, 153 (1970): "The Application of Nuclear Magnetic Resonance to Pharmacological Problems"

36 Rowe J. J. M., Hinton J. and Rowe K. L., Chem. Rev. *70*, 1 (1970): "Nuclear Magnetic Resonance Studies on the Biochemistry of Biopolymers"

Monographs, Proceedings of Meetings on NMR in Biology, and Review Articles after 1970

37 Bovey F. A.: "High Resolution NMR of Macromolecules", Academic Press, New York, 1972

38 Dwek R. A.: "Nuclear Magnetic Resonance in Biochemistry. Applications to Enzyme Systems", Clarendon Press, Oxford 1973

39 Franconi C. (ed.): "Magnetic Resonances in Biological Research", Gordon and Breach, New York, 1971

40 Lasker S. E. and Milvy P. (eds.): "Electron Spin Resonance and Nuclear Magnetic Resonance in Biology and Medicine and Magnetic Resonance in Biological Systems", Ann. New York Acad. Sci., Vol. 222 (1973)
41 Jardetzky O. and Wade-Jardetzky N. G., Ann. Rev. Biochem. *40*, 605 (1971): "Applications of Nuclear Magnetic Resonance Spectroscopy to the Study of Macromolecules"
42 Cohn M. and Reuben J., Accounts Chem. Res. *4*, 214 (1971): "Paramagnetic Probes in Magnetic Resonance Studies of Phosphoryl Transfer Enzymes"
43 Sykes B. D. and Scott M. D., Ann. Rev. Biophys. Bioeng. *1*, 27 (1972): "Nuclear Magnetic Resonance Studies of the Dynamic Aspects of Molecular Structure and Interactions in Biological Systems"
44 Cohen J. S. In Experimental Methods of Biophysical Chemistry *6*, 521 (1972): "Nuclear Magnetic Resonance Investigation of the Interactions of Biomolecules"
45 Bovey F. A., Brewster A. I., Patel D. J., Tonelli A. E. and Torchia D. A., Acc. Chem. Res. *5*, 193 (1972): "Determination of the Solution Conformations of Cyclic Polypeptides"
46 Horwitz A. F. In Membrane Molecular Biology, C. F. Fox and A. D. Keith (Eds.), Sinauer Publishers, Stamford, Conn., 1972, p. 164: "Nuclear Magnetic Resonance Studies on Phospholipids and Membranes"
47 Wüthrich K., Naturwissenschaften *60*, 221 (1973): "Magnetische Kernresonanzspektroskopie in der Biologischen Forschung"
48 Phillips W. D. In NMR of Paramagnetic Molecules, G. N. LaMar, W. DeW. Horrocks and R. H. Holm (Eds.), Academic Press, New York 1973, p. 421: "Biological Applications"
49 Kearns D. R. and Shulman R. G., Acc. Chem. Res. *7*, 33 (1974): "High Resolution Nuclear Magnetic Resonance Studies of the Structure of Transfer Ribonucleic Acid and other Polynucleotides in Solution"
50 Wüthrich K., Experientia *30*, 577 (1974): "Nuclear Magnetic Resonance in Protein Research"

CHAPTER II

PROTON NMR OF AMINO ACIDS—AN INTRODUCTION TO HIGH RESOLUTION NMR SPECTROSCOPY

The proton NMR spectra of amino acids and small peptides are generally well resolved and in this sense, contain many of the typical NMR features of organic molecules. In this chapter they were therefore chosen as an illustrative background for a brief introduction to high resolution proton NMR experiments. The reader already acquainted with the technique will probably go directly to Sections II.6 and II.7 which concentrate on specific aspects of the amino acid NMR spectra.

Principal NMR Features in Amino Acids

The typical NMR-spectral features of amino acids which we will have to discuss here follow from their chemical structures. Since there are different types of protons in an amino acid molecule, the NMR spectrum will consist of several resonances with different chemical shifts. These resonances can further be split up into multiplets by spin–spin coupling. This is illustrated in Fig. II-1 with the spectrum of L-alanine. This molecule contains two types of chemically and magnetically non-equivalent protons which are non-exchanging in protic solvents, i.e. the three methyl protons and the C^{α}-proton. The methyl resonance is split into a doublet by spin–spin coupling with the C^{α}-proton, and the latter appears as a quartet due to the coupling with the three methyl protons. From this spectrum the chemical shifts and the spin–spin coupling constant J are readily obtained, as indicated in Fig. II-1. Independent of the fine structure of the resonances, their integrated intensities provide additional support for the assignments given in Fig. II-1. With increasing complexity of the chemical structures, and in particular with increasing number of protons in the molecules studied, more complicated NMR spectra may be obtained from which the spectral parameters can no longer be evaluated by mere inspection. The general basis for the interpretation of such complex spin systems will be discussed in Section II.3, and in the Appendix II.

In addition to the protons bound to carbon atoms, which are non-

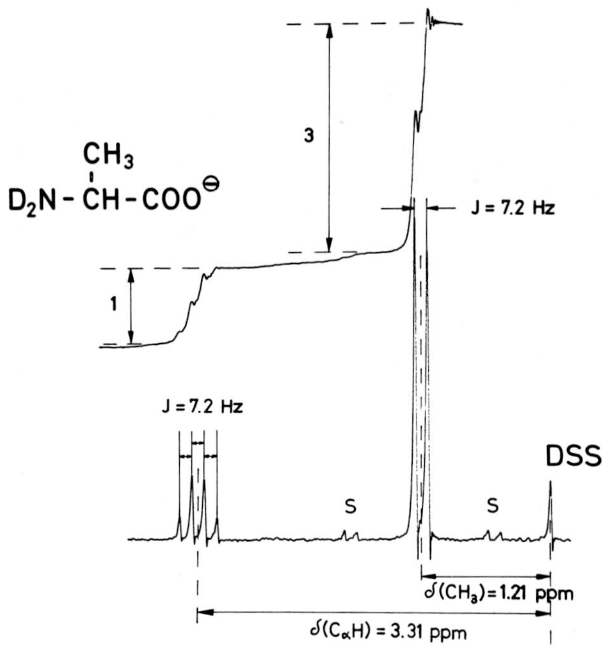

Fig. II-1. Proton NMR spectrum at 60 MHz of a 0.5-M solution of L-alanine in D_2O, pD $\simeq$ 13.0, $T = 35°$. DSS is used as an internal reference. The resonance assignments can be derived from the relative signal intensities in the integrated spectrum (upper trace), and from the multiplet structures. In this example the NMR parameters δ and J can be directly read from the line positions, as indicated by the arrows. The weak lines labelled s are the spinning side bands of the methyl doublet resonance.

exchanging in protic solvents, the amino acids contain several potentially exchangeable protons, of which the exact number depends on the solvent environment. In aqueous solution it varies with pH according to

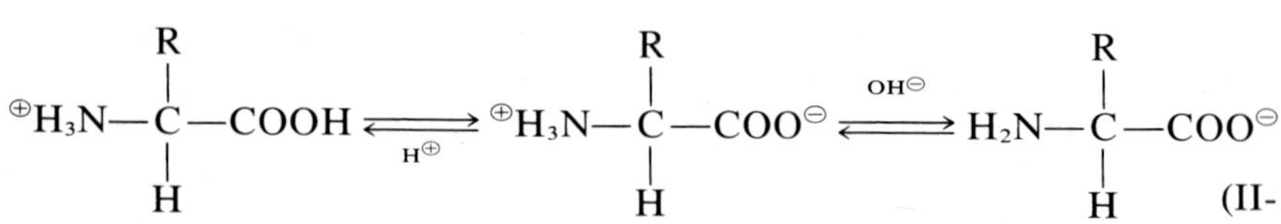

(II-1)

Because of the reactions II-1, essentially all the resonances in the individual amino acids vary with pH, as will be discussed in more detail in Section II.6. The exchangeable protons in the individual amino acids can generally not be observed as separate NMR signals in aqueous media because there is rapid exchange of the amino and carboxyl protons with protons of the solvent. This and other manifestations of rate processes in the NMR spectra will be the subject of Section II.5.

1. PROTON CHEMICAL SHIFTS

The position of the nuclear resonance of a particular proton depends on a number of factors. These include the covalent structure of the molecule, and non-bonding intramolecular and intermolecular interactions which can contribute to the shielding of the nucleus against the external magnetic field. The theoretical basis of nuclear shielding by the electronic environment is discussed in the textbooks on NMR mentioned in Section I.4. Here we will base the interpretation of the chemical shifts in amino acids on a set of empirical resonance positions in diamagnetic organic molecules (Table II-1).

TABLE II-1

Proton Chemical Shifts in some diamagnetic organic structures*[1].

Molecular structure — Proton chemical shift δ in ppm from TMS

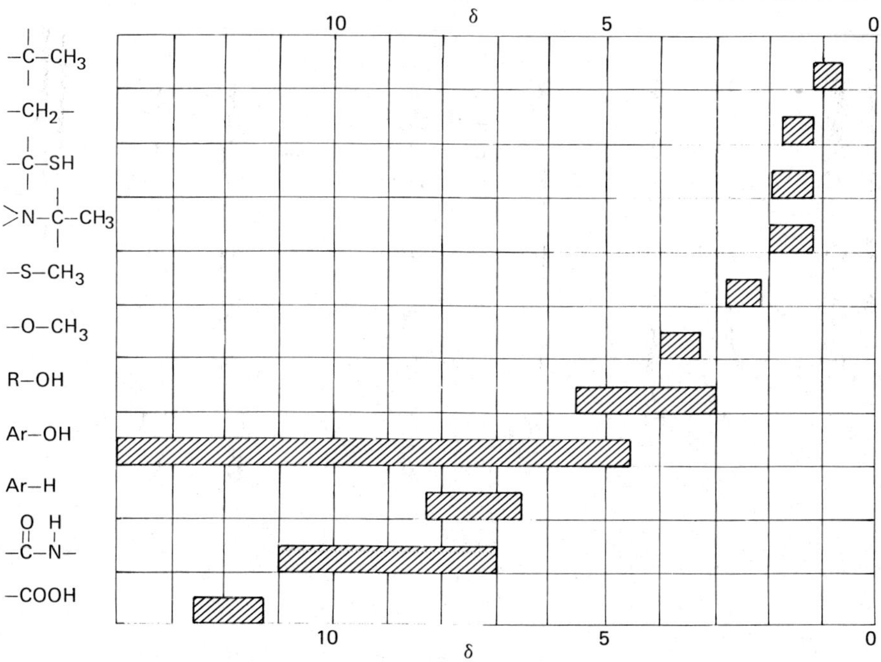

*The bars indicate the spectral regions where the resonances can occur under different conditions of solvent and temperature.

Standard Reference Compounds

In Table II-1 the resonance positions for protons in a variety of organic compounds are given with respect to the reference compound tetramethyl silane, $Si(CH_3)_4$, which is usually abbreviated as "TMS". Different derivatives of TMS, which have similar resonance positions, are also in use. For biological work, the water soluble sodium salt of 2,2-dimethyl-2-silapentane-5-sulfonate ("DSS"), $(CH_3)_3-Si-CH_2-CH_2-CH_2-SO_3^{\ominus}Na^{\oplus}$, is often employed. The chemical shifts δ are given in parts per million (ppm) of the resonance frequency, where positive numbers indicate shifts to lower field at constant frequency*.

Deshielding by Electronegative Groups

The table reveals an obvious correlation between chemical constitution and NMR chemical shifts, i.e. that electronegative atoms and groups cause deshielding of nuclei attached to or near them. This is usually referred to as the inductive effects in shielding. As is seen in Table II-1, the electrons surrounding the protons in aliphatic hydrocarbons provide an effective shield so that these protons are observed at high field strength of H_0. Electronegative groups such as $-S-$, $-O-$, $-COO^{\ominus}$, reduce the electron density on nearby protons, and hence reduce the shielding against the applied magnetic field. Therefore protons attached to or near electronegative groups appear in positions at lower field H_0. This deshielding effect falls off rapidly with increasing number of bonds between the observed protons and the electronegative group.

Ring Current Shifts

All chemical shifts in Table II-1 can qualitatively be accounted for by inductive effects, with the exception of the low field position for aromatic protons. Additional deshielding of aromatic protons† arises because the diamagnetic susceptibility of an aromatic ring is greater in the direction perpendicular to the plane of the ring than parallel to it. An interpretation of these diamagnetic anisotropies of aromatic molecules is that the application of the magnetic field H_0 induces a circulation of the π-electrons in the plane

---

*Other units in use include the τ-scale, where τ (in ppm) = 10—δ (in ppm from TMS), and the indication of low field shifts at constant frequency by negative numbers[1].

†Shielding effects similar to those in aromatic rings have been observed or postulated for different classes of compounds, e.g. acetylenes, olefinic and carbonyl double bonds[1].

of the ring[2]. In its consequences, this electron flow would be equivalent to a superconducting current in a ring of radius equal to that of the aromatic ring. This ring current produces a local magnetic field H_R, often referred to as the ring current field, which opposes the externally applied field H_0 in the areas above and below the plane of the aromatic molecule, and reinforces it in the ring plane outside the contours of the ring (Fig. II-2). Several different

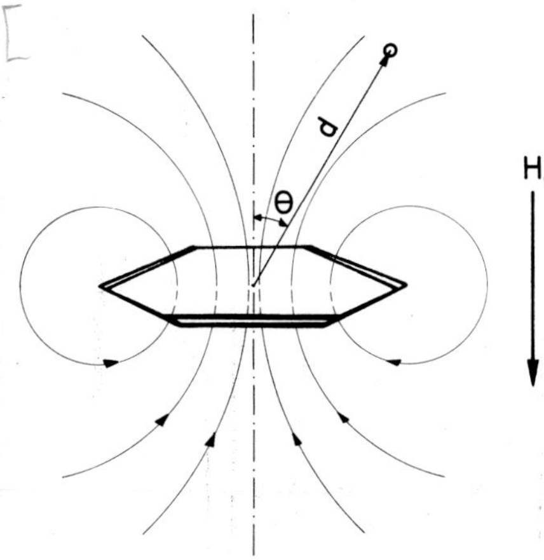

Fig. II-2. Local magnetic ring current field H_R induced in aromatic rings by an external field H_0 applied perpendicular to the ring plane.

models have been proposed for a quantitative assessment of the shielding zone about aromatic rings. Pople[3] replaced the circular magnetic shell produced by the electronic ring current by an equivalent dipole μ at the center of the ring. With this approximation, the ring current shielding in ppm for a nucleus at a distance d from the center of the rapidly tumbling ring is

$$\delta_R = 10^6 \frac{ne^2a^2}{6\pi mc^2d^3}(3\cos^2\theta - 1) \qquad \text{(II-2)}$$

where θ is the angle between d and the normal to the ring plane (Fig. II-2), n is the number of circulating electrons with charge e and mass m, a is the radius of the ring, and c the velocity of light. In this simple model the ring current field increases with the number of electrons, and with the square of the radius of the ring. For any given set of a and n, the ring current shielding decreases rapidly with increasing distance d. Upfield shifts will result for nuclei lying within a cone with its axis perpendicular to the plane of the ring and through its center, and with an apex angle of ca. 55°, downfield shifts for nuclei located outside of this cone. These principal qualitative features of

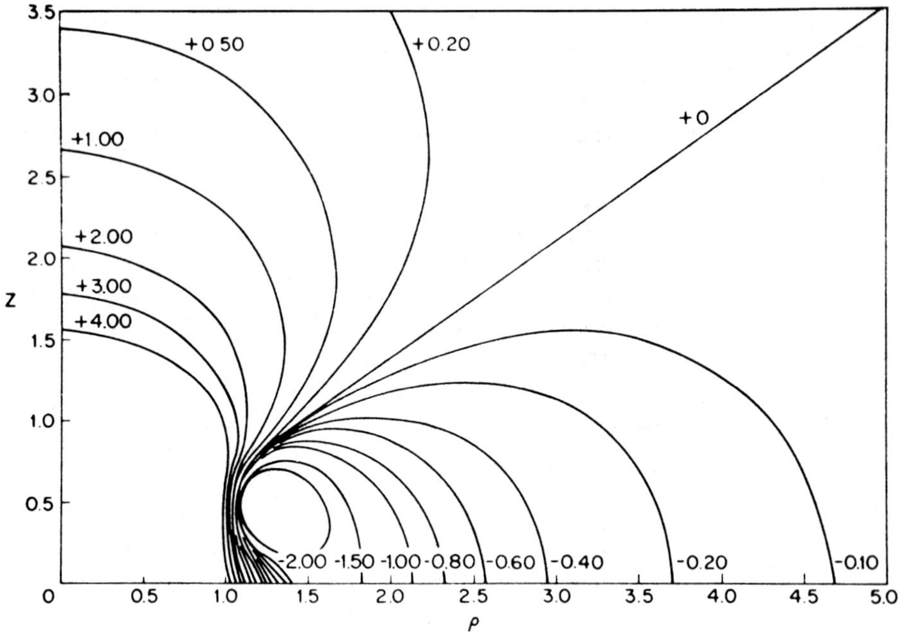

Fig. II-3. The shielding zone about a rapidly tumbling benzene ring. The plot represents one quadrant of a plane through the center of the ring and normal to the ring plane. The origin of the plot is at the ring center, and the ρ-axis is in the ring plane. ρ and z are in units of 1.39 Å. The isoshielding lines are in ppm, where positive numbers indicate shifts to higher field. (From Johnson and Bovey[4]).

the ring current field are preserved in more elaborate models. As an illustration, Fig. II-3 shows the shielding zone about a benzene ring as computed by Johnson and Bovey[4] from a model in which the circular path of the electrons was retained.

Proton Chemical Shifts in Amino Acids

Taking the Table II-1 as a guide, the following typical resonance positions for the common amino acids are to be expected. The methyl protons of Ala, Val, Leu, Ile, and Thr should be at 0.8 to 1.5 ppm, the methyl protons of Met and most of the methylene protons of the aliphatic amino acid side chains between 1.5 and 3.0 ppm, the C^{α}-protons between 3.5 and 4.5 ppm, and the ring protons of the aromatic side chains at around 7.0 ppm. The tables in Section II.7, where the proton NMR parameters for the different amino acids are given for different solvents, will show that the expected chemical shifts are actually observed.

2. SPIN–SPIN COUPLING

The Spin–Spin Coupling Constant J

We have seen in Fig. II-1 that the NMR signals may be split into multiplets by spin–spin coupling. These interactions between the protons are mediated through the intervening covalent bonds, and therefore spin–spin coupling between the different protons of an organic molecule is quite generally observed. The extent of these interactions is described by the spin–spin coupling constants J, which are usually given in Hz. In contrast to the chemical shift, J is independent of the applied magnetic field H_0. Proton–proton spin–spin coupling constants from 0 to 20 Hz are typically observed in organic molecules.

Coupling constants for nuclei separated by n chemical bonds are customarily designated as ^nJ. The coupling between two methyl protons of alanine (Fig. II-1) would thus be represented by ^2J, and is referred to as a "geminal" proton–proton coupling. The coupling between the methyl protons and the C^α-proton of alanine would accordingly be ^3J, and is called "vicinal" proton–proton coupling. Subscripts may be added to further relate the spin–spin coupling with the molecular structure. The vicinal coupling in alanine may thus be designated $^3J_{HC^\alpha HC^\beta}$ to indicate that the ^3J-coupling is between protons bonded to the α- and β-carbon atoms. Throughout this chapter, we shall use an abridged notation, where the vicinal proton–proton coupling in alanine is represented by $^3J_{\alpha\beta}$. In subsequent chapters, where we shall be concerned exclusively with vicinal spin–spin couplings, the prefix ^{3} will also be omitted, and we will write $^3J_{HC^\alpha HC^\beta} \equiv J_{\alpha\beta}$.

Proton–proton couplings through two and three bonds are almost always the largest. Couplings transmitted only by the σ bond system usually decrease by at least an order of magnitude with each additional intervening bond. The resulting so-called "long range couplings" are usually too small to be resolved in the spectra of biological macromolecules. The same applies to the couplings transmitted through four or more bonds in aromatic systems, even though the interactions via the π-electrons can cause some quite appreciable long range couplings, as will be seen below.

Spin–spin couplings between protons and other nuclei may also be manifested in the ^{1}H-NMR spectra. For example, the couplings between directly bonded ^{13}C and ^{1}H, $^1J_{^{13}CH}$, are of the order of 150 Hz (see Chapter V). However, because of the low natural abundance of ^{13}C (Table I-1), the resulting doublet structures in the ^{1}H NMR spectra have only 1.1% of the total signal intensity. The analysis of these "^{13}C satellites" in the ^{1}H NMR spectra has been of considerable interest for certain NMR studies in organic chemistry[1], but has so far had little impact on the investigations of biopolymers with natural isotope distribution.

First Order Multiplet Patterns

The spin–spin coupling phenomenon may be explained by the following simple model: The internuclear coupling arises through mutual polarization of the spins, which depends on the spin states m_I. For a macroscopic sample of spins $I = 1/2$, there is an almost equal probability of finding spins in the α-($m_I = 1/2$) or β-($m_I = -1/2$) state (Equation I-1). Therefore the resonance of the methyl protons in Fig. II-1 is split into two lines of equal intensity by the coupling with the single C^α-proton. On the other hand the C^α-proton in the polarizing field of the three methyl protons sees the following 9 spin states with essentially identical probability.

$F_n = 3/2$	$F_n = 1/2$	$F_n = -1/2$	$F_n = -3/2$
α α α	α α β	α β β	β β β
	α β α	β α β	
	β α α	β β α	

Since the extent of the polarization by n equivalent neighboring spins depends on $F_n = \sum_{i=1}^{n} m_i$, the resulting multiplet is a quartet with relative line intensities of 1:3:3:1. Generalizing, one has that if a proton is coupled with n equivalent spins 1/2, its resonance will be split into $(n + 1)$ peaks, and the relative intensities of the multiplet components are given by the Pascal triangle:

```
n = 0          1
    1        1   1
    2      1   2   1
    3    1   3   3   1
```

As we will see in Section II.3, the multiplet patterns can in certain cases also deviate appreciably from the simple "first order" appearance described here.

Vicinal Proton–Proton Couplings

Besides the chemical shifts, the vicinal spin–spin coupling constants ^3J are the most important source of structural information in the high resolution proton NMR spectra of biopolymers. We shall see later on, that vicinal spin–spin coupling is particularly important for structural studies of peptides. This prominent role of vicinal couplings is based on the dependence of ^3J on the torsion angle θ that defines the rotational states about the bond linking the two atoms to which the coupled spins are attached:

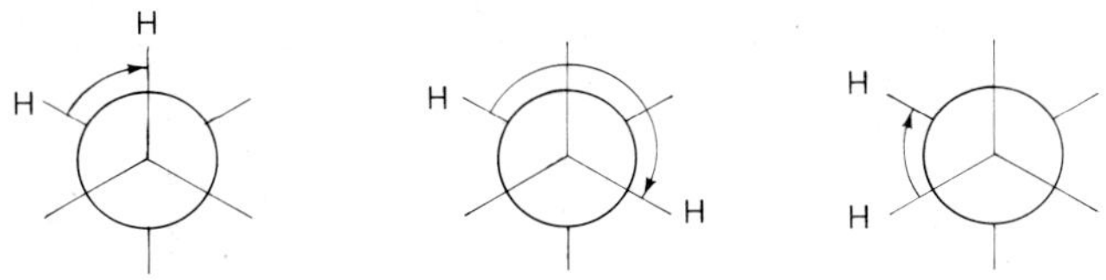

gauche $\theta = 60°$ *trans* $\theta = 180°$ *gauche* $\theta = 60°$

For the vicinal coupling constants between protons on adjacent sp^3 carbon atoms, Karplus[5,6] derived the following relations:

$$\left.\begin{array}{ll} {}^3J = 8.5\cos^2\theta - 0.28 & 0° \leqslant \theta \leqslant 90° \\ {}^3J = 9.5\cos^2\theta - 0.28 & 90° \leqslant \theta \leqslant 180° \end{array}\right\} \quad \text{(II-3)}$$

These functions are plotted in Fig. II-4. It can be seen that ^3J is ca. 1.8 Hz for

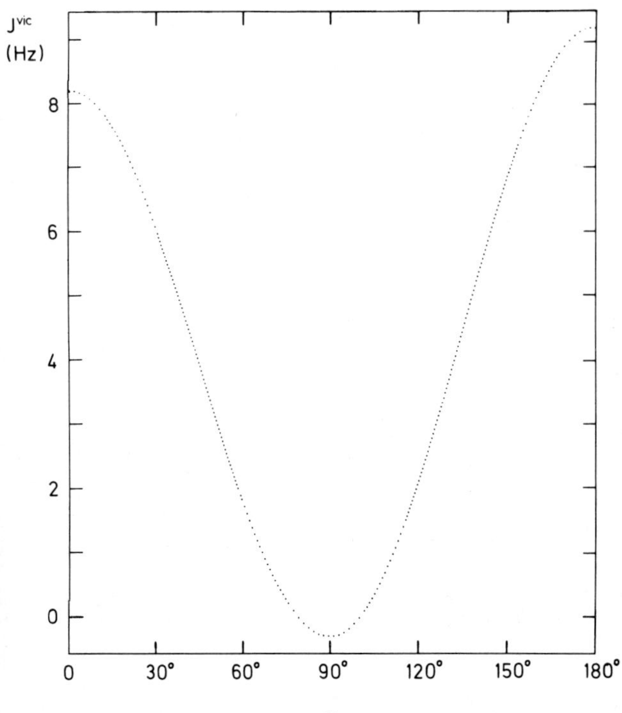

Fig. II-4. The Karplus function describing the dependence of vicinal proton–proton coupling on the dihedral angle θ in the H–C–C–H bond system.

$\theta = 60°$, and ca. 9.2 Hz for $\theta = 180°$. The magnitude of ^3J is further strongly influenced by substituent effects. However, as will be further discussed in subsequent chapters, the qualitative features of the function in Fig. II-4 are also observed for the angular dependence of ^3J between nuclei bound to non-equivalent atoms.

Geminal Proton–Proton Couplings

Geminal couplings ^2J for most sp^3-hybridized groups are of the order (−) 10 to (−)15 Hz*. However, the geminal couplings between the three methyl protons of alanine are clearly not manifested in the spectrum of Fig. II-1. It can be shown from symmetry arguments[7] that J-couplings between chemical-shift-equivalent nuclei generally do not produce a multiplet splitting of the NMR-signal.

Proton–Proton Couplings in Aromatic and Heterocyclic Rings

Proton–proton spin–spin couplings likely to be encountered in experiments with biopolymers include those in aromatic rings and heterocyclic molecules. Typical J-values for aromatic rings are[8]

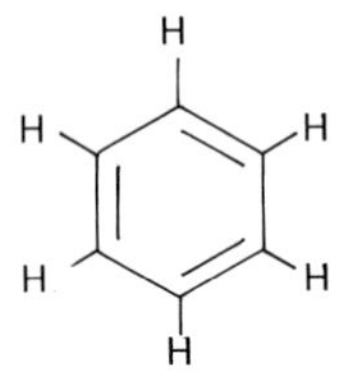

J_{ortho} : 6.5–9.4 Hz
J_{meta} : 0.8–3.0 Hz
J_{para} : 0.4–1.0 Hz

In pyridine as a representative heterocyclic structure type, one has the following couplings:

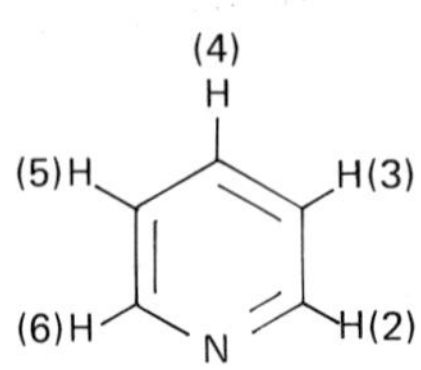

$J_{2,3}$: 4.5–5.5 Hz
$J_{3,4}$: 7.5–8.2 Hz
$J_{2,4}$: 0.7–1.9 Hz
$J_{3,5}$: 0.3–1.6 Hz
$J_{2,6}$: 0.0–0.4 Hz
$J_{2,5}$: 0.5–0.9 Hz

---

*Spin–spin coupling constants can be positive or negative. Generally, geminal couplings are found to be negative, vicinal couplings positive. The sign of the coupling constants is most clearly manifested in certain multiple resonance experiments[1].

3. SPECTRAL ANALYSIS OF COMPLEX SPIN SYSTEMS

Weak and Strong Spin–Spin Coupling in Nuclear Spin Systems

In the spectrum of alanine (Fig. II-1), the relative chemical shift $\Delta\delta$ between the methyl protons and the C^{α} proton resonance is 126 Hz, which is large compared to the spin–spin coupling $^{3}J_{\alpha\beta}$ of 7.2 Hz. Correspondingly, Fig. II-1 represents the spectrum of a typical "weakly coupled" spin system. It is a general feature of weakly coupled spin systems with $J \ll \Delta\delta$ (in Hz), that the NMR spectra consist of well separated first order multiplets with regular spacings between the constituent lines, and a binomial distribution of the line intensities. Spin systems characterized by relatively small values of the quotient $\Delta\delta/J$ are called "strongly coupled". The corresponding NMR spectra often appear as complex systems of lines, with spacings which exhibit no obvious regularity, and line intensities which deviate widely from binomial. The NMR parameters can then no longer be obtained by mere inspection of the spectrum, and hence the calculation of J values and chemical shifts for systems of strongly coupled spins is an important aspect of high resolution NMR spectroscopy.

The analysis of a strongly coupled spectrum requires the computation by quantum mechanical methods of the energy levels and stationary state wave functions of the coupled spin systems in a static magnetic field, and then to use perturbation methods and selection rules to give the probabilities of transitions occurring between these levels when a resonant radio frequency field is applied. Fig. II-5 represents the result of such a calculation for a system consisting of two spins $I = \frac{1}{2}$ (see Appendix II). It is seen that for $\Delta\delta/J = 1000$, one gets a clean first order spectrum with two well separated doublets. One notices further that the spectra with $\Delta\delta/J \geqslant 5$ can approximately be treated according to the rules for weakly coupled spin systems. On the other hand, it is obvious from Fig. II-5 that one would obtain erroneous values for $\Delta\delta$, when trying to solve the spectra characterized by $\Delta\delta/J < 5$ with the use of first order approximations. The situation is quite analogous for more complex spin systems, and the limit of $\Delta\delta/J \geqslant 5$ may quite generally be employed as a rule of thumb to appraise whether a first order approximation would yield meaningful parameters. However, with increasing number of spins the number of resonance lines, and hence the complexity of the spectrum, may increase quite dramatically.

Systematic Notation for Spin Systems

A decisive step at the outset of the analysis of a complex spectrum is the characterization of the spin systems in terms of a now generally adopted

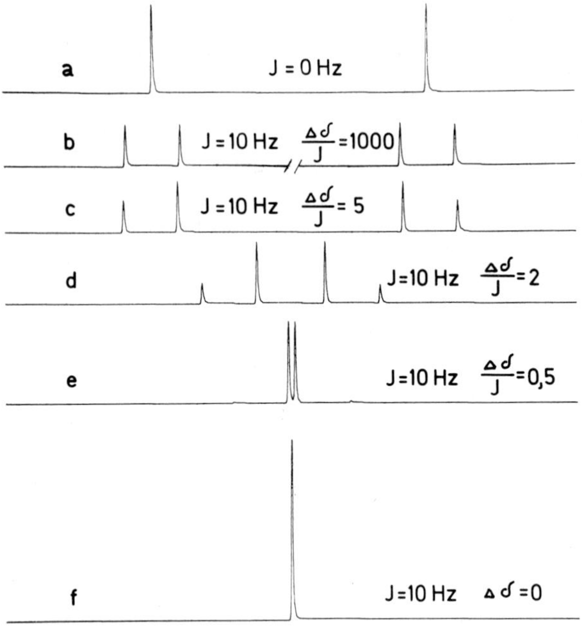

Fig. II-5. Calculated spectra for a spin system of two spins 1/2 with various ratios of the relative chemical shift $\Delta\delta = (1/2\pi)\gamma H_0(\sigma_1 - \sigma_2)$ to the spin–spin coupling constant J.

alphabetical notation for spin systems within molecules[9]. Two chemical shift nonequivalent nuclei are denoted A and B if $\Delta\delta \approx J$, and A and X if $\Delta\delta \gg J$. For the two coupled spins of Fig. II-5e, one would thus have an AB spin system, in Fig. II-5b an AX spin system. Correspondingly, three nonequivalent spins are denoted ABC, or AMX, respectively. n equivalent spins, i.e. spins which have identical chemical shifts, and identical spin–spin couplings with all the other spins in the system, are denoted A_n. Two spins with identical chemical shifts yet different spin–spin couplings with other nuclei in the system, are denoted AA′.

Practical Aspects of the Analysis of Complex NMR Spectra

For practical purposes, computer programs are available for the computation of spectra for systems with up to 8 spins[10], and many representative strongly coupled spectra have also been tabulated[11]. These tables, and the input for the NMR spectral simulation programs are usually arranged according to the above alphabetical notation. Therefore, once a new spectrum has been recorded, the first step in the analysis will be to find the systematic notation for the spin system. This can usually be achieved on the basis of the covalent

TABLE II-2
Description of Spin Systems in Amino Acid Residues*.

Molecular structure	Spin system	Occurrence in amino acids
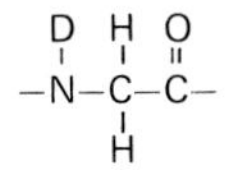	A_2 or AB	Gly
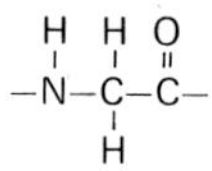	A_2X or ABX	Gly
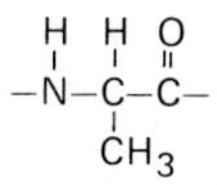	A_3MX	Ala
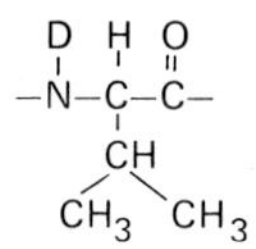	A_3B_3MX	Val
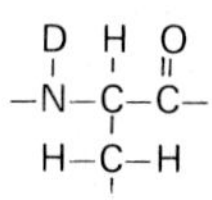	ABX	Ser, Asp, Asn, Cys, Trp, Phe, Tyr, His
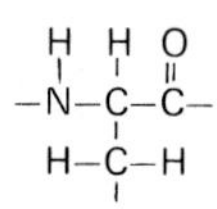	ABMX	Ser, Asp, Asn, Cys, Trp, Phe, Tyr, His
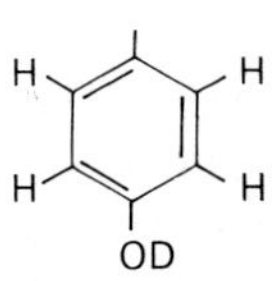	AA′BB′	Tyr

*Only the proton spins are taken into consideration.

structure of the molecule, provided it is known, and from comparison with the spectra of structurally related compounds or with tabulated spectra of classified spin systems. The next step is to obtain approximate values for the NMR parameters, e.g. from inspection of the spectrum, or from comparison with tabulated spectra (see the Appendix II). These tentative NMR parameters are used as the initial input for the computer program, and are then optimized either by a trial-and-error search for a good fit between computed and experimental spectrum, or by a least squares fit of the experimental data[10]. The chemical shifts and spin–spin couplings thus obtained may then be used to unravel the structural and conformational features of the molecule under investigation.

The Appendix to this chapter presents an introduction to the quantum mechanical background of the analysis of complex spin systems. In addition, a concise description of the spin systems which are most frequently encountered in peptides and proteins will also be given. For a more thorough treatment of the quantum mechanical background and the practical aspects of the interpretation of complex NMR spectra, the reader is referred to the textbooks mentioned in Section I.4.

Spin Systems in Amino Acid Residues

The description of some typical spin systems in amino acid residues is given in Table II-2. Special attention is called to the observation that while the β-methyl protons in Ala are equivalent because of the rotation about the $C^\alpha - C^\beta$ carbon–carbon bond, β-methylene protons in amino acids are non-equivalent even if there is rapid rotation about $C^\alpha - C^\beta$. This structural nonequivalence of C^β-substituents is also manifested in the different proton NMR parameters for the two γ-methyl groups in valine.

4. EXPERIMENTAL ASPECTS

At the present time commercially available instruments operate at proton NMR frequencies between 60 and 360 MHz (Table II-3). They are routinely equipped with a field-frequency lock to assure short term and long term stability, attachments for multiple irradiation experiments, and for variable temperature operation. As was already discussed in Chapter I, the main emphasis in instrumentation for biological NMR work has to be on high sensitivity and spectral resolution. These two aspects will now be discussed in some more detail.

TABLE II-3

Magnetic fields H_0 and proton NMR frequencies ν_0 in some common commercial spectrometers*.

H_0[KG]	ν_0[MHz]
14.1	60
21.1	90
23.5	100
51.7	220
63.4	270
70.5	300
84.6	360

*At resonance frequencies ν_0 of 220 MHz and higher, the magnetic field H_0 is produced in a superconducting solenoid.

Sensitivity–Spectrum Accumulation in Digital Computers

The sensitivity for detection of the NMR signals of a given nucleus is improved at higher fields H_0, because the magnetisation of a macroscopic sample, and hence the net absorption of energy in the NMR experiment increase (Equation I-1). Theory predicts that the sensitivity increases with $\nu_0^{3/2}$.

The sensitivity is further determined by the mode of operation, and can be greatly improved by spectrum accumulation in a computer. In the continuous wave (CW) or "conventional" mode of operation, where the resonance condition (Equation I-5) is attained by slowly sweeping the frequency ν_0 at constant field H_0, the spectra obtained in n successive sweeps are stored in a computer of average transients. The signal-to-noise ratio will thus be improved as the square root of the number of traces, $(S/N)\alpha\sqrt{n}$ (Fig. II-6). Until 1971 essentially all the high resolution proton NMR spectrometers were operated in the CW mode, and most of the ^{1}H NMR spectra discussed in this book had been recorded with this technique. With the more recently introduced Fourier transform (FT) mode of operation, the principles of which will be introduced in Chapter IV, an improvement of the sensitivity by a factor 10 over that of the CW experiment can be obtained[12].

As one is often interested in investigations of rather dilute solutions of biological materials, it is also of importance that the sample size in modern spectrometers has been increased over the conventional 5 mm tubes to diameters of 10 to 20 mm. Finally it should be mentioned that the sensitivity

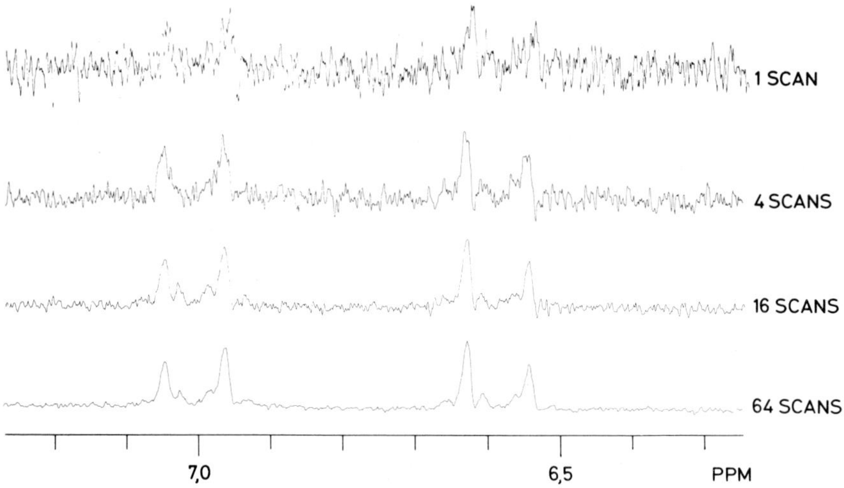

Fig. II-6. Aromatic region of the ^{1}H-NMR spectrum at 100 MHz of a 0.001-M solution of L-tyrosine in D_2O, pD = 7.0, recorded with a computer of average transients. Sample size 5 mm, sweep time 100 sec, sweep width 100 Hz, filter time constant 1 sec.

for CW experiments at fixed field and fixed sample size has been substantially increased over the last decade through improved design of the NMR probes.

Spectral Resolution at High Fields

The spectral resolution increases rather dramatically at higher fields H_0 because the chemical shifts (in Hz)* increase with H_0, whereas the line widths (Fig. I-2) are for most practical purposes either independent of H_0, or decrease with H_0. This is of particular relevance for work with high molecular weight compounds where the proton NMR spectra are very crowded with a large number of rather broad resonances. From the discussion in Section II.3 it is evident that the spectra of small molecules can also be greatly simplified at higher fields H_0. Since the chemical shifts $\Delta\delta$ increase with H_0, and the spin–spin coupling constants J are independent of the field, a molecule which yields a complex, strongly coupled spectrum at 60 MHz can display almost pure first order NMR features at higher fields (Fig. II-7).

*It is one of the advantages of the δ- and τ-scales for chemical shifts that measurements at different fields H_0 can be directly compared, even though the shifts increase proportional to H_0 if given in frequency units (Equation I-2).

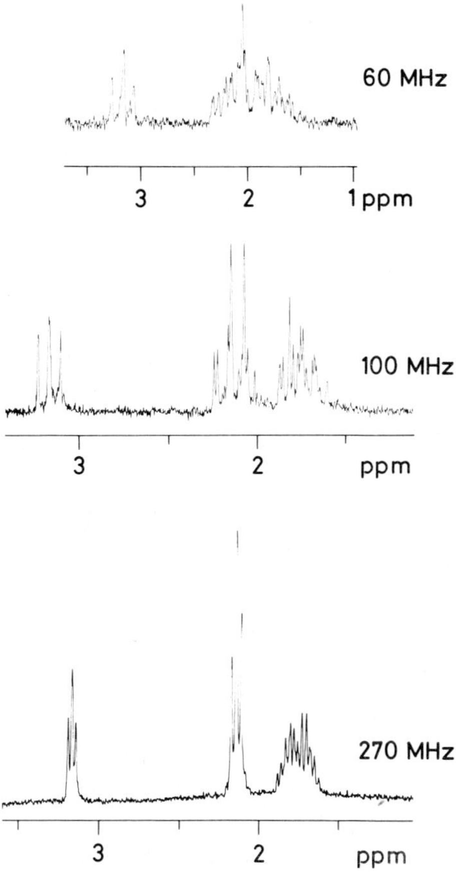

Fig. II-7. Proton NMR spectra of a solution of glutamic acid in D_2O at 60, 100, and 270 MHz.

Sample Preparation

In most biological NMR experiments, aqueous solutions are preferred. Because of the interference caused by the strong resonance of H_2O, deuterium oxide is used in most instances. The acidity of D_2O solutions can be measured with the glass electrode used for H_2O when the following correction is made[13]

$$pD = pH \text{ (meter reading)} + 0.40. \quad \text{(II-4)}$$

In D_2O solution all the exchangeable protons of the biological material will in principle be replaced by deuterium, and hence valuable NMR spectral information can be lost. This is the main reason why one often tries to

TABLE II-4
Properties of some common NMR solvents*.

Solvent	Structure	Chemical shift δ (rel. to TMS or DSS)	Exchange of labile protons with deuterium	Dielectric constant ϵ (25°)[e]	pK_{a1}[e]	pK_{a2}[e]
Water	H_2O	~ 4.8 (20°)[a]	no[b]	78.5	− 1.8	15.3
Deuterium oxide	D_2O	~ 4.8 (20°)[a, c]	yes			
Methanol	CH_3OH	3.35	no[b]	32.6	− 2.0	16.0
d_4-Methanol	CD_3OD	3.35[c, d]	yes			
Trifluoroacetic acid	CF_3COOH	~ 12.5	no[b]	8.4	− 10.6	0.2
d-Trifluoroacetic acid	CF_3COOD	~ 12.5[c]	yes			
d_6-Dimethylsulfoxide	$(CD_3)_2—S{=}O$	2.58[c, d]	no	46.4	0.0	
d-Chloroform	$CDCl_3$	7.27[c]	no	4.9	− 6.0	24.0
d_7-N-dimethylformamide	$(CD_3)_2NCD{=}O$	2.88, 2.97, 8.02[c]	no	36.7	0.0	
d_5-Pyridine	C_5D_5N	7.0, 7.6, 8.6[c, d]	no	12.3	5.2	
Trifluoroethanol	CF_3CH_2OH	~ 6.5, ~ 12.0	no[b]	26.7	− 8.2	12.4
d_3-Trifluoroethanol	CF_3CD_2OD	~ 6.5, ~ 12.0[c, d]	yes			

*Indicated are the solvent structure, the ^{1}H-NMR spectral properties, the observability of labile protons in the dissolved materials, the dielectric constant, and the acid-base character. The latter three properties are of particular interest for studies of peptides.

[a]The water resonance moves upfield at higher temperature (at 50°, $\delta_{H_2O} \approx 4.5$ ppm).

[b]Labile protons of the solute may in certain cases not be observable as separate NMR signals because of rapid proton exchange with the solvent.

[c]NMR of residual protons in the deuterated solvent.

[d]The solvent spectrum may contain multiplet resonances arising from hetero-nuclear proton-deuterium spin–spin coupling.

[e]The dielectric constants ϵ, pK_{a1}, which indicate the pH for half protonation of the neutral bases, and pK_{a2}, which indicate the pH for half protonation of the Brönsted acid, are from Ref. 14. The pK_a values are meant only as a general indication of the acid-base character, which may be different in different solvent mixtures.

complement experiments in D_2O with additional spectra taken in H_2O. For work with peptides, non-aqueous solvents are often chosen by necessity, because peptides can exhibit small solubility in water. In many instances when it is felt that a certain molecule might experience a non-aqueous environment under physiological conditions, different solvents can also be chosen for the NMR studies in an attempt to mimic the biological system. An additional important aspect of the choice of solvent is that quite often the structural interpretation of polypeptide NMR data can greatly be aided when the spectra are recorded in various solvents with different dielectric constants, and different acid-base character (Table II-4).

Quite often the amount of substance required for NMR is one of the critical aspects for the applicability of the method. From the point of view of the NMR sensitivity, the sample concentration for ^{1}H-NMR has to be of the order 10^{-2} to 10^{-4}-molar, or higher, depending on the instrument. To present a specific example, single proton resonances in a 0.01-M solution can be recorded with a proton NMR spectrometer operating in the CW-mode at 100 MHz or higher frequency in a 5 mm sample tube with a single sweep of 5 to 30 minutes. The sample volume needed is ca. 0.3 ml. Hence when working with a protein of molecular weight 20000, the amount of dry material needed would be ca. 60 mg. The minimal concentration can then be appreciably reduced by increasing the sensitivity through time-averaging devices or operation in the FT mode. The sample volume can also be reduced by the use of special microcells (Fig. II-8). With the investment of sufficient instrument time, it is thus in principle possible to obtain a proton NMR spectrum from ca. 1 to 3 mg of a substance with molecular weight 20000.

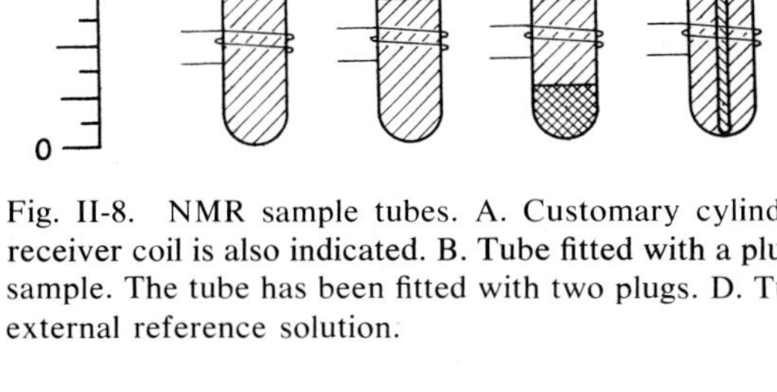

Fig. II-8. NMR sample tubes. A. Customary cylindrical sample tube. The position of the receiver coil is also indicated. B. Tube fitted with a plug to prevent vortexing. C. Simple micro sample. The tube has been fitted with two plugs. D. Tube fitted with a coaxial capillary for an external reference solution.

Sample Spinning

In high resolution NMR it is established practice to use cylindrical sample tubes of 5 mm (or larger) outside diameter and ca. 150 mm length (Fig. II-8). During operation the sample tube is rotated about its long axis at a frequency of ca. 20 Hz to 120 Hz, depending on the instrument, to obtain the high resolution needed. One of the side effects of this sample spinning is that it causes vortexing of the solution, which may reduce the effective sample volume in the receiver coil. Therefore one uses rather large sample volumes for routine NMR (Fig. II-8A). A smaller sample volume, i.e. ca. 0.3 ml in a 5 mm tube, can be used if vortexing is suppressed through a plug fitted into the tube in contact with the solution (Fig. II-8). Of course, this plug has to be made of NMR—inert material, e.g. glass or teflon[15].

Another side-effect of sample spinning is the appearance of spinning side bands at integer multiples of the spinning frequency from the parent NMR signal (Fig. II-1). These spinning side bands, which arise from the modulation of the magnetic field inhomogeneities by the mechanical spinning can cause problems whenever there is a strong solvent resonance in a dilute solution. The first, second, and sometimes even the third spinning side bands of the solvent resonance can then have intensities equal to or greater than those of the resonances of the dissolved material.

Internal and External Referencing

Chemical shifts can in principle be measured with respect to a reference compound dissolved in the sample, or with respect to an external reference which can be introduced into the sample in a capillary (Fig. II-8D). When using internal standards, one has to take into account that the NMR properties of the molecules investigated could be altered by the reference compound, e.g. through complex formation. However, with the commonly used standards TMS and DSS, this appears not to be a common difficulty, even though it is advisable in little known systems to check carefully for such potential pitfalls. With the use of external standards intermolecular effects are excluded. However this advantage of external referencing is more than counterbalanced by the disadvantage that the sample and the reference do not experience the same effective magnetic field because of their different magnetic susceptibilities. Therefore in order to compare chemical shifts measured in different solutions vs. the same external reference, one would have to make a correction for the different susceptibilities. The same is true if a given sample is studied under different conditions of pH, temperature, or ionic strength. As is discussed in more detail in textbooks on high resolution NMR[1,16], these susceptibility corrections are by no means straightforward. It is therefore advisable to use internal referencing whenever it is at all possible.

Resonance Assignments from Double Resonance Experiments

As was mentioned above, modern NMR spectrometers are equipped with attachments for multiple irradiation experiments. At this point two types of double resonance experiments are briefly introduced which are frequently employed for the identification of the NMR of coupled groups of protons in complex spin systems. Further discussions of double resonance phenomena will be given in subsequent chapters.

In a spin decoupling experiment the sample is simultaneously irradiated with two radiofrequency fields. The "observing field" H_1 is the field used in all CW experiments to attain the resonance conditions for the different protons (Equation I-5) by sweeping slowly through the range of resonance frequencies (ν_1). The "decoupling field" H_2, on the other hand, has a fixed frequency ν_2 which corresponds to the resonance frequency of one of the multiplets in the spectrum. If the amplitude of the H_2-field is adequately chosen, this double resonance irradiation will selectively eliminate the spin–spin couplings with the nuclei which resonate at the frequency ν_2.

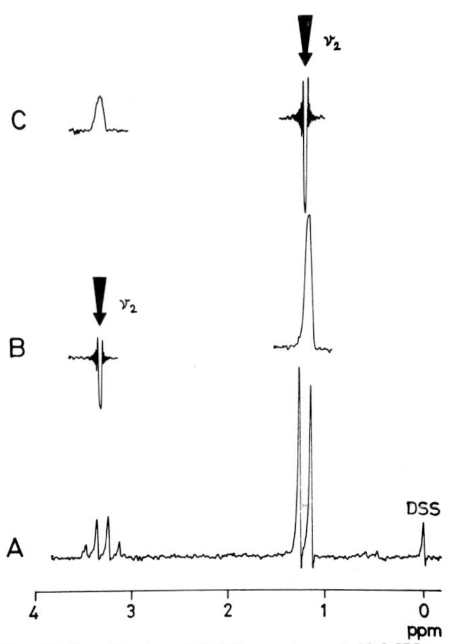

Fig. II-9. Proton NMR spectra at 60 MHz of a solution of L-alanine in D_2O, pD ≃ 13.0. A. CW spectrum. B. Spectrum recorded with double resonance irradiation of the quartet at 3.31 ppm. C. Spectrum recorded with double resonance irradiation of the doublet at 1.21 ppm.

Hence in the spectrum observed by sweeping ν_1, all the multiplet structures arising from spin–spin coupling with the ν_2-nuclei will disappear. This is illustrated in Fig. II-9. Because sizeable proton–proton spin–spin couplings occur only between protons separated by two or three chemical bonds, the neighboring groups of protons in a molecule can thus in principle be identified through homonuclear spin decoupling. Certain regions of a spectrum can be considerably simplified by spin decoupling, which may also be a great help for the spectral analysis.

In an INDOR experiment, the observing field H_1 is kept fixed at ν_1, the resonance frequency of one of the nuclei, i.e. "one sits with the H_1 field on an individual NMR line". The double irradiation frequency ν_2 is then swept through the spectrum, and one records the effect of this operation on the observed line. Again, as in the spin decoupling experiments, neighboring groups of protons can in principle be identified. The INDOR technique is often particularly helpful for resonance assignments in spectral regions which are crowded with a large number of resonances.

5. MOLECULAR RATE PROCESSES—THE TIME SCALE OF NMR EXPERIMENTS

In the amino acid spectra discussed so far we have noticed that the three methyl protons in alanine are magnetically equivalent because of the rapid intramolecular rotation about the $C^\alpha - C^\beta$ bond (Fig. II-1), and that the amino and carboxyl protons are not observed as separate peaks because of rapid exchange with protons or deuterium of the aqueous solvent. Other rate processes which can affect the NMR spectra include interconversions between different molecular conformations, ligand exchange reactions in intermolecular complexes, spin–spin coupling with rapidly relaxing nuclei and electron exchange in paramagnetic systems. In this section a qualitative discussion of these phenomena is presented. A more rigorous treatment of rate processes will be given in subsequent chapters.

It is a common feature of the rate processes accessible to NMR studies that the spin under observation finds itself jumping forth and back at random between different magnetically not equivalent environments. For simplicity of presentation we choose the case where a spin is jumping between two equally populated sites A and B where its resonance positions are respectively ν_A and ν_B, and the life times with respect to the exchange reaction are $\tau_A = \tau_B = \tau$. One then has

$$A(\tau_A) \xrightleftharpoons{\nu_e = \frac{1}{2\pi\tau}} B(\tau_B) \qquad \text{(II-5)}$$

the exchange frequency in Hz, ν_e, given in Equation II-5. To assess the influence of the exchange process on the NMR spectrum, one compares the jumping frequency ν_e with the separation of the resonance positions $|\nu_A - \nu_B|$, where the latter can be due to different chemical shifts δ_A and δ_B, different spin–spin couplings in the two environments, or a combination of different chemical shifts and coupling constants.

Fast and Slow Exchange Limits

There are two cases which are relatively easy to handle, called fast and slow exchange. In the fast exchange limit $\nu_e \gg |\nu_A - \nu_B|$, i.e. the spin is jumping so rapidly compared with the difference in resonance frequencies that it experiences on the NMR time scale the average of the two environments A and B. Therefore a single line at the average resonance position is observed. The influence of the exchange reaction is then often dominant over the spin relaxation in determining the line width $\delta\nu_{1/2}$ of the average resonance according to

$$\delta\nu_{1/2} \approx \frac{(\nu_A - \nu_B)^2}{\nu_e} \quad \text{fast exchange} \qquad \text{(II-6)}$$

This line broadening effect is illustrated in Fig. II-10 for different ratios of $|\nu_A - \nu_B| : \nu_e$.

In the slow exchange limit $\nu_e \ll |\nu_A - \nu_B|$, i.e. on the NMR time scale the spin sees only one of the two environments A and B, and hence two separate resonances are observed in the positions ν_A and ν_B. Again, the line widths $\delta_{1/2}$ of the two resonances can be determined predominantly by the exchange reaction according to

$$\delta\nu_{1/2} \approx \nu_e \quad \text{slow exchange} \qquad \text{(II-7)}$$

The slow exchange case is represented in Fig. II-10, which also shows some spectra of the intermediate exchange region, where $\nu_e \approx |\nu_A - \nu_B|$.

With these basic concepts on rate phenomena in NMR in mind, one can readily explain the magnetic equivalence of the methyl protons in alanine (Fig. II-1). On the one hand the differences between the resonance frequencies of the individual protons for any given rotation angle, which come from different chemical shifts and different spin–spin coupling with the C^α-proton (Equation II-3), are expected to be of the order of several Hz, while on the other hand the rotation rate about the $C^\alpha - C^\beta$ bond is much faster (see Appendix V). Hence a rotating methyl group represents a typical fast exchange case. The proton exchange between the amino and carboxyl groups of an amino acid and the solvent water molecules is another example of fast exchange. Because the water protons are present in a large excess,

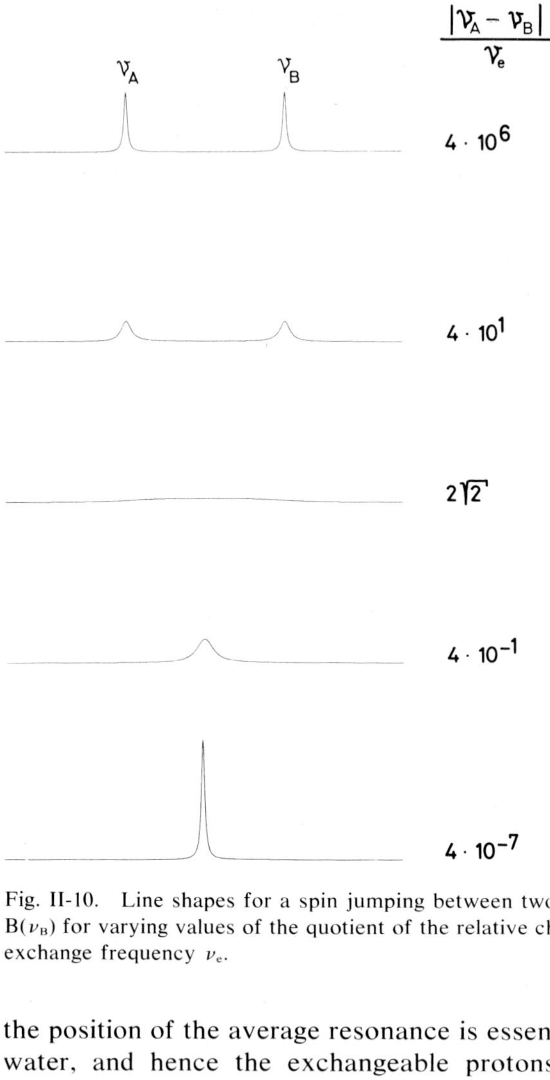

Fig. II-10. Line shapes for a spin jumping between two equally populated states A(ν_A) and B(ν_B) for varying values of the quotient of the relative chemical shift in Hz, $|\nu_A - \nu_B|$, and the exchange frequency ν_e.

the position of the average resonance is essentially identical to that of pure water, and hence the exchangeable protons of the amino acid do not manifest themselves noticeably in the NMR*.

*In a 0.1-M solution of an amino acid, there are ca. 1000 water protons for every carboxyl proton. With $\delta_{H_2O} = 4.8$ ppm and $\delta_{-COOH} = 12.0$ ppm, the position of the average resonance resulting from proton exchange between $-COOH$ and H_2O is

$$\delta_{av} = \frac{1000 \times 4.8 + 12.0}{1001} \text{ ppm} \approx 4.8 \text{ ppm}.$$

6. pH-DEPENDENCE OF THE ^{1}H-NMR IN AMINO ACIDS

The proton NMR spectra of the three differently protonated species in Equation II-1 are shown for alanine in Fig. II-11. It is seen that successive deprotonation of the carboxylic acid and the amino group cause upfield shifts of the C^α- and C^β-proton resonances. The approximate extents of these shifts for amino acids with non-ionizable side chains are given in Table II-5.

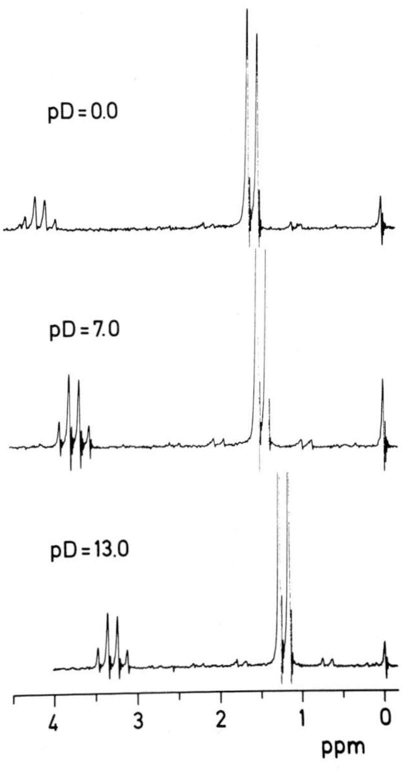

Fig. II-11. Proton NMR spectra at 60 MHz of a solution of L-alanine in D_2O at different pD.

pH-Titration with NMR

At all presently available magnetic field strengths H_0, the chemical shifts arising from different pH (Table II-5) are small compared to the rate of the proton exchange reactions II-1. During the entire pH-titration of an amino acid, one therefore observes only one spectrum, which corresponds to the

TABLE II-5

Approximate chemical shifts arising from the deprotonation reactions II-1 in an amino acid with a non-ionizeable aliphatic side chain.

	$^{\oplus}H_3N-CH(R)-COOH \rightarrow {}^{\oplus}H_3N-CH(R)-COO^{\ominus}$	$^{\oplus}H_3N-CH(R)-COO^{\ominus} \rightarrow H_2N-CH(R)-COO^{\ominus}$
αCH	-0.4 ppm	-0.5 ppm
βCH	-0.1 ppm	-0.3 ppm
γCH	-0.05 ppm	-0.1 ppm
δCH	-0.02 ppm	-0.04 ppm

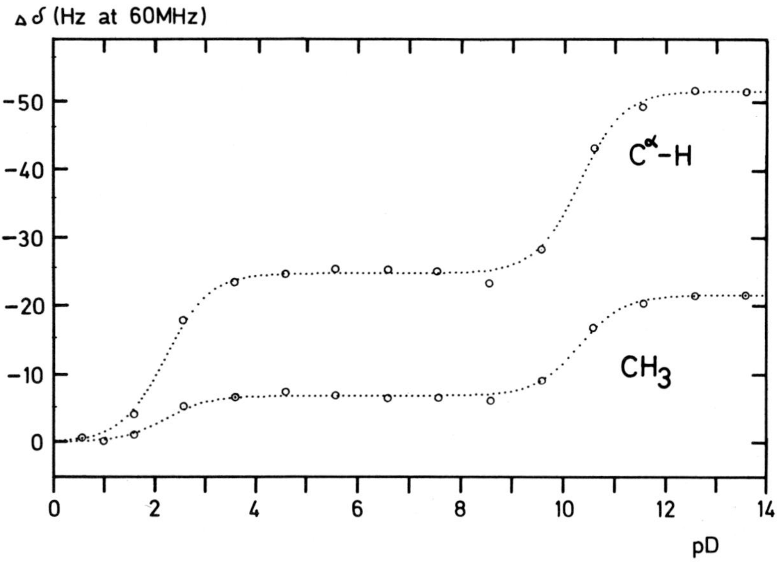

Fig. II-12. Variation of the chemical shifts of the α- and β-protons in alanine with pD.

weighted average of the species present in the solution, and the titration can thus be followed in the NMR spectrum (Fig. II-12).

Protonation of Amino Acid Side Chains

Additional pH-dependences are observed in the spectral features of amimo acids with ionizeable side chains. An important example in practice is the pH-dependence of the imidazole ring proton resonances. It may be recalled that the pK_a-value for the imidazole ring falls into the physiological pH-range (Fig. II-13) and that protonation–deprotonation reactions of histidine appear to be essential for numerous enzyme reactions.

Practical applications in protein studies have also been made of the manifestation of the deprotonation of tyrosine in the proton NMR (Equation II-8).

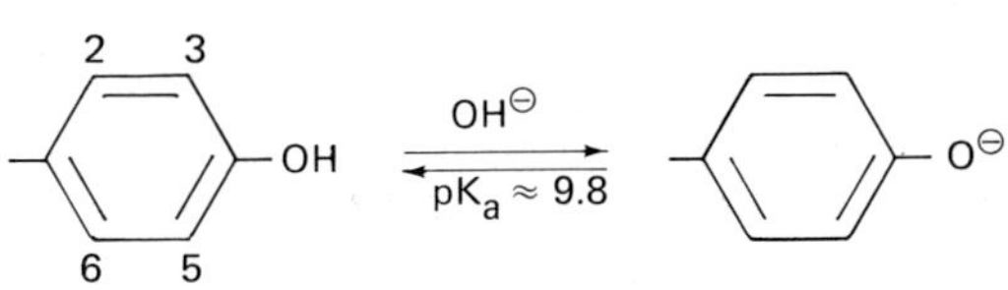

(II-8)

δ(3,5) 6.90 6.65

δ(2,6) 7.20 7.05

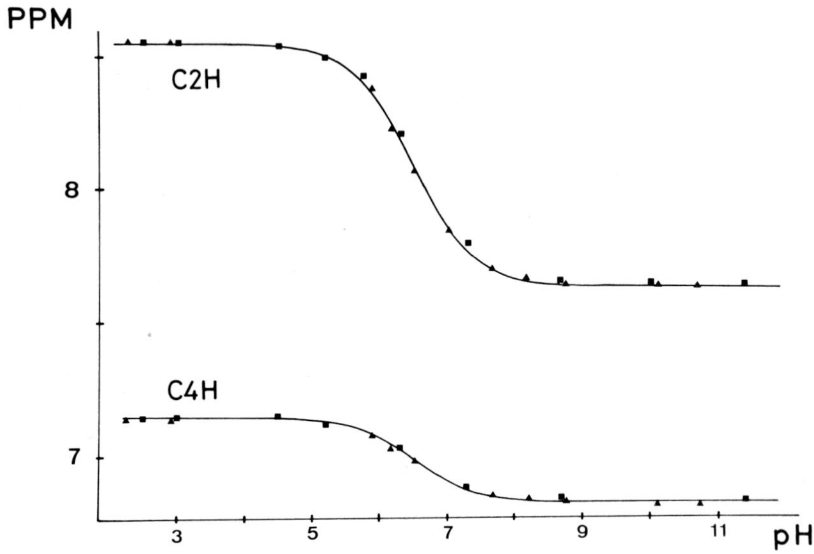

Fig. II-13. Chemical shift variation of the C-2 and C-4 imidazole ring protons of histidine during pH-titration. The data were obtained for the single histidyl residue in calcitonin M (from Masson[17]).

The protonation–deprotonation reactions of the side chain carboxylic acid groups in aspartic acid and glutamic acid, and the amino group in lysine have similar effects on the neighboring non-exchanging protons as do the corresponding reactions of the backbone groups. As in all the other cases mentioned, one has fast exchange between the different species present during titration, so that a single average spectrum is observed throughout.

In addition to the pH-dependence of the chemical shifts, small variations with pH have also been observed for some of the spin–spin coupling constants in amino acids. These observations have been tabulated by Roberts and Jardetzky[18].

7. PROTON NMR PARAMETERS FOR THE COMMON AMINO ACIDS

The foregoing sections have shown that the proton NMR spectra of free amino acids display in themselves a wealth of interesting features. However, starting with the first NMR studies of amino acids, the principal interest has been to obtain the basic spectral data needed for the investigation of peptides and proteins. The data tabulated in this section have also been selected mainly with the application to studies of peptides and proteins in mind.

Spectral parameters in the three solvents D_2O, TFA, and DMSO are included in the Tables II-6 to II-9. D_2O is the solvent of choice for most biological studies. Trifluoroacetic acid has the advantage of dissolving many peptides which are essentially insoluble in aqueous media, and is in certain cases also a suitable solvent for studies of the denatured forms of peptides and proteins. Dimethylsulfoxide is frequently used for NMR investigations of peptides, and is in certain cases a suitable system for studies of the denatured forms of biopolymers.

For work with peptides and proteins one is really more interested in the spectra of the amino acid residues than in those of the free amino acids. The latter are indeed more nearly models for the end groups rather than for the bulk of a polypeptide chain. As will be seen in the following, two different methods have in the past been chosen by different authors in attempts to obtain the NMR data of the amino acid residues. In one of these procedures certain empirical rules are used to predict the NMR chemical shifts for the amino acid residues from the corresponding data on the free amino acids. The second procedure, which is in principle more straightforward but also much more demanding from the experimental point of view, is to study the amino acid residues directly in oligopeptides.

Free Amino Acids in D_2O

The chemical shift data for the common amino acids in the different ionization states presented in Table II-6 are essentially those compiled by Roberts and Jardetzky[18]. Earlier data on amino acids in D_2O solution have been published by different authors[19,20], and in general they agree quite well with those given in Table II-6. An atlas of the amino acid spectra at 300 MHz has been assembled by Gelan and Anteunis, and can be obtained on request from these authors*.

In D_2O solution all the exchangeable protons are replaced by deuterium, which simplifies the spectra to some extent. Yet a complete spectral analysis as outlined in the Appendix II is still not easily obtained for all the amino acids. Thus in proline, and in the amino acids with long aliphatic side chains, i.e. in particular leucine, isoleucine, lysine and arginine, all the spectral details are not at present unambiguously understood. In as far as the proton–proton spin–spin coupling constants have been determined, they seem to vary only little with the ionization state[18], and are nearly identical in D_2O solution and in TFA (Table II-8).

---

*J. Gelan and M. Anteunis, Dept. of Organic Chemistry, State University of Gent, Gent 9000, Belgium.

TABLE II-6

Proton chemical shifts for the common α-amino acids in the different ionization states (II-1)*.

Amino acid	$^{\oplus}D_3N{-}C(R)(H){-}COOD$				$^{+}D_3N{-}C(R)(H){-}COO^{\ominus}$				$D_2N{-}C(R)(H){-}COO^{\ominus}$			
	αCH	βCH	Others		αCH	βCH	Others		αCH	βCH	Others	
Gly	3.94				3.55				3.22			
Ala	4.20	1.59			4.11				3.22	1.22		
Val	3.98	2.37	γCH_3	1.05	3.59	2.27	γCH_3	0.98	3.05	1.92	γCH_3	0.86
				1.07				1.33				0.92
Ile	4.05	2.06	γCH_2	2.42	3.65	1.95	γCH_2	1.29	3.10	1.64	γCH_2	1.24
				2.42				1.29				1.24
			γCH_3	1.04			γCH_3	1.00			γCH_3	0.93
			δCH_3	0.96			δCH_3	0.94			δCH_3	0.89
Leu	4.07	1.80	γCH	1.80	3.72	1.70	γCH	1.70	3.27	1.46	γCH	1.68
		1.80	δCH_3	0.96		1.70	δCH_3	0.96		1.42	δCH_3	0.92
				0.96				0.94				0.90
Ser	4.22	4.07			3.84	3.95			3.35	3.75		
		4.03				3.95				3.72		
Thr	4.42	4.03	γCH_3	1.36	3.58	4.25	γCH_3	1.32	3.10	3.95	γCH_3	1.20
Pro	4.45	2.41	γCH_2	2.06	4.11	2.31	γCH_2	2.00	3.46	2.12	γCH_2	1.72
		2.17	δCH_2	3.42		2.11	δCH_2	3.40		1.72	δCH_2	2.74
								3.33				3.02
Asp	4.39	3.14			4.09	3.01			3.56	2.64		
		3.14				3.00				2.29		
Glu	4.16	2.28	γCH_2	2.64	3.81	2.15	γCH_2	2.55	3.23	1.84	γCH_2	2.21
		2.24				2.15				1.84		
Lys	4.08	1.94	γCH_2	1.52	3.46	1.69	γCH_2	1.43	3.24	1.53	γCH_2	1.28
		1.94	δCH_2	1.71		1.69	δCH_2	1.69		1.53	δCH_2	1.40
			ϵCH_2	3.06			ϵCH_2	2.99			ϵCH_2	2.60

Arg	4.13	2.02	γCH_2	1.88	3.21	1.63	γCH_2	1.63	3.19	1.61	γCH_2	1.61
		2.02	δCH_2	3.27		1.63	δCH_2	3.21		1.61	δCH_2	3.19
Asn	4.38	3.06			4.00	2.92			3.59	2.67		
		3.06				2.87				2.44		
Gln	4.18	2.27	γCH_2	2.55	3.77	2.14	γCH_2	2.45	3.27	1.86	γCH_2	2.31
		2.27				2.14				1.86		
Met	4.24	2.28	γCH_2	2.69	3.85	2.15	γCH_2	2.97	3.23	1.88	γCH_2	2.56
		2.28	SCH_3	2.12		2.15	SCH_3	2.13		1.88	SCH_3	2.12
Cys	4.31	3.16			3.98	3.08			3.56	2.98		
		3.11				3.05				2.80		
$(CyS)_2$	4.55	3.47			—	—			3.58	3.13		
		3.38								2.92		
			Ring protons				Ring protons				Ring protons	
Trp[a]	4.43	3.45	C2H	7.38	4.05	3.45	C2H	7.31	3.59	3.02	C2H	6.72
		3.50	C4H	7.73		3.31	C4H	7.72		3.19	C4H	7.26
			C5,6H	7.28			C5,6H	7.28			C5,6H	6.72
			C7H	7.59			C7H	7.52			C7H	7.02
Phe[a]	4.38	3.33	Ring	7.37	3.98	3.26	Ring	7.37	3.51	2.99	Ring	7.33
		3.26				3.13				2.86		
Tyr[a]	3.32	3.24	C3,5H	6.90	3.93	3.05	C3,5H	6.90	3.44	2.72	C3,5H	6.67
		3.19	C2,6H	7.18		3.17	C2,6H	7.19		2.88	C2,6H	7.04
His[a]	4.45	3.49	C4H	7.49	3.98	3.19	C4H	7.05	3.51	2.97	C4H	6.93
		3.49	C2H	8.70		3.14	C2H	7.75		2.84	C2H	7.68

*The solvents used were DCl in D_2O, D_2O, and NaOD in D_2O. The data are in ppm from internal DSS. (For comparison with the chemical shifts vs. TMS measured in different solvents, add 0.47 ppm to the shifts given in this table).
[a]Numeration of the aromatic amino acids used throughout this book.

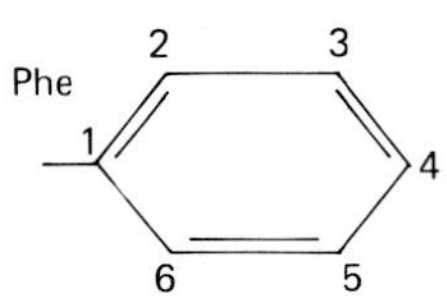

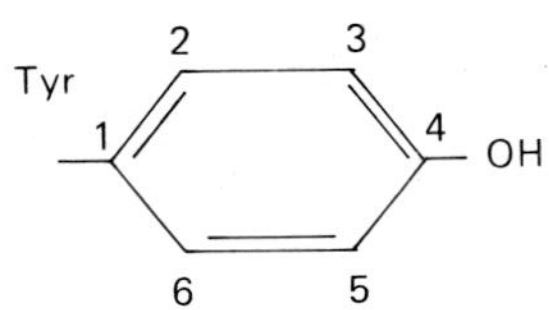

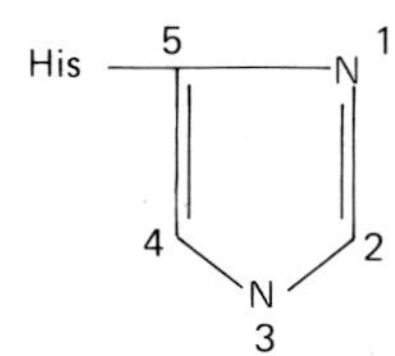

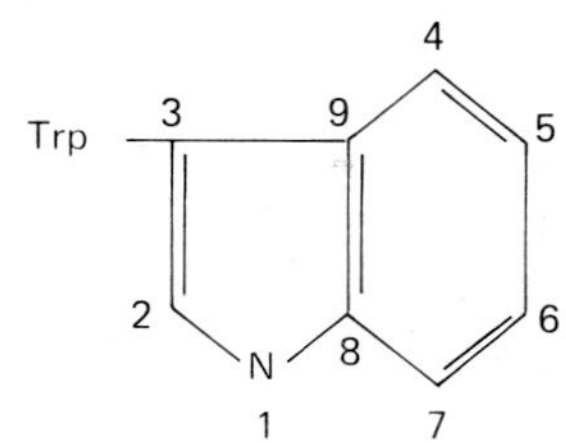

TABLE II-7

Proton NMR chemical shifts from internal DSS and line widths of amino acid residues used for computing the NMR spectra of random coil proteins in D_2O*.

Amino acid	Proton type	Number of protons per residue	δ (ppm from DSS)	$\delta\nu_{1/2}$ (Hz at 200 MHz)
Gly	CH_2	2	4.0	10
Ala	CH	1	4.6	15
	CH_3	3	1.4	10
Val	αCH	1	4.1	10
	βCH	1	2.2	15
	CH_3	6	0.9	10
Ile	αCH	1	4.1	10
	βCH	1	1.9	10
	γCH_3	1	1.1	15
	γCH_2	1	1.4	15
	δCH_3	6	0.9	10
Leu	αCH	1	4.2	15
	$CH_2 + \gamma CH$	3	1.6	10
	CH_3	6	0.9	10
Ser	αCH	1	4.3	15
	CH_2	2	3.9	10
Thr	αCH	1	4.1	10
	βCH	1	4.3	15
	CH_3	3	1.2	10
Pro	CH	1	4.6	10
	$\beta CH_2 + \gamma CH_2$	4	2.1	20
	δCH_2	2	3.3	15
Asp	CH	1	4.6[a]	15
	CH_2	2	2.7[a]	30
Glu	CH	1	4.3	15
	βCH_2	2	2.0[a]	15
	γCH_2	2	2.3[a]	10
Lys	CH	1	4.0	15
	$\beta CH_2 + \delta CH_2$	4	1.7	20
	γCH_2	2	1.4[a]	15
	ϵCH_2	2	3.0[a]	10
Arg	CH	1	3.7	15
	βCH_2	2	1.8	20
	γCH_2	2	1.7	15
	δCH_2	2	3.2	10

TABLE II-7 (*contd.*)

Amino acid	Proton type	Number of protons per residue	δ (ppm from DSS)	$\delta\nu_{1/2}$ (Hz at 200 MHz)
Asn	CH	1	4.5	15
	CH_2	2	2.8	20
Gln	CH	1	4.3	15
	βCH_2	2	2.1	15
	γCH_2	2	2.3	10
Met*	CH	1	4.4	15
	βCH_2	2	2.1	15
	γCH_2	2	2.7	10
	SCH_3	3	2.1	3
Cys	CH	1	(4.5)	15
	CH_2	2	3.0[(a)]	15
Trp	αCH	1	4.6	15
	CH_2	2	3.4	15
	Indole C2	1	7.2	5
	C5,C6	2	7.1	10
	C4,C7	1 + 1	7.4/7.5	10
Phe	CH	1	4.5	15
	CH_2	2	3.1	20
	Ring	5	7.3	15
Tyr	CH	1	4.4	15
	CH_2	2	3.0	20
	Ring C3,C5	2	6.8[(a)]	10
	C2,C4	2	7.1[(a)]	10
His	CH	1	4.5	15
	CH_2	2	3.2	20
	Ring C2	1	7.1[(a)]	5
	C6	1	7.9[(a)]	5

*The resonance positions marked (a) are pH-dependent (see Table II-6).

Amino Acid Residues in D_2O

On the basis of systematic studies with glycyl-containing peptides, Nakamura and Jardetzky[21] found that mainly only the C^{α}-proton resonances are sizeably affected by the incorporation of an amino acid into a peptide chain. The C^{α}-proton resonance shift arising from the replacement of the free amino and carboxylic acid groups in the zwitterionic form of the amino acids by peptide bonds in a random coil peptide chain is ≈ 0.5 ppm.

TABLE II-8

Proton chemical shifts and proton–proton coupling constants in the common amino acids in solution in TFA or d-TFA*.

Amino acid	Chemical shifts δ (ppm from TMS)					Spin–spin coupling constants J (Hz)[a]				
	$NH_3^{\oplus}$	αCH	βCH	Others		$^3J_{HN\alpha}$	$^3J_{\alpha\beta}$	$^2J_{\beta\beta}$	Others	
Gly	7.47	4.28				5.7				
Ala	7.41	4.49	1.86			5.8	7.3			
Val	7.33	4.32	2.60	γCH_3	1.25	5.7	4.2		$^3J_{\beta\gamma}$	6.9
					1.25					
Ile	7.35	4.41	2.28	γCH_2	1.70	5.5	3.6	13.6	$^3J_{\beta\gamma CH_2}$	7.2
					1.55					
				γCH_3	1.21				$^3J_{\beta\gamma CH_3}$	7.0
				δCH_3	1.10				$^3J_{\gamma\delta}$	7.0
Leu	7.38	4.48	(2.00)	γCH	(2.00)	5.5	6.7	....	$^3J_{\gamma\delta}$	5.8
			(2.00)	δCH_3	1.11					
					1.10					
Ser	7.70	4.65	4.56			6.0	4.0	13.5		
			4.51							
Thr	7.63	4.44	4.82	γCH_3	1.67	5.5	4.5		$^3J_{\beta\gamma}$	6.5
Pro	8.08	4.78	2.74	γCH_2	2.35	5.5	8.0	14.0	$^3J_{\beta\gamma}$	6.8
	7.71		2.50		2.29					
				δCH_2	3.82					
					3.76					
Asp	7.73	4.76	3.54			5.1	4.7	....		
			3.54							
Glu	7.71	4.60	2.63	γCH_2	3.01	5.5	5.6	15.5	$^3J_{\beta\gamma}$	6.2
			2.55							
Lys	7.60	4.52	2.35	γCH_2	(1.82)	5.8	5.6	15.0	$^3J_{\beta\gamma}$	(6.0)
			2.26	δCH_2	2.00					
				ϵCH_2	3.38				$^3J_{\delta\epsilon}$	(6.0)
				ϵ-$NH_3^{\oplus}$	6.97				$^3J_{\epsilon NH_3^{\oplus}}$	(6.0)

Arg	7.60	4.46	2.34	γCH_2	(2.06)	5.5	5.3	15.0	$^3J_{\beta\gamma}$	(6.0)
			2.25	δCH_2	3.43				$^3J_{\gamma\delta}$	6.5
				ϵNH	6.50				$^3J_{\delta\epsilon}$	5.3
				$NH_2CNH_2^{\oplus}$	6.19					
Asn	7.79	4.72	3.48	$CONH_2$	7.39	5.5	5.1	....		
			3.48							
Gln	7.90	4.55	2.64	γCH_2	2.99	5.0	5.9	15.5	$^3J_{\beta\gamma}$	6.2
			2.55	$CONH_2$	7.58					
Met	7.73	4.67	2.64	γCH_2	2.96	5.5	6.0	15.7	$^3J_{\beta\gamma}$	6.5
			2.50	SCH_3	2.27				$^4J_{\gamma SCH_3}$	(0.0)
Cys	7.58	4.68	3.41	SH	1.84	5.3	5.0	15.5	$^3J_{\beta SH}$	9.1
			3.36							
				Ring protons					Ring protons	
Trp	7.05	4.71	3.79	C2H	7.24	....	6.0	15.5		
			3.57	C4,7H	7.58					
					7.42					
				C5,6H	7.29					
					7.19					
				NH	....					
Phe	7.33	4.68	3.64	Ring	7.43	....	6.5	14.5		
			3.37							
Tyr	7.37	4.64	3.60	C3,5H	7.03	....	6.5	15.0	J_{ortho}	8.5
			3.34	C2,6H	7.27					
His	7.82	4.91	3.87	C2H	8.73	....	6.4	....	$J_{2,4}$	1.4
			3.87	C4H	7.66					
				NH	....					

*The *J*-values given for the couplings with methylene groups correspond to the average of the couplings with the individual methylene protons. Numbers in parenthesis are uncertain.

[a]The following abbreviated notations are used:

$^3J_{H_N H_C^\alpha} \equiv {}^3J_{HN\alpha}$ $\quad$ $^3J_{H_C^\alpha H_C^\beta} \equiv {}^3J_{\alpha\beta}$

$^2J_{H_C^\beta H_C^\beta} \equiv {}^2J_{\beta\beta}$ $\quad$ $^3J_{H_C^\beta H_C^\gamma} \equiv {}^3J_{\beta\gamma}$

$^3J_{H_C^\epsilon H_N^\zeta} \equiv {}^3J_{\epsilon NH_3^\oplus}$ (in lysine) $\quad$ $^4J_{H_C^\gamma H_C^\epsilon} \equiv {}^4J_{\gamma SCH_3}$ (in methionine)

Additional chemical shifts of the C^{α}-proton resonances which can arise from nearest neighbor influences other than protonation–deprotonation reactions at the end groups, were found to be of the order 0.0 to 0.2 ppm. In the proton NMR of the side chain resonances, the overall chemical shifts which can result from the incorporation of the amino acid into a peptide chain are much smaller, i.e. from 0.0 to ca. 0.3 ppm, whereby the larger shifts occur only in certain C^{β}-proton resonances.

These rules are implicit in a very useful compilation of chemical shift data by McDonald and Phillips[22]. On the basis of NMR studies in the free amino acids and in a number of dipeptides, these authors proposed a set of proton NMR parameters for computing NMR spectra of random coil proteins, and refined these data through actual comparison with a considerable number of random coil proteins. As is seen in Table II-7, McDonald and Phillips made the realistic assumption that different lines in the spectrum of an amino acid which are separated by less than ca. 15 Hz either by their different chemical shifts or by spin–spin couplings would not be resolved in a protein ^{1}H NMR spectrum at 220 MHz. The presence of unresolved fine structure is then accounted for in the resonance line widths $\delta\nu_{1/2}$. To complement the data of McDonald and Phillips, chemical shifts for the C^{α}-protons were taken from Table II-6, and the above mentioned correction of 0.5 ppm applied to allow for the replacement of the free terminal groups by peptide bonds.

Free Amino Acids in Trifluoroacetic Acid

Since the proton NMR line of TFA is located outside the spectral region of the amino acid resonances (Table II-4), studies are readily practicable in protonated TFA, where the exchangeable protons can also be observed, and in deuterated TFA, where the spectra are somewhat simplified because of the replacement of the exchangeable protons by deuterium. A combination of NMR experiments at 60 MHz and 220 MHz in TFA and *d*-TFA has resulted in the data given in Table II-8, which were taken from a paper by Bak et al.[23]. An earlier compilation of corresponding data by Bovey and Tiers[24] had been based on measurements at 40 MHz. Both groups of authors pointed out that special care had to be taken in the experiments with cysteine, serine, threonine, and tryptophan. When these amino acids were dissolved in anhydrous TFA, the NMR spectra revealed the occurrence of chemical reactions with the solvent.

The data of Table II-8 were used by Bak et al.[23] for the interpretation of the NMR spectra of insulin in TFA, which is assumed to be in a random coil form. They found a satisfactory agreement between the experimental and the predicted spectrum when the chemical shifts of the free amino acids

TABLE II-9

Proton chemical shifts and proton–proton coupling constants of the common amino acid residues measured in d_6-DMSO solutions of the protected linear tetrapeptides F_3C–CO–Gly(1)–Gly(2)–X(3)–L–Ala(4)–OCH_3, where X stands for the different amino acids*.

Amino acid residue	Chemical shifts δ (ppm from TMS)					Spin–spin coupling constants J (Hz)[c]			
	NH	αCH	βCH	Others		$^3J_{HN\alpha}$	$^3J_{\alpha\beta}$	Others	
Gly	8.15	3.76				6.0			
		3.76							
Ala	8.04	4.34	1.22			7.5	7.5		
Val	7.88	4.26	1.97	γCH_3	0.98	9.0	7.0	$^3J_{\beta\gamma}$	4.5
					0.92				
Ile	8.26	4.25	( )	γCH_2	( )	9.0			
					( )				
				γCH_3 }	0.86				
				δCH_3 }	0.84				
Leu	7.95	4.37	( )	γCH	( )	8.0		$^3J_{\gamma\delta}$	5.5
			( )	δCH_3	0.89				
					0.85				
Ser	7.98	4.34	3.60	OH	5.92	8.0	3.5	($^3J_{\beta OH}$	5.5)
			3.60						
Thr	7.83	4.22	3.93	γCH_3	1.07	8.0	5.0	$^3J_{\beta\gamma}$	6.0
				OH	4.79			$^3J_{\beta OH}$	5.0
Asp	8.21	4.63	(2.70)	COOH	12.23	8.0	(5.0)		
			(2.44)				(8.0)		
Glu	8.05	4.34	1.87	γCH_2	2.26	8.0	7.5	$^3J_{\beta\gamma}$	8.0
			1.87		2.26				
				COOH	12.08				

TABLE II-9 (*contd.*)

Amino acid residue	Chemical shifts δ (ppm from TMS)					Spin–spin coupling constants J (Hz)[c]			
	NH	αCH	βCH	Others		$^3J_{HN\alpha}$	$^3J_{\alpha\beta}$	Others	
Lys	8.06	4.29	(1.57)	γCH_2	(1.40)	8.0			
			(1.57)		(1.40)				
				δCH_2	(1.57)				
					(1.57)				
				ϵCH_2	2.73				
					2.73				
				$NH_3^{\oplus}$	8.03				
Arg[a]	8.07	4.38	(1.56)	γCH_2	(1.56)				
			(1.56)		(1.56)				
				δCH_2	3.17				
					3.17				
Asn	8.14	4.61	(2.60)	$CONH_2$	7.28	7.0	(5.0)		
			(2.38)		6.88		(8.0)		
Gln	8.06	4.32	(1.89)	γCH_2	(2.13)	7.5	(5.0)	($^3J_{\beta\gamma}$	8.0)
			(1.80)		(2.13)		(8.0)		
				$CONH_2$	7.24				
					6.76				
Met	8.07	4.41	1.87	γCH_2	2.47	8.0	6.5	$^3J_{\beta\gamma}$	8.0
			1.87		2.47			($^4J_{\gamma SCH_3}$	0.0)
				SCH_3	2.05				
Cys_2	8.29	4.64	3.16			7.0	5.0		
			2.88				9.5		

				Ring protons				Ring protons	
Trp	8.09	4.58	3.16	C2H	7.17	8.0	4.5	$J_{4,5}$	8.0
			2.93	C4H	7.65		8.5	$J_{5,6}$	8.5
				C5H	7.01			$J_{6,7}$	9.0
				C6H	7.10				
				C7H	7.35				
				NH	10.76				
Phe[b]	8.14	4.56	3.03	ØH	7.26	8.5	4.0		
			2.75				10.0		
Tyr	8.04	4.46	2.91	C3,5H	6.64	8.0	4.0	J_{ortho}	8.5
			2.64	C2,6H	7.02		9.0		
				OH	9.13				
His	8.31	4.66	3.22	C2H	8.35	8.5	5.0		
			2.96	C4H	7.35		7.5		
				NH	14.20				

*No data are given for Pro because there are two different conformations of the tetrapeptide, with the Gly(2)–Pro(3) bond in the cis and trans form, respectively. For the parameters given in brackets, the limited spectral resolution did not allow an unambiguous confirmation by spectrum simulation.

(a) Arg was studied in the form of the N^{ω}-nitro derivative.

(b) The aromatic protons of Phe have slightly different chemical shifts. The number given corresponds to the single line observed at 100 MHz.

(c) The same notation is used as in Table II-8.

were used with the appropriate corrections for end group effects, as discussed above.

Amino Acid Residues in d_6-DMSO

The spectra of the twenty common amino acid residues given in Table II-9 were obtained from investigations at 100 MHz and 270 MHz of the protected linear tetrapeptides

$$F_3C-\overset{\overset{\displaystyle O}{\|}}{C}-\text{Gly(1)}-\text{Gly(2)}-\text{X(3)}-\text{L-Ala(4)}-OCH_3$$

where X stands for the 20 common amino acids.[25] These data can be used without further modifications as a basis for the interpretation of the proton NMR spectra of peptides and proteins in DMSO-solution.

The experiments with the tetrapeptides (II-9) yielded also data on the effects of the amino acid sequence on the proton NMR in random coil peptides[25]. The chemical shifts observed for the C^α-protons of Gly(2) next to the 20 different amino acid residues X are all between 3.76 and 3.86 ppm. The corresponding range of chemical shifts for the C^α-proton of Ala(4) is 4.21 to 4.32 ppm, and for the C^β-protons of Ala(4) it is 1.28 to 1.31 ppm. Hence as in the above mentioned investigations[21,22], it was found that specific amino acid sequence effects on the chemical shifts of the side chain proton NMR are rather small.

8. BIBLIOGRAPHY

1 Bovey F. A., Nuclear Magnetic Resonance Spectroscopy. Academic Press, New York, 1969
2 Pauling L., J. Chem. Phys. *4*, 673 (1936)
3 Pople J. A., J. Chem. Phys. *24*, 1111 (1956)
4 Johnson C. E. and Bovey F. A., J. Chem. Phys. *29*, 1012 (1958)
5 Karplus M., J. Chem. Phys. *30*, 11 (1959)
6 Karplus M., J. Amer. Chem. Soc. *85*, 2870 (1963)
7 Pople J. A., Schneider W. G. and Bernstein H. J., High Resolution Nuclear Magnetic Resonance. McGraw-Hill, New York, 1959, p. 117
8 See Ref. 1, p. 368
9 See Ref. 7, p. 98
10 Swalen J. D., Progr. Nucl. Magnetic Resonance Spectroscopy *1*, 205 (1966)
11 E.g. see Ref. 1, Appendix D, pp. 275–349
12 Ernst R. R., Adv. Magnetic Resonance *2*, 1 (1966)
13 Glasoe P. K. and Long F. A., J. Phys. Chem. *64*, 188 (1960)
14 Llinás M. and Klein M. P., J. Amer. Chem. Soc. 1975 (submitted)
15 Hewitt R. C., Rev. Sci. Instr. *38*, 831 (1967)

16 See Ref. 7, p. 80
17 Masson A., "Konformationsstudien an Calcitonin M und am Trypsin Inhibitor BPTI". Ph.D. Thesis, ETH, Zürich (1974)
18 Roberts G. C. K. and Jardetzky O., Adv. Protein Chem. *24*, 447 (1970)
19 Jardetzky O. and Jardetzky C. D., J. Biol. Chem. *233*, 383 (1958)
20 Mandel M., J. Biol. Chem. *240*, 1586 (1965)
21 Nakamura A. and Jardetzky O., Proc. Nat. Acad. Sci. *58*, 2212 (1967)
22 McDonald C. C. and Phillips W. D., J. Amer. Chem. Soc. *91*, 1513 (1969)
23 Bak B., Dambmann C., Nicolaisen F., Pedersen E. J. and Bhacca N. S., J. Mol. Spectr. *26*, 78 (1968)
24 Bovey F. A. and Tiers V. D., J. Amer. Chem. Soc. *81*, 2870 (1959)
25 Bundi A., Grathwohl Ch., Hochmann J., Keller R. M., Wagner G. and Wüthrich K., J. Magn. Res. *18*, 191 (1975)
26 Slichter C. P., "Principles of Magnetic Resonance". Harper and Row, New York, 1963, p. 13
27 Emsley J. W., Feeney J. and Sutcliffe L. H., "High Resolution Nuclear Magnetic Resonance Spectroscopy". Pergamon Press, New York, 1967, p. 362

APPENDIX II

COMPUTATION OF STRONGLY COUPLED NMR SPECTRA. AB AND ABX SPIN SYSTEMS

In Section II.3, it was mentioned that the NMR spectra of strongly coupled spin systems cannot usually be analyzed by mere inspection, but require comparison with spectra computed with a suitable set of chemical shifts and spin–spin coupling constants. Here, we shall briefly survey the quantum mechanical background of these spectral simulations.

First, the complete computation will be presented for a system of two spins 1/2. This is a suitable system to illustrate the different aspects of the calculation, and it is also of considerable interest for work with peptides. The second part of the Appendix will be devoted to three-spin systems. Table II-2 has shown that three-spin systems are quite common in amino acids, and in the following chapter we will see that the analysis of ABX spin systems is an important aspect of NMR work with peptides. Even though proper solutions of strongly coupled systems with three or more spins require in all cases a complete quantum mechanical calculation, we will see that certain patterns of the ABX spectra become recognizable with experience.

A spin system may be described by a spin Hamilton operator. Using a suitable set of basis functions for the system treated, one evaluates the energies and eigenfunctions corresponding to the eigenstates of the operator. These stationary states define all the resonance frequencies which may occur in the system. The probabilities for the individual resonance frequencies to be manifested in the spectrum, i.e. the intensities of the individual resonance lines, are given by the probabilities of the occurrence of transitions between the different pairs of stationary states in the system.

Two Spin-Systems

The spin Hamiltonian for a system of two non-equivalent protons is

$$\mathcal{H} = -\gamma \frac{h}{2\pi} H_0[(1-\sigma_1)I_{z1} + (1-\sigma_2)I_{z2}] + J\boldsymbol{I}_1 \cdot \boldsymbol{I}_2 \qquad \text{(II-10)}$$

where γ is the gyromagnetic ratio for protons, σ_1 and σ_2 are the shielding constants for the two spins, I_{z1} and I_{z2} are the z-components of the nuclear spin operators $\boldsymbol{I}_1$ and $\boldsymbol{I}_2$, and J is the spin–spin coupling constant. A convenient set of basis functions is obtained by taking all the possible combinations of α- and β-states for the two spins

$$\phi_1 = |\alpha\alpha\rangle \quad \phi_2 = |\alpha\beta\rangle \quad \phi_3 = |\beta\alpha\rangle \quad \phi_4 = |\beta\beta\rangle \tag{II-11}$$

To compute the matrix elements of the spin Hamiltonian, it is useful to rewrite the operator $\boldsymbol{I}_1 \cdot \boldsymbol{I}_2$ as

$$\boldsymbol{I}_1 \cdot \boldsymbol{I}_2 = I_{z1}I_{z2} + \tfrac{1}{2}(I_1^+I_2^- + I_1^-I_2^+) \tag{II-12}$$

where I^+ and I^- are the shift operators[26].

For the operators on the right hand side of Equation II-12 one has

$$\begin{aligned} I_z|\alpha\rangle &= \tfrac{1}{2}|\alpha\rangle \\ I_z|\beta\rangle &= \tfrac{1}{2}|\beta\rangle \\ I^+|\alpha\rangle &= 0 \\ I^+|\beta\rangle &= |\alpha\rangle \\ I^-|\alpha\rangle &= |\beta\rangle \\ I^-|\beta\rangle &= 0 \end{aligned} \tag{II-13}$$

Using the relations II-13, one obtains for the matrix elements of the spin Hamiltonian II-10

$$\begin{array}{c|cccc} & |\alpha\alpha\rangle & |\alpha\beta\rangle & |\beta\alpha\rangle & |\beta\beta\rangle \\ \hline |\alpha\alpha\rangle & \frac{1}{2\pi}\gamma H_0\left(-1+\frac{1}{2}\sigma_1+\frac{1}{2}\sigma_2\right)+\frac{1}{4}J & 0 & 0 & 0 \\ |\alpha\beta\rangle & 0 & \frac{1}{4\pi}\gamma H_0(\sigma_1-\sigma_2)\frac{1}{4}J & \frac{1}{2}J & 0 \\ |\beta\alpha\rangle & 0 & \frac{1}{2}J & -\frac{1}{4\pi}\gamma H_0(\sigma_1-\sigma_2)\frac{1}{4}J & 0 \\ |\beta\beta\rangle & 0 & 0 & 0 & \frac{1}{2\pi}\gamma H_0\left(1-\frac{1}{2}\sigma_1-\frac{1}{2}\sigma_2\right)+\frac{1}{4}J \end{array} \tag{II-14}$$

From the energy matrix II-14 one sees then that ϕ_1 and ϕ_4 are already eigenfunctions of the Hamiltonian II-10, with the energies given in the matrix. Because ϕ_2 and ϕ_3 are mixed by the spin Hamiltonian, one has to solve the secular equation II-15, where $\Delta\delta = (1/2\pi)\gamma H_0(\sigma_1 - \sigma_2)$ is the relative chemical shift between the two protons in Hz, to obtain the other two energies and eigenfunctions.

$$\begin{vmatrix} \frac{1}{2}\Delta\delta - \frac{1}{4}J - E & \frac{1}{2}J \\ \frac{1}{2}J & -\frac{1}{2}\Delta\delta - \frac{1}{4}J - E \end{vmatrix} = 0 \qquad \text{(II-15)}$$

Using the conventional expressions II-16

$$\begin{aligned} C &= \tfrac{1}{2}(J^2 + (\Delta\delta)^2)^{1/2} \\ C \cos 2\theta &= \tfrac{1}{2}\Delta\delta \\ C \sin 2\theta &= \tfrac{1}{2}J \end{aligned} \qquad \text{(II-16)}$$

one has then the energies E (in units of the Planck constant h) and eigenfunctions ψ for the four states given in II-17

$$\begin{aligned} E_1 &= \frac{1}{2\pi}\gamma H_0\left(-1 + \frac{1}{2}\sigma_1 + \frac{1}{2}\sigma_2\right) + \frac{1}{4}J & \psi_1 &= |\alpha\alpha\rangle \\ E_2 &= -C - \frac{1}{4}J & \psi_2 &= \cos\theta|\alpha\beta\rangle - \sin\theta|\beta\alpha\rangle \\ E_3 &= C - \frac{1}{4}J & \psi_3 &= \sin\theta|\alpha\beta\rangle + \cos\theta|\beta\alpha\rangle \\ E_4 &= \frac{1}{2\pi}\gamma H_0\left(1 - \frac{1}{2}\sigma_1 - \frac{1}{2}\sigma_2\right) + \frac{1}{4}J & \psi_4 &= |\beta\beta\rangle \end{aligned} \qquad \text{(II-17)}$$

For a complete description of the spectrum one further needs the transition probabilities, and hence the line intensities, for the transitions between the different stationary states. Since the radiofrequency field is applied perpendicular to H_0, the intensity $I_{\psi\to\psi'}$ of the transition between the states ψ and ψ' is given by

$$\begin{aligned} I_{\psi\to\psi'} &\propto |\langle\psi|I_{x1} + I_{x2}|\psi'\rangle|^2 \\ &\propto |\langle\psi|I_1^+ + I_2^+ + I_1^- + I_2^-|\psi'\rangle|^2 \end{aligned} \qquad \text{(II-18)}$$

Using the operator Equations II-13 in Equation II-18, one finds that there are four "allowed" transitions in an AB system, with the transition probabilities listed in Table II-10. The corresponding transition frequencies are obtained as the energy differences between corresponding pairs of the energy levels II-17. In the Table II-10, the transition frequencies are given relative to the mean chemical shift, $(\gamma/2\pi)H_0(1 - \frac{1}{2}\sigma_1 - \frac{1}{2}\sigma_2)$.

The schematic AB spectrum in Fig. II-14, where it was assumed that the spin A is more shielded, shows that once an AB system is recognized as such, the parameters δ_A, δ_B, and J can readily be obtained from the following features: (i) There are two pairs of lines with spacing J. (ii) The spacing of the center peaks is $2C - |J|$. (iii) The spacing of the outer peaks is

TABLE II-10
Transition Frequencies and Relative Intensities for the AB System.

Transition	Transiton frequency (rel. to the center of the spectrum) $C = \frac{1}{2}\sqrt{J^2 + \Delta\delta^2}$	Intensity $\sin 2\theta = \frac{J}{2C}$
$4 \rightarrow 2$	$C + \frac{1}{2}J$	$1 - \sin 2\theta$
$3 \rightarrow 1$	$C - \frac{1}{2}J$	$1 + \sin 2\theta$
$4 \rightarrow 3$	$-C + \frac{1}{2}J$	$1 + \sin 2\theta$
$2 \rightarrow 1$	$-C - \frac{1}{2}J$	$1 - \sin 2\theta$

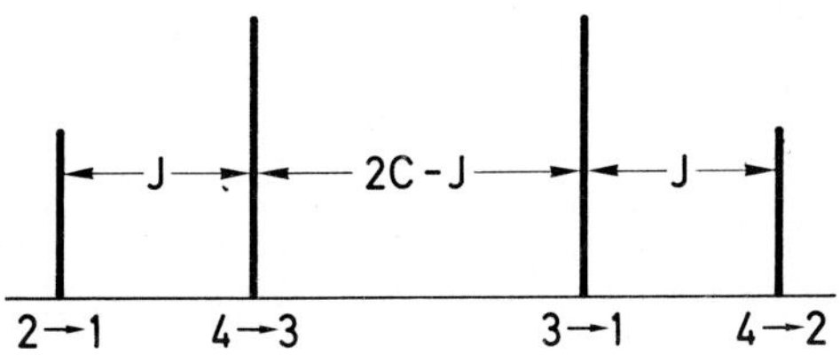

Fig. II-14. Schematic AB spectrum.

$2C + |J|$. (iv) The pattern is symmetrical about the position corresponding to the mean chemical shift, $\delta_{mean} = (\delta_A + \delta_B)/2$.

The expressions for the transition frequencies and intensities in Table II-10 provide a ready explanation for the spectral features observed in Fig. II-5, where the results of the above calculation are shown for AB systems with different quotients $\Delta\delta/J$. (i) If $J = 0$, the pair of transitions $4 \rightarrow 2$ and $3 \rightarrow 1$ are degenerate and give a single spectral line, and the same holds for the pair of transitions $4 \rightarrow 3$ and $2 \rightarrow 1$ (Fig. II-5a). (ii) If $\Delta\delta \gg J$, we have that $C \approx \Delta\delta$, and $\sin 2\theta \approx 0$. The resulting spectrum, which is of the AX type, consists of two doublets with line spacings equal to J. The spacing between the doublet centers is $\Delta\delta$, and the four lines have equal intensity (Fig. II-5b). (iii) When $\Delta\delta \approx J$, $\sin 2\theta$ becomes > 0, and we have that the transitions corresponding to the two center lines are more strongly allowed than those of the outer lines. With decreasing value of the quotient $\Delta\delta/J$, we have more and more a two line spectrum near the mean chemical shift position, with the forbidden lines appearing as weak satellites in the positions $\pm(C + \frac{1}{2}|J|)$ (Fig. II-5, d and e). (iv) With $\Delta\delta = 0$, the spectrum reduces to a single line. We have here another illustration of the general rule that spin–spin coupling between chemical shift equivalent nuclei is not manifested in the NMR spectra[7]. Finally, it might also be pointed out that for all possible values of $\Delta\delta/J$, the transition frequencies and intensities depend only on the absolute value of J, and hence the sign of the coupling constant is not manifested in the spectral features of an AB spin system.

Asymmetric Three Spin Systems: *AMX, ABX*

An AMX system is comprised of three spins with chemical shifts δ_A, δ_M, and δ_X, and with spin–spin couplings J_{AM}, J_{AX}, and J_{MX}. The chemical shift differences in Hz, $\Delta\delta_{AM}$, $\Delta\delta_{AX}$, and $\Delta\delta_{MX}$, are large compared to the spin–spin couplings. The corresponding spectrum has typical first order features, and contains three well separated four line patterns with equal intensity for all the lines (Fig. II-15). It is seen that each group of four lines is symmetrical

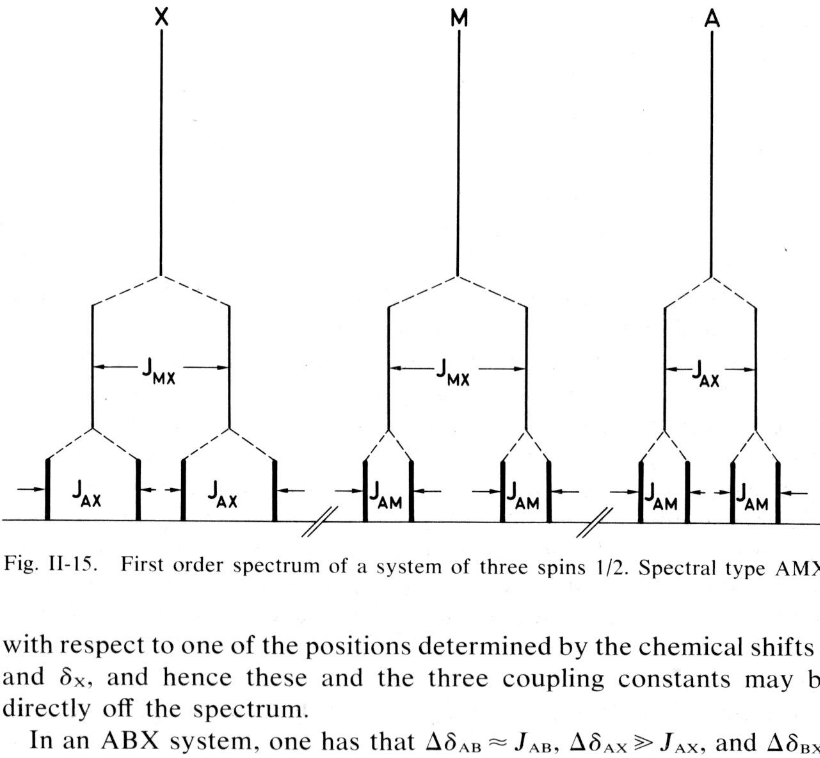

Fig. II-15. First order spectrum of a system of three spins 1/2. Spectral type AMX.

with respect to one of the positions determined by the chemical shifts δ_A, δ_M, and δ_X, and hence these and the three coupling constants may be read directly off the spectrum.

In an ABX system, one has that $\Delta\delta_{AB} \approx J_{AB}$, $\Delta\delta_{AX} \gg J_{AX}$, and $\Delta\delta_{BX} \gg J_{BX}$. In analogy to Equation II-10, the spin Hamiltonian for an ABX system may be written

$$\mathcal{H} = -\gamma\frac{h}{2\pi}H_0[(1-\sigma_A)I_{zA} + (1-\sigma_B)I_{zB} + (1-\sigma_X)I_{zX}]$$
$$+ J_{AB}\boldsymbol{I}_A\cdot\boldsymbol{I}_B + J_{AX}\boldsymbol{I}_A\cdot\boldsymbol{I}_X + J_{BX}\boldsymbol{I}_B\cdot\boldsymbol{I}_X \qquad \text{(II-19)}$$

A convenient set of basis functions includes the eight combinations of α- and β-states for the three spins: $|\alpha\alpha\alpha>$, $|\alpha\alpha\beta>$, $|\alpha\beta\alpha>$, $|\beta\alpha\alpha>$, $|\alpha\beta\beta>$, $|\beta\alpha\beta>$, $|\beta\beta\alpha>$, $|\beta\beta\beta>$. From an analogous computation to that for the AB-system, the transition frequencies and intensities in Table II-11 are obtained.

Table II-11 lists 15 lines for the ABX type spectrum. One line has intensity zero, so that an ABX spectrum may exhibit up to 14 lines. Twelve of these originate from the three spins, which is reminiscent of the AMX type spectrum (Fig. II-15), and two are "combination lines" in the sense that they decrease to vanishing intensity as $\Delta\delta_{AB}$ increases. Fig. II-16 represents a theoretical ABX spectrum computed with the assumptions that $J_{AX} > J_{BX} > 0$, and $\delta_A > \delta_B$. This spectrum, as one would anticipate from Table II-11, contains 8 lines in the AB part, and 6 lines in the X part. The 6 lines of the X

TABLE II-11

Transition Frequencies and Relative Intensities for the ABX System.

Transition number	Origin	Transition frequency	Intensity
1	B	$\frac{1}{2}(\nu_A+\nu_B)-\frac{1}{4}(2J_{AB}+J_{AX}+J_{BX})-D_-$	$(\cos\theta_- - \sin\theta_-)^2$
2	B	$\frac{1}{2}(\nu_A+\nu_B)-\frac{1}{4}(2J_{AB}-J_{AX}-J_{BX})-D_+$	$(\cos\theta_+ - \sin\theta_+)^2$
3	B	$\frac{1}{2}(\nu_A+\nu_B)+\frac{1}{4}(2J_{AB}-J_{AX}-J_{BX})-D_-$	$(\cos\theta_- + \sin\theta_-)^2$
4	B	$\frac{1}{2}(\nu_A+\nu_B)+\frac{1}{4}(2J_{AB}+J_{AX}+J_{BX})-D_+$	$(\cos\theta_+ + \sin\theta_+)^2$
5	A	$\frac{1}{2}(\nu_A+\nu_B)-\frac{1}{4}(2J_{AB}+J_{AX}+J_{BX})+D_-$	$(\cos\theta_- + \sin\theta_-)^2$
6	A	$\frac{1}{2}(\nu_A+\nu_B)-\frac{1}{4}(2J_{AB}-J_{AX}-J_{BX})+D_+$	$(\cos\theta_+ + \sin\theta_+)^2$
7	A	$\frac{1}{2}(\nu_A+\nu_B)+\frac{1}{4}(2J_{AB}-J_{AX}-J_{BX})+D_-$	$(\cos\theta_- - \sin\theta_-)^2$
8	A	$\frac{1}{2}(\nu_A+\nu_B)+\frac{1}{4}(2J_{AB}+J_{AX}+J_{BX})+D_+$	$(\cos\theta_+ - \sin\theta_+)^2$
9	X	$\nu_X-\frac{1}{2}(J_{AX}+J_{BX})$	1
10	X	$\nu_X+D_+-D_-$	$\cos^2(\theta_+-\theta_-)$
11	X	$\nu_X-D_++D_-$	$\cos^2(\theta_+-\theta_-)$
12	X	$\nu_X+\frac{1}{2}(J_{AX}+J_{BX})$	1
13	comb.	$2\nu_0-\nu_X$	0
14	comb.	$\nu_X-D_+-D_-$	$\sin^2(\theta_+-\theta_-)$
15	comb.	$\nu_X+D_++D_-$	$\sin^2(\theta_+-\theta_-)$

The following conventional expressions have been used in Table II-11:
ν_A, ν_B, and ν_X are the respective chemical shifts in Hz.
$\Delta\delta_{AB}=(\nu_A-\nu_B)$, $\Delta\delta_{AX}=(\nu_A-\nu_X)$, $\Delta\delta_{BX}=(\nu_B-\nu_X)$
$D_+\cos 2\theta_+ = \frac{1}{2}\Delta\delta_{AB}+\frac{1}{4}(J_{AX}-J_{BX})$
$D_+\sin 2\theta_+ = \frac{1}{2}J_{AB}$
$D_-\cos 2\theta_- = \frac{1}{2}\Delta\delta_{AB}-\frac{1}{4}(J_{AX}-J_{BX})$
$D_-\sin 2\theta_- = \frac{1}{2}J_{AB}$
comb. stands for "combination line" (see text)
D_+ and D_- may alternatively be expressed by
$D_+ = \{[\frac{1}{2}\Delta\delta_{AB}+\frac{1}{4}(J_{AX}-J_{BX})]^2+\frac{1}{4}J^2_{AB}\}^{1/2}$
$D_- = \{[\frac{1}{2}\Delta\delta_{AB}-\frac{1}{4}(J_{AX}-J_{BX})]^2+\frac{1}{4}J^2_{AB}\}^{1/2}$

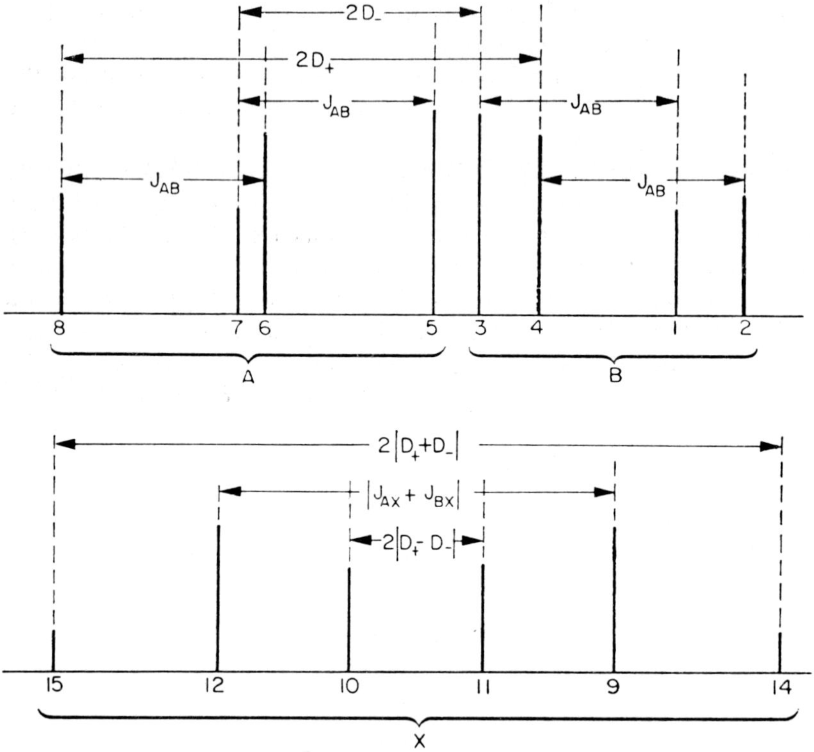

Fig. II-16. Theoretical ABX type spectrum showing the separate AB (upper trace) and X (lower trace) parts. The lines are numbered assuming $J_{AX} > J_{BX} > 0$, and $\nu_A > \nu_B$. Note that the transitions 2,4,6,8 and 1,3,5,7 form two symmetrical quartet patterns. The X part forms a symmetrical pattern consisting of six lines (from Emsley, Feeney and Sutcliffe[27]).

part are arranged symmetrically with respect to the position δ_X. As indicated in Fig. II-16, certain parameters may readily be read off an ABX spectrum, provided the individual lines can be attributed to specific transitions. These resonance line assignments to the theoretical transition energies (Table II-11) depend upon the relative signs of the coupling constants involved, and can thus in principle be made in different ways in the spectrum of an unknown spin system. The following equalities among the spacings between individual pairs of lines, which must always hold for all ABX spectra, are helpful in avoiding errors in the assignment of transitions:

$$\begin{aligned} 3-1 &= 4-2 = 7-5 = 8-6 \\ 2-1 &= 4-3 = 11-9 = 12-10 \\ 10-9 &= 12-11 = 6-5 = 8-7 \\ 10-11 &= (8-4)-(5-1) \end{aligned} \qquad \text{(II-20)}$$

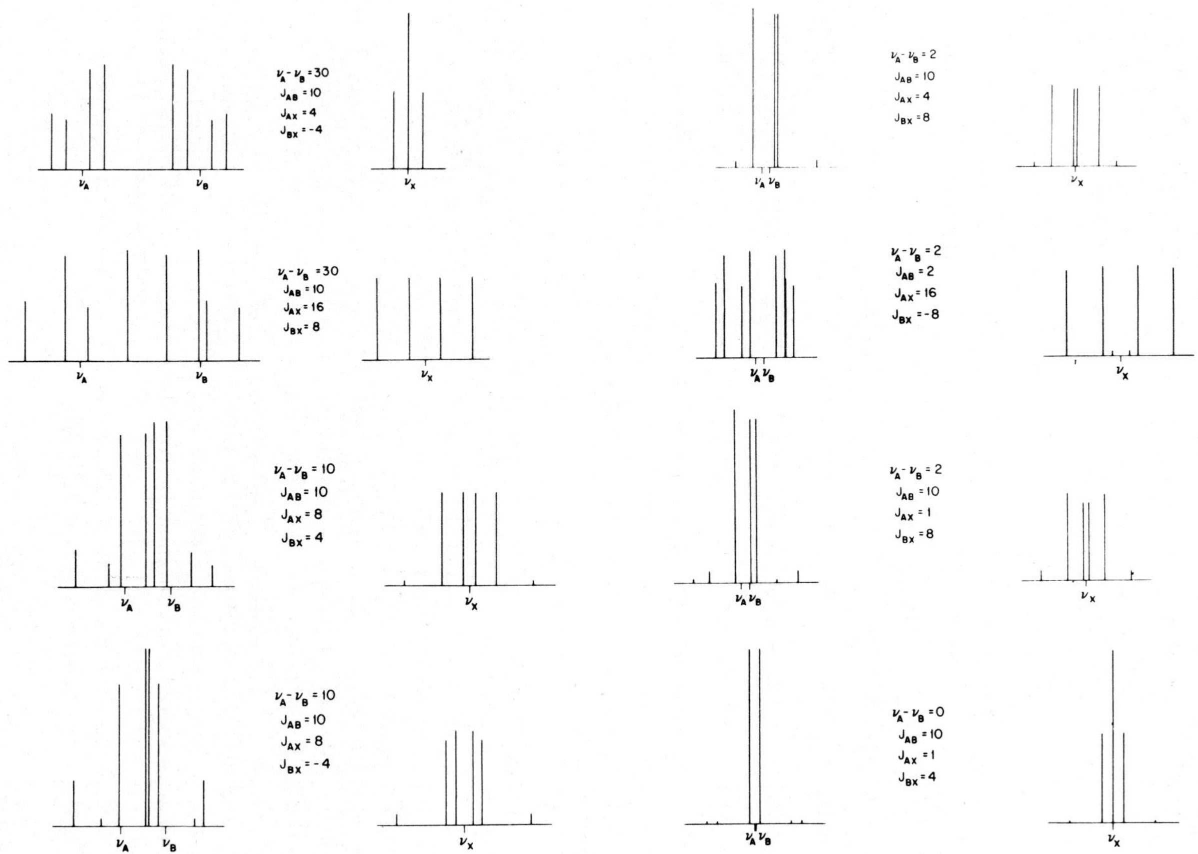

Fig. II-17. Selected ABX type spectra for different combinations of the parameters $(\nu_A - \nu_B) \equiv \Delta\delta_{AB}$, J_{AB}, J_{AX}, and J_{BX} (from Bovey[11]).

Once the resonance lines are correctly assigned to the theoretical transition energies, one may then use that the spacings between some of the latter are directly related to the NMR parameters. Examination of Table II-11 shows that the following pairs of resonance lines are separated by J_{AB}

$$|J_{AB}| = |3-1| = |4-2| = |7-5| = |8-6|. \tag{II-21}$$

Similarly, two lines are separated by $(J_{AX}+J_{BX})$

$$|(J_{AX}+J_{BX})| = |12-9|. \tag{II-22}$$

One may thus readily obtain $|J_{AB}|$, $|J_{AX}+J_{BX}|$, and δ_X, but neither $\Delta\delta_{AB}$ nor $|J_{AX}-J_{BX}|$ can be directly read off the spectrum. From the line spacings II-23 to II-26, one may obtain D_+ and D_-

$$2|D_+| = |6-2| = |8-4| \tag{II-23}$$

$$2|D_-| = |5-1| = |7-3| \tag{II-24}$$

$$2|D_+ + D_-| = |15-14| \tag{II-25}$$

$$2|D_+ - D_-| = |11-10| \tag{II-26}$$

However, in contrast to the spectrum of Fig. II-16, where specific assumptions were made concerning δ_A and δ_B, and J_{AX} and J_{BX}, one will not a priori be able to differentiate between the line spacings D_+ and D_- in an unknown spectrum.

The spectral features described above may help to classify the spectrum, and yield approximate NMR parameters for the initial input of the computer simulation. In the comparison of experimental and simulated theoretical spectra, one will also consider the influence of the relative signs of J_{AX} and J_{BX} on the transition frequencies and intensities (Table II-11).

Depending on the relative magnitudes and signs of the NMR parameters, ABX spectra can be deceptively simple. Fewer resonance lines than predicted are then observed in the spectra because some resonance lines coalesce and others decrease to vanishing intensities for certain relationships between the parameters. Quite generally, less information is available from a simplified spectrum of this kind in that only average values of coupling constants can be determined. Moreover, it is in some cases all too easy to draw erroneous conclusions from these simplified ABX type spectra. A few selected examples of ABX spectra are represented in Fig. II-17. Note that closely spaced lines in the line diagrams may not necessarily be resolved in a real spectrum with a real line width.

CHAPTER III

HIGH RESOLUTION PROTON NMR STUDIES OF STRUCTURE AND CONFORMATION IN PEPTIDES AND PROTEINS

Since the NMR spectra of a polypeptide chain depend primarily on the amino acid composition, and can also be affected by the amino acid sequence, NMR can be employed as a complimentary technique to the chemical investigations of the covalent structures in polypeptide chains. Yet, an even more important role for high resolution proton NMR in peptide and protein research is its application for studies of the molecular conformations, and most of this chapter will be devoted to this aspect. Conformational studies of oligopeptides and proteins will be discussed in separate sections, because at the present stage the choice of the appropriate NMR approach for conformational studies will in general depend to a large extent on the size of the molecule. It is to be expected, however, that with the further improvements of the NMR techniques, and with the use of isotopically labelled polypeptide chains, many of the experiments described here for oligopeptides will also become applicable to larger molecules in the near future.

The experiments included in this chapter were chosen to illustrate the basic principles of the method. Additional examples of ^{1}H-NMR studies of polypeptide chain conformations will be included in subsequent chapters. There, they will also be considered in the light of complementary data obtained from different NMR techniques, such as the application of paramagnetic shift reagents, and observation of other nuclei than protons.

1. ANALYSIS OF AMINO ACID COMPOSITION AND AMINO ACID SEQUENCE

High Resolution ^{1}H-NMR Spectra of Random Coil Polypeptides Chains

In the context of the present discussion, a protein molecule is taken to be in a "random coil form" when all the amino acid side chains extend freely

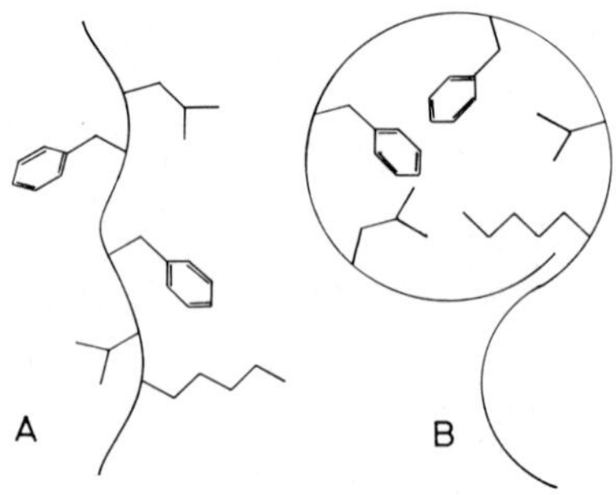

Fig. III-1. Schematic representation of a segment of the polypeptide chain in an extended random coil form (A) and a globular conformation (B) of a protein molecule.

into the solvent (Fig. III-1A)*. Since the chemical shifts are primarily determined by the covalent structure of the peptide and the magnetic susceptibility of the solvent environment, one would expect that the NMR of the amino acid side chains in such a random coil polypeptide chain correspond essentially to the sum of the resonances of the constituent amino acid residues (Tables II-7 and II-9). This is illustrated in Fig. III-2, which shows the proton NMR spectrum of a human polypeptide hormone, calcitonin M. This molecule (Fig. III-3) has been synthesized by chemical methods, and its physiological role in the regulation of the blood calcium level has been extensively investigated. NMR studies have shown that calcitonin M in aqueous solution is predominantly in an extended random coil form[1]. In the spectral regions of the aliphatic amino acid side chains between 0 and 3 ppm, and the aromatic side chains between 6 and 9 ppm, the experimental spectrum A and the computed spectrum B in Fig. III-2 are nearly identical. This shows that all the labile hydrogen atoms, i.e. those bonded to O, N, or S, had been replaced by deuterium atoms of D_2O, and that the different amino acid residues of one kind, e.g. the three phenylalanines in positions 16, 19, and 22 (Fig. III-3), are magnetically essentially equivalent. Here the methyl resonances of the five threonyl residues 6, 11, 13, 21, and 25 (Fig. III-3) at approximately 1.1 ppm (the second line from the right in Fig. III-2) are the only apparent exception, in that they are slightly shifted with respect to each other in the spectrum A of Fig. III-2.

Because the amino acid side chain resonances are essentially identical

---

*It should be noted that this is a rather loose definition of the term "random coil form". It encompasses quite generally non-globular extended solution conformations of polypeptide chains, including denatured forms of protein molecules with intact disulfide bridges (see Section III.4), even if they contain energetically favored spatial arrangements of the polypeptide backbone (see Section V.4).

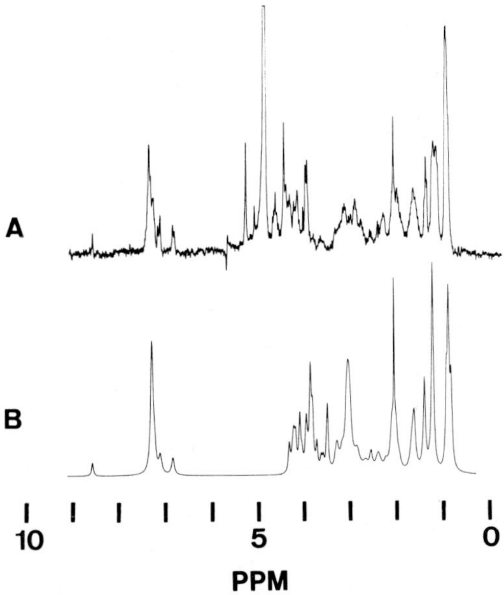

Fig. III-2. ¹H-NMR spectrum at 220 MHz of calcitonin M in a D_2O-solution, pD = 5.0. A. Experimental spectrum. B. Hypothetical spectrum computed on the basis of the amino acid composition (Fig. III-3) as the sum of the resonances of the individual amino acid residues (Table II-7). The strong narrow lines between 4 and 6 ppm in the spectrum A correspond to the resonance of the residual protons of the solvent, and its spinning side bands (from Masson[1]).

(1) (5) (10)
H–Cys–Gly–Asn–Leu–Ser–Thr–Cys–Met–Leu–Gly–Thr–Tyr–Thr–Gln–

(15) (20) (25)
–Asp–Phe–Asn–Lys–Phe–His–Thr–Phe–Pro–Gln–Thr–Ala–Ile–Gly–Val–

(30)
–Gly–Ala–Pro–NH_2

Fig. III-3. Amino acid sequence in calcitonin M (R. Neher et al., Helv. Chim. Acta *51*, 1900 (1968)).

with those of the individual amino acid residues (Table II-7), certain resonance lines in Fig. III-2A, can readily be identified from their chemical shifts. Thus the line at 0.85 ppm contains the methyl resonances of Leu(4), Leu(9), Ile(27), and Val(29) (Fig. III-3), the resonances between 1.0 and 1.1 ppm come from the methyl groups of the five threonyl residues, the line at 1.3 ppm from the methyls of the two alanyl residues in positions 26 and 31, and the sharp line at 2.0 ppm from the methyl protons of Met-8. The resonance assignments in the aromatic region are given in Fig. III-4B. The small line width of the resonance at 2.0 ppm is quite typical for methionine,

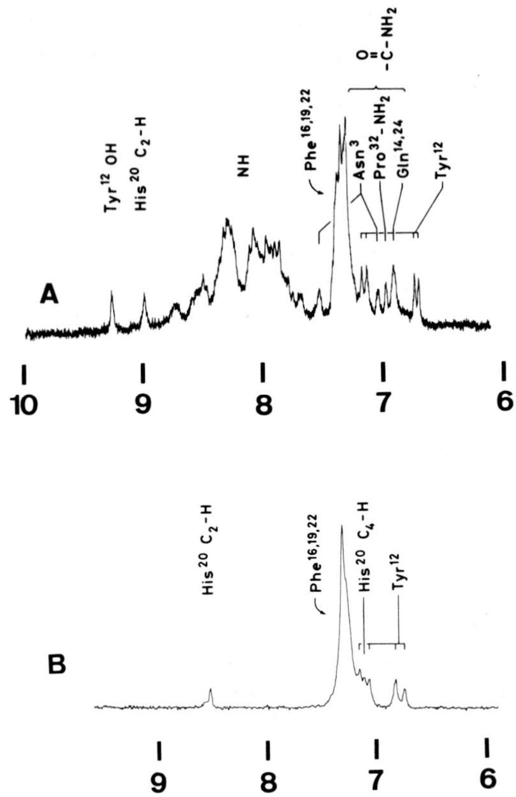

Fig. III-4. Spectral region from 6 to 10 ppm of the ^{1}H-NMR spectra of calcitonin M. A. Solvent DMSO-d$_6$, $T = 20°$, $\nu_0 = 220$ MHz. B. Solvent D_2O, $T = 28°$, $\nu_0 = 100$ MHz. The resonance assignments are also given. "NH" in spectrum A identifies the overlapping resonances of the approx. 30 backbone amide protons (from Masson[1]).

because its methyl protons experience no spin–spin coupling with other nuclei.

When working with polypeptides which had been synthesized by classical chemical methods, one is often in the fortunate position to obtain partial sequences for the purpose of comparative NMR studies. These experiments can then yield particularly reliable information on the resonance assignments. Furthermore, because a "random coil spectrum" of the amino acid side chains is by the above definition identical to the sum of the spectra of the constituent fragments, such comparative studies can reveal whether subtle deviations from the hypothetical spectrum computed with the NMR parameters of the individual amino acid residues (Tables II-7, II-9) are really due to the occurrence of preferred molecular conformations, or whether they are, as appears to be the case with the slight nonequivalence of the five threonyl methyl resonances in calcitonin M (Fig. III-2A)[1], a consequence of the amino acid sequence.

To illustrate the use of peptide fragments for the identification of individual resonances in the random coil spectra of polypeptide molecules, some observations in dimethylsulfoxide solutions of calcitonin M, where the labile protons can also be observed, shall be recounted[1]. Between 0 and 4 ppm, the spectrum in DMSO-d_6 is essentially identical to that in D_2O (Fig. III-2A). At around 5.0 ppm, the hydroxyl protons of Thr and Ser are observed, and between 6 and 10 ppm there are in addition to the resonances of the non-labile aromatic protons (Fig. III-4B) those of the backbone amide protons, the side chain amide protons, and the hydroxyl proton of tyrosine (Fig. III-4A). While the latter could readily be identified from its typical chemical shift (Table II-9), the specific assignments of the side chain amide protons of Asn(3), Gln(14), and Gln(24), and of the chain-terminal amide protons of Pro(32) given in Fig. III-4 were based on comparison with suitable peptide fragments (Figs. III-5 and III-6). Between 10 and 15 ppm, the spectrum of calcitonin M in DMSO-d_6 (Fig. III-7) contains the resonances of the carboxylic acid proton of Asp(15), and the labile imidazole ring protons of His(20). Fig. III-7 shows further that the fragment H-1-28-OH, which has a free carboxylic acid group at the chain terminal, contains an additional line in this spectral region. The essentially identical chemical shifts of the resonances of corresponding labile protons in calcitonin M and its constituent peptide fragments, can also be taken as additional evidence for the random coil character of the observed NMR spectrum in DMSO-d_6 solution.

Additional examples of proton NMR spectra of proteins in random coil forms will be presented in Section III.4.

Amino Acid Composition of Polypeptide Chains

The foregoing discussion of the spectral properties of random coil polypeptide chains shows that NMR can in principle be used for studies of the amino acid composition. The reliability of the method will of course depend very much on the systems under investigation, e.g. it will only be applicable when a random coil spectrum of the molecule can at all be obtained. It will quite generally rather be useful for the detection of certain specific amino acid residues than for a complete analysis of the amino acid composition. Thus in our example calcitonin M, the presence or absence of His, Tyr, Phe, Met, Ala, or Thr would readily be apparent in the ^{1}H-NMR spectra (Figs. III-2 and III-4), whereas it might be difficult, if not impossible, to determine the number of, say, glycyl, leucyl, or isoleucyl residues.

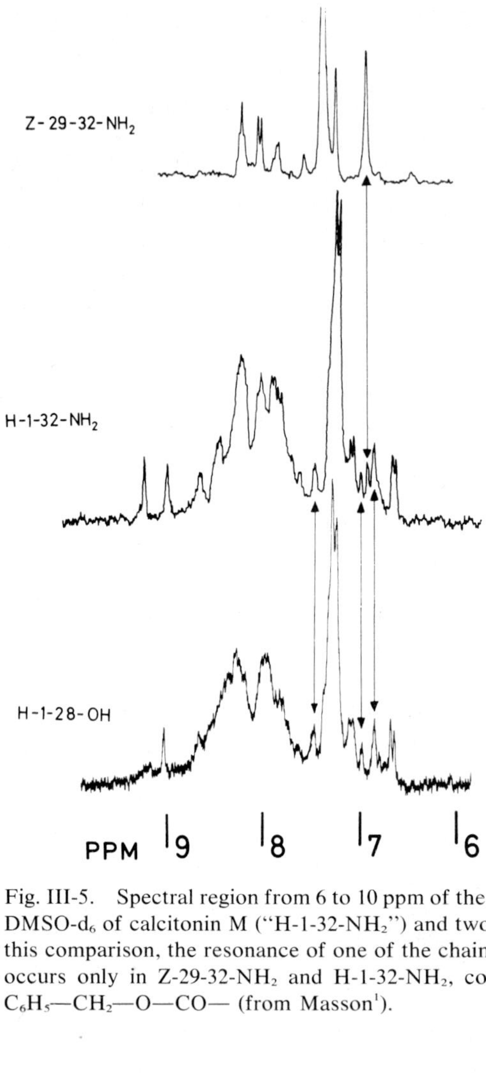

Fig. III-5. Spectral region from 6 to 10 ppm of the ^{1}H-NMR spectra at 220 MHz in solutions in DMSO-d_6 of calcitonin M ("H-1-32-NH_2") and two of its constituent peptide fragments. From this comparison, the resonance of one of the chain terminal amide protons at 6.95 ppm, which occurs only in Z-29-32-NH_2 and H-1-32-NH_2, could be identified[1]. Z = Benzyloxycarbonyl, C_6H_5—CH_2—O—CO— (from Masson[1]).

Overall, it would appear that random coil polypeptide NMR spectra could for certain specific problems be a valuable complementary technique to the chemical methods for the determination of the amino acid composition. It is particularly attractive as a non-destructive and rapid alternative method for product control in polypeptide synthesis.

Instead of the examination of the random coil polypeptide chains, acid hydrolysis and subsequent NMR analysis of the resulting mixture of amino acids has also been considered as a technique for studies of the amino acid

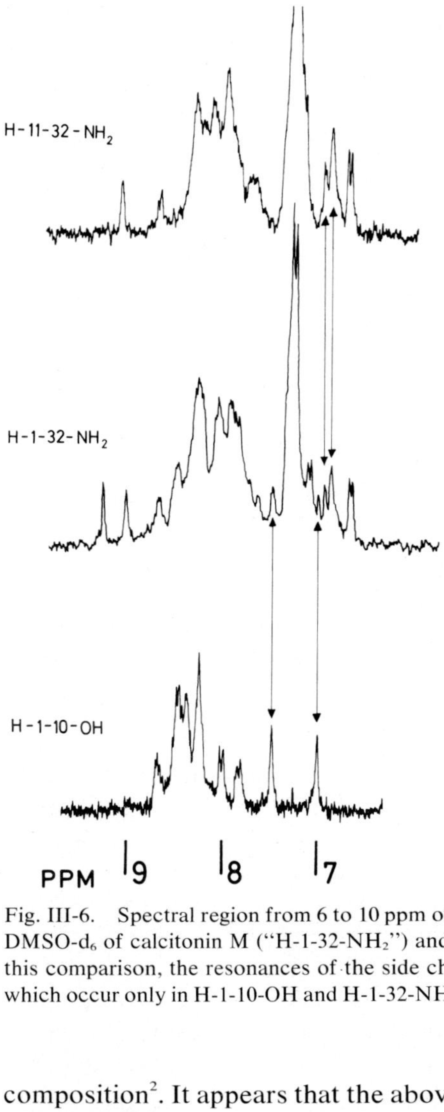

Fig. III-6. Spectral region from 6 to 10 ppm of the ^{1}H-NMR spectra at 220 MHz in solutions in DMSO-d_6 of calcitonin M ("H-1-32-NH_2") and two of its constituent peptide fragments. From this comparison, the resonances of the side chain amide protons of Asn(3) at 7.0 and 7.5 ppm, which occur only in H-1-10-OH and H-1-32-NH_2 (see Fig. III-3) were identified (from Masson[1]).

composition[2]. It appears that the above-mentioned limitations of the analysis of the random coil polypeptide spectra are also inherent in this experiment.

Amino Acid Sequence in Linear Oligopeptides

The discussion of the proton NMR in random coil polypeptide chains has shown that the readily assigned proton resonances, i.e. those of the amino acid side chains, are very little affected by the amino acid sequence. Yet the

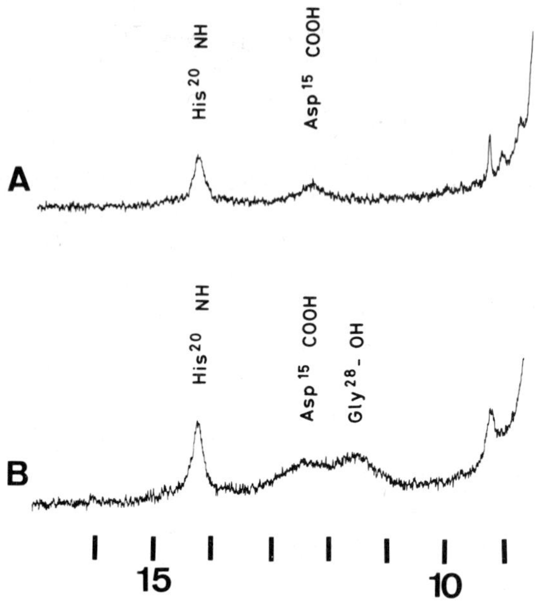

Fig. III-7. Spectral region from 9 to 16 ppm of the ^{1}H-NMR spectra at 220 MHz in solutions in DMSO-d_6 of A, calcitonin M (H-1-32-NH_2), and B, the calcitonin M peptide fragment H-1-28-OH. The resonance assignments are also indicated (from Masson[1]).

pH-dependence of the amino acid spectra described in Section II.6 provides a means to identify the amino- and carboxy-terminal amino acids in linear unprotected peptides. Thus, even with the use of a routine ^{1}H-NMR spectrometer, the complete sequence of linear di- and tri-peptides can be obtained from a pH-titration[3]. Since small chemical shifts upon variation of the electric charges localized on the chain terminal groups arise also in the penultimate amino acid residues, the spectral resolution obtained with the modern spectrometers operating at around 300 MHz makes even a complete sequence analysis of linear tetra- and pentapeptides practicable[4]. In favorable cases, i.e. in particular in the absence of ionizeable amino acid side chains, it should in principle be possible to identify at least the amino-terminal and the carboxy-terminal amino acids in longer polypeptide chains. Since the NMR experiment is non-destructive, it could in certain instances be a valuable complementary technique for amino acid sequence analysis. Just as with the NMR determination of the amino acid composition, its most attractive potential application is for product control in polypeptide synthesis.

2. MANIFESTATION OF DIFFERENT MOLECULAR CONFORMATIONS IN THE ^{1}H-NMR PARAMETERS

Polypeptide Backbone Conformations

The polypeptide backbone consists of a chain of peptide groups linked together through single bonds via the α-carbon atoms. As a consequence of the partial double bond character of the N—C′ bond, the peptide groups are preferentially in a near-planar form. A widely used model for the description of polypeptide chain conformations, which has also been the starting point for numerous NMR investigations, is therefore based on the assumption that all the peptide groups are rigid planes with standard dimensions. While the trans form (Fig. III-8) is common for unsubstituted amide groups, cis peptide groups (Fig. III-8) are frequently encountered in N-substituted amino acids, i.e. in proline, and in the N-methyl-amino acids which occur in certain microbial peptides. With the assumption of standard trans and cis peptide groups, the backbone conformations can unambiguously be described by the torsion angles ϕ_i and ψ_i about the N—C$^\alpha$ and C$^\alpha$—C′ bonds (Figs. III-9 and III-10). In this "rigid model", only certain combinations of ϕ and ψ are sterically allowed for each individual dipeptide fragment, the extent of the allowed regions in the ϕ-ψ plane depending on the amino acid side chains, and on the dimensions chosen for the peptide groups[5].

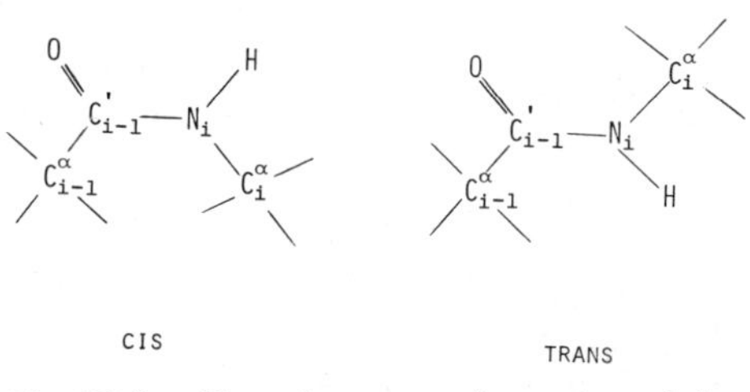

Fig. III-8. Cis and trans conformation of the peptide group.

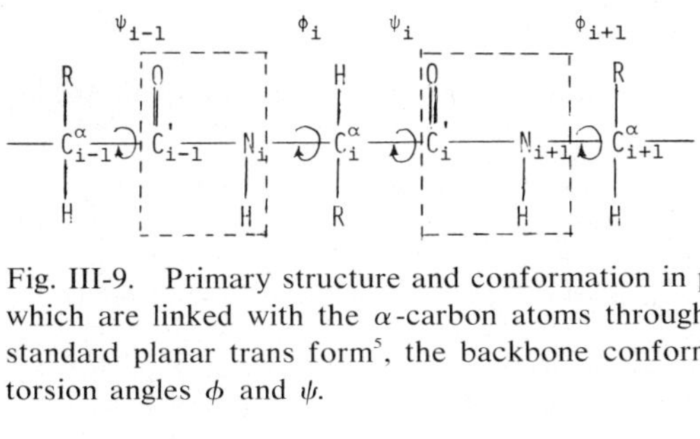

Fig. III-9. Primary structure and conformation in polypeptide chains. If the peptide groups, which are linked with the α-carbon atoms through single bonds, are assumed to be in the standard planar trans form[5], the backbone conformation can be uniquely described by the torsion angles ϕ and ψ.

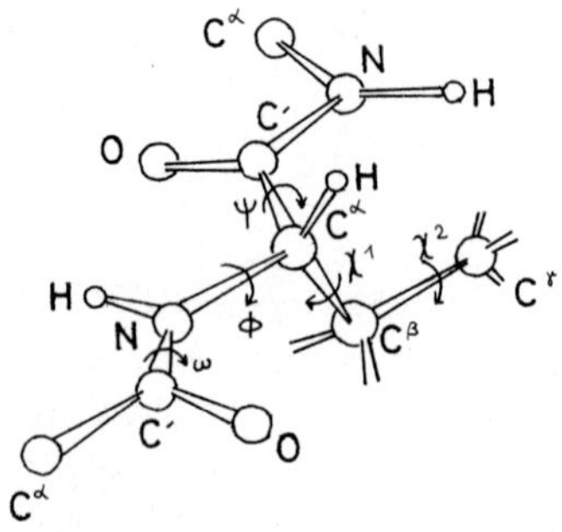

Fig. III-10. Same peptide fragment as in Fig. III-9. In addition to ϕ and ψ, the torsion angles ω, which describe deviations of the peptide group from the trans planar form, and χ^j, which characterize the spatial arrangement of the amino acid side chains[5], are also indicated.

Angular Dependence of the Vicinal Spin–Spin Coupling Constants between Atoms of the Peptide Backbone

Experimental and theoretical investigations have shown that a Karplus-type angular dependence (Fig. II-4) is a general feature of vicinal spin–spin coupling constants. Therefore, a complete rigid model description of the peptide conformations in terms of the torsion angles ϕ_i and ψ_i may in principle be derived from the ^{1}H-NMR data: The proton–proton coupling constant $J_{HN\alpha}$* is related to the torsion angle ϕ_i (Figs. III-9 and III-10). In peptides with ^{13}C-enriched carbonyl carbon positions, the torsion angles ϕ_i can e.g. also be studied via the ^{13}C—^{1}H coupling $J_{^{13}C'H^\alpha}$*. In ^{15}N-enriched peptides, the torsion angles ψ_i are accessible to NMR investigation through the ^{15}N—^{1}H coupling constant $J_{^{15}NH^\alpha}$*. The practical aspects of using these vicinal coupling constants for conformational studies will now be discussed in some more detail.

The formalism for describing the angular dependence of $J_{HN\alpha}$ follows closely that used originally by Karplus for hydrocarbon segments. $J_{HN\alpha}$ can be expressed by a relation of the general form[6–13]

$$J_{HN\alpha} = A\cos^2\theta + B\cos\theta + C\sin^2\theta \qquad \text{(III-1)}$$

where A, B, and C are empirical constants, and θ is the dihedral angle

---

*Since throughout Chapter III we shall encounter exclusively vicinal spin–spin couplings, the notation for the spin–spin coupling constants introduced in Table II-8 has been further abbreviated by omitting the prefix[3]. The proton–proton coupling between the amide proton i and the C_i^α-proton (Fig. III-9) then becomes $J_{HN\alpha}$, and the vicinal couplings between protons on adjacent carbon atoms of the amino acid side chains are $J_{\alpha\beta}$, $J_{\beta\gamma}$, etc. The heteronuclear ^{13}C – ^{1}H spin–spin coupling between the carbonyl carbon atom (i – 1) and the C_i^α-proton (Fig. III-9) is $J_{^{13}C'H^\alpha}$, and the ^{15}N—^{1}H coupling between the C_i^α-proton and the (i + 1)th amide nitrogen is $J_{^{15}NH^\alpha}$.

between the planes fixed by H—C$^{\alpha}$—N, and C$^{\alpha}$—N—H, respectively. Practical appiications of Equation III-1 for conformational studies of peptides and proteins are limited by various factors which can be traced to three basically different sources. The first one has to do with the uncertainties in the numerical values for the coefficients A, B, and C to be used for the fragment H—N—C$^{\alpha}$—H in Equation III-1. This problem arises because these coefficients must be determined empirically from studies of model amide compounds, and it seems to be a common feature of the models used so far that there remained some ambiguities about the rotational states corresponding to the observed spin–spin coupling constants. As a consequence, the coefficients A, B, and C proposed by different authors on the basis of experiments with different model compounds are somewhat different (Table III-1). In addition, one has to include that the vicinal coupling

TABLE III-1

Coefficients proposed for equation III-1 on the basis of experimental observations by different authors*.

Reference	Coefficients (in Hz)		
	A	B	C
Bystrov et al.[7]	9.3	−0.5	0.9
Bystrov et al.[10]	9.8	−1.1	0.4
Ramachandran et al.[8]	8.6	−1.7	1.5
Donzel[11]	9.7	−0.4	0.1
Thong et al.[12]	9.3	−3.5	0.3

*The coefficients are valid for the fragment —NH—CH_3. For the fragments —NH—CH_2C and $\text{—NH—CH}\left\langle\begin{smallmatrix}C\\C\end{smallmatrix}\right.$ the respective electronegativity corrections are to be applied (from Bystrov et al.[10]).

constants are decreased by the introduction of electronegative substituents R_1 and R_2 in the fragment —NH—CHR_1R_2[6]. The relation III-2 has been proposed to allow for these electronegativity effects[6]

$$J_{HN\alpha} = J^{obs}_{HN\alpha}\left(1 - \alpha \sum_i \Delta E_i\right)^{-1} \qquad \text{(III-2)}$$

where $J_{HN\alpha}$ is the coupling constant to be used in Equation III-1, $J^{obs}_{HN\alpha}$ the measured vicinal coupling constant, α a proportionality factor for which numerical values between 0.05 and 0.1 have been proposed[6, 10], and ΔE_i the electronegativity difference between hydrogen and the i-th C^{α}-substituent. Following a suggestion by Ramachandran et al.[8], this leads to the following approximate electronegativity corrections in peptides:

If Equation III-3 applies to the fragment $—NH—CH\langle^{C}_{C}$,

$$J_{HN\alpha} = 7.9 \cos^2\theta - 1.5 \cos\theta + 1.3 \sin^2\theta \qquad \text{(III-3)}$$

$J^{obs}_{HN\alpha}$ for $—NH—CH_3$ would have to be reduced by 6%, and $J^{obs}_{HN\alpha}$ for $—NH—CH_2C—$ by 3% prior to its use in Equation III-3 for determining the corresponding dihedral angle θ. Finally, as theoretical investigations have shown[6, 13], $J_{HN\alpha}$ is in addition to its dependence on θ (Equations III-1 and III-3) also slightly influenced by the rotational states about the adjoining $C^{\alpha}—C'$ bond (ψ in Fig. III-10), and by deviations of the peptide groups from their standard planar cis or trans forms. Since these small effects are rather difficult to include in detail into the analysis of $J_{HN\alpha}$, it appears that they could for practical purposes best be allowed for in a presentation of the angular dependence of $J_{HN\alpha}$ in terms of "permissible ranges". Fig. III-11

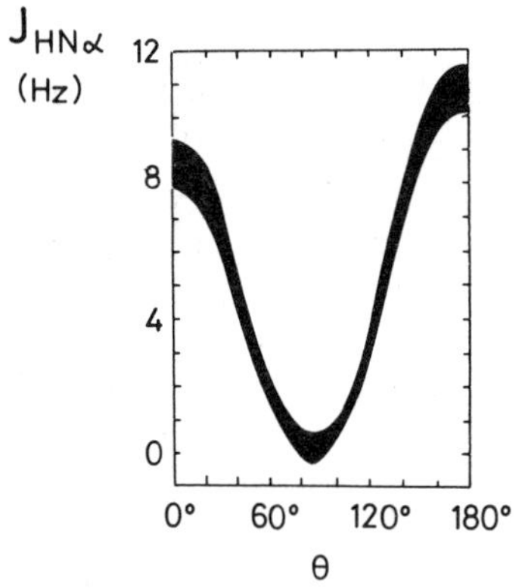

Fig. III-11. Empirical angular dependence of $J_{HN\alpha}$ in the fragment $—NH—CH_3$ presented in the form of "permissible ranges" (see text) for each value of the dihedral angle θ (H—N—C^{α}, N—C^{α}—H) (from Bystrov et al.[10]).

shows the permissible ranges for $J_{HN\alpha}$ of the fragment $—NH—CH_3$ which correspond to the coefficients A, B, and C of Bystrov et al[10].

The second source of ambiguities in the analysis of $J_{HN\alpha}$ stems from the dynamic character of the solution conformations in polypeptide chains. Thus the observed spin–spin coupling constants will in general correspond to the average over a certain range of values for θ, rather than to a rigid angle

θ. For each molecule studied, one then has to decide on the range of θ-values to be included, and on the weighing factors to be attached to the different θ's. The simplest model would assume that all the values of θ in the allowed range from θ_1 to θ_2 have the same weighing factor. Using Equation III-3[8] for the angular dependence of $J_{HN\alpha}$, one then has for the observed average of the vicinal spin–spin coupling constant in a fragment

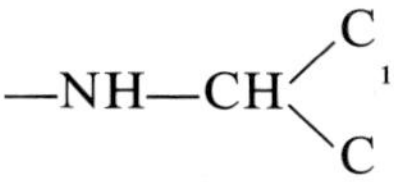
[14]

$$J^{av}_{HN\alpha} = \frac{\int_{\theta_1}^{\theta_2} (7.9 \cos^2 \theta - 1.5 \cos \theta + 1.3 \sin^2 \theta)\, d\theta}{\int_{\theta_1}^{\theta_2} d\theta} \tag{III-4}$$

Integration yields

$$J^{av}_{HN\alpha} = \frac{f(\theta_2) - f(\theta_1)}{\theta_2 - \theta_1} \tag{III-5}$$

with

$$f(\theta) = 1.6 \sin 2\theta - 1.5 \sin \theta + 4.6\theta$$

The third factor which can restrict the use of Equation III-1 for studies of peptide conformations is the limited spectral resolution of the ^{1}H-NMR spectra, and certain other experimental limitations. Quite generally, the spectral resolution of even the most advanced spectrometers will at present not allow the determination of all the vicinal coupling constants $J_{HN\alpha}$ when the size of the peptide chain exceeds 10 to 15 amino acid residues. In smaller peptides, the application of Equation III-1 will depend on the solvent. As can be seen from Table II-4, the amide protons are in general replaced by deuterons in many of the common solvents, including D_2O, and the coupling constants $J_{DN\alpha}$ have not yet been observed in peptides.

The dihedral angle θ of Equation III-1 is related to the standard torsion angle ϕ used for the description of the peptide conformation (Figs. III-9 and III-10)[5] by $\theta = |\phi - 60°|$ for a L-amino acid, and by $\theta = |\phi + 60°|$ for a D-amino acid. These correspondences are given in Table III-2 for a number of θ-values. In general there exists an ambiguity of two possible values of ϕ for each θ corresponding to a given value of $J_{HN\alpha}$. Since there are also two possible angles θ for most of the $J_{HN\alpha}$-values (Fig. III-11), one can thus have as many as four possible torsion angles ϕ corresponding to an observed value of the vicinal spin–spin coupling constant $J_{HN\alpha}$.

An additional comment is called for in the case of glycine. In the ABX spin system of the C^α-methylene protons and the amide proton in this

TABLE III-2

Relation between the dihedral angle θ and the torsion angle ϕ used for the description of peptide conformations according to the IUPAC-IUB nomenclature[5d]*.

	θ	0	30	60	90	120	150	180
L-amino acid residues	ϕ	60	90 30	120 0	150 −30	180 −60	−150 −90	−120
D-amino acid residues	ϕ	−60	−30 −90	0 −120	30 −150	60 −180	90 150	120

*The angles θ and ϕ are given in degrees.

residue, a more reliable experimental value can often be obtained for the sum of the two vicinal coupling constants, $\Sigma J_{HN\alpha_2}$, than for the individual constants J_{AX} and J_{BX}. In terms of the torsion angle ϕ (Figs. III-9 and III-10)[5], this sum can be written as

$$\sum J_{HN\alpha_2} = J_{HN\alpha}(\phi + 60°) + J_{HN\alpha}(\phi - 60°) \qquad \text{(III-6)}$$

Combining Equation III-3 for the angular dependence of $J_{HN\alpha}$ with Equation III-6 then yields for the angular dependence of $\Sigma J_{HN\alpha_2}$[14, 15]

$$\sum J_{HN\alpha_2} = 6.0 \cos^2 \phi - 1.5 \cos \phi + 12.5 \sin^2 \phi \qquad \text{(III-7)}$$

Relatively few practical applications of the heteronuclear spin–spin coupling constants $J_{^{13}C'H^\alpha}$ and $J_{^{15}NH^\alpha}$ for conformational studies of peptides have so far been reported. This is mainly so because suitable isotope-labelled peptides are not yet readily available. On the other hand, a potential advantage over work with $J_{HN\alpha}$ would be that the limited spectral resolution of the ^{1}H-NMR spectra in larger peptides can here in principle be overcome by measurements of $J_{^{13}C'H^\alpha}$ and $J_{^{15}NH^\alpha}$ in the ^{13}C- and ^{15}N-NMR spectra. Of the two parameters in question, $J_{^{15}NH^\alpha}$ has attracted particular interest because it appears at present to be the only measurable quantity which would be directly related to the torsion angle ψ (Fig. III-10) in the solution conformations of peptide molecules.[16, 17] This is further discussed in Chapter VII.

Backbone Conformations with Intramolecular Hydrogen Bonds

When the amide protons in a polypeptide chain are freely accessible to the solvent molecules, they have typical chemical shifts for a given solvent (Table II-9), and the temperature dependence of the chemical shifts is also a characteristic quantity for a given solvent system. Furthermore, the solvent-accessible amide protons are rapidly exchanged with hydrogen atoms (or deuterium atoms) of protic solvents (Table II-4). In molecules where some of the amide protons are involved in intramolecular hydrogen bonds, or otherwise shielded from the solvent, the chemical shifts and the temperature coefficients for these protons will often be clearly different from those of the solvent-exposed amide protons, and the rate of the exchange reactions with the solvent will in general be considerably slower. These observations on the labile protons can in many cases yield valuable information on the peptide conformations, as will be further illustrated in Section III.3.

Trans and Cis Peptide Groups

As was mentioned at the outset of this section, only trans peptide groups (Fig. III-8) are commonly found in amino acid residues of linear polypeptides, whereas in prolyl peptides, and other N-substituted peptide groups, the cis-form is also quite frequently encountered. ^{1}H-NMR has greatly contributed to the characterization of the equilibria between the cis and trans forms of peptide groups through studies of both model amides, and a variety of imino acid residues. These data imply that in simple linear imino acid residues, the trans form is generally more stable, with $\Delta G°$ of the order 0.1 to 2.0 Kcal·mol^{-1} [18,19], and the free energy of activation for the cis-trans isomerization was found to be of the order $\Delta G^{\ddagger} \approx 20$ Kcal·mol^{-1}.[19–22] These numbers show that in solutions at ambient temperature, the cis and trans forms of the imino acid residues can occur simultaneously, and the cis-trans interconversion will be slow on the NMR time scale. Hence the NMR of the two species can in principle be observed simultaneously. As will be illustrated in Section III.3, the cis-trans isomerism of peptide groups is often manifested in different chemical shifts of certain proton resonances. It has also been suggested that long range spin–spin coupling could in principle be employed to differentiate between the cis and trans forms of certain imino acid residues[23].

Recently evidence has been accumulating that non-planar peptide groups might be quite common[24,25], and that the rigid model might not yield an adequate description of the conformations in many peptide molecules. If the rigid model were thus abandoned, a complete characterization of the backbone conformations would have to include all the torsion angles ω_i (Fig. III-10)[5], or a corresponding quantity which would describe the deviations of the peptide groups from the standard planar dimensions[25]. As was mentioned above, the deviations from the standard peptide group dimensions appear to be manifested in small variations of the spin–spin coupling constants $J_{HN\alpha}$. It seems rather unlikely, however, that this observation will find much application for the investigation of non-planar peptide groups. More promising NMR approaches for studies of the deviations of the peptide groups from their standard dimensions will be discussed in the sections on ^{13}C-NMR and ^{15}N-NMR of peptides.

Spatial Arrangement of the Amino Acid Side Chains

As indicated in Fig. III-10, the spatial arrangement of the amino acid side chains can be characterized by the torsion angles χ^j (Ref. 5). Different rotational states about the side chain bonds will be manifested in different vicinal coupling constants between protons on adjoining carbon atoms. For the coupling constant $J_{\alpha\beta}$ between the C$^\alpha$- and C$^\beta$-protons in the fragment

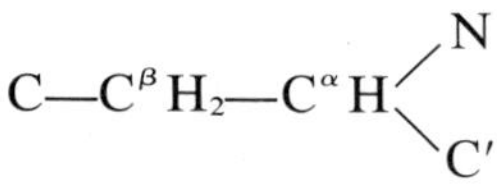

, the following Karplus-type relation has been proposed[26]

$$J_{\alpha\beta} = 11.0 \cos^2 \theta - 1.4 \cos \theta + 1.6 \sin^2 \theta \qquad \text{(III-8)}$$

where θ is the dihedral angle between the planes H—C$^\alpha$—C$^\beta$, and C$^\alpha$—C$^\beta$—H. The factors limiting the applications of $J_{HN\alpha}$ for studies of the peptide backbone apply also to $J_{\alpha\beta}$.[6] Thus, to account for the electronegativity corrections, the coupling constants observed for the fragments

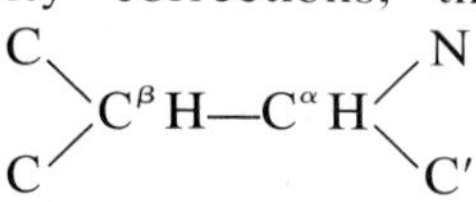

, and

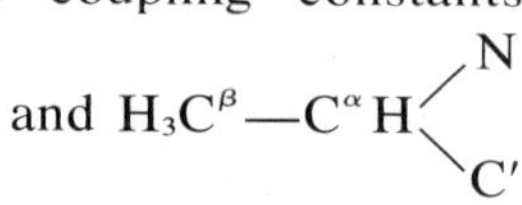

would have to be increased by 3%, or reduced by 3%, respectively, prior to their use for the determination of θ through Equation III-8, and correspondingly larger corrections of $J_{\alpha\beta}^{obs}$ would account for the oxygen atoms bound to C$^\beta$ in threonine and serine[26]. The coefficients A, B, and C of Equation III-8 were determined from experiments with a series of cyclic model compounds[26]. They were found to be in satisfactory agreement with the vicinal coupling constants observed in cyclo-L-cystine, where the H—C$^\alpha$—C$^\beta$—H angles are likely to be rigidly held near 30° and 90°, and the values for $J_{\alpha\beta}$ were found to be 7.0 and ≈ 0 Hz, respectively[27]. It has been suggested that in addition to $J_{\alpha\beta}$, the vicinal coupling constant between ^{13}C′ and C$^\beta$-protons, and between the backbone amide nitrogen atoms in ^{15}N-enriched peptides and the C$^\beta$-protons might also be used for the experimental determination of χ^j (Fig. III-10)[17].

In practice it is often difficult even in small peptides to obtain reliable data on the vicinal coupling constants $J_{\alpha\beta}$, $J_{\beta\gamma}$ etc., which would determine the conformations of the amino acid side chains. One therefore attempts to complement the measurements of the vicinal coupling constants by studies of intramolecular interactions between different segments of the polypeptide chain, which can arise as a consequence of a given set of torsion angles in the backbone, and in the amino acid side chains. Intramolecular interactions in natural peptide chains which are likely to be manifested in the ^{1}H-NMR spectra, are further discussed below. For NMR studies of peptides which can be synthesized by chemical methods, the introduction of covalently bound marker groups, which would amplify the NMR manifestations of intramolecular conformation-dependent interactions has also been suggested[28]. These synthetic probes could e.g. be aromatic rings which would then act similarly to the behavior of the aromatic amino acid side chains discussed in the following paragraph.

Natural ^{1}H-NMR Probes in Proteins

The description of the molecular conformations through the torsion angles ϕ, ψ, ω, and χ^j (Fig. III-10) can of course be applied to any polypeptide chain, independent of the size, and complete sets of the torsion angles for globular proteins have been obtained from single crystal X-ray studies[5c]. With high resolution ^{1}H-NMR studies, however, it is at present not usually possible to investigate the complete set of vicinal spin–spin coupling constants in a protein molecule, in part because of the limited spectral resolution available even with the most advanced instrumentation. Instead one has so far mainly concentrated on the investigation of the chemical shifts which arise in the spectra of proteins as a consequence of the spatial folding of the polypeptide chain (Refs. 28–30 of Chapter I).

Manifestations of the protein conformation in the ^{1}H-NMR can be due to interactions through space or via weak non-covalent bonds between different segments of the polypeptide chain. As is shown schematically in Fig. III-1B, some of the amino acid side chains in the interior of a globular protein are no longer freely accessible to the solvent, and are instead in close contact with other amino acids. Across the protein molecule, the different amino acid residues of a given kind are thus in different micro-environments characterized by different local magnetic susceptibilities, and hence their chemical shifts may be different. As a consequence of this dispersion of the chemical shifts of structurally identical protons, the ^{1}H-NMR spectra of native globular proteins contain rather broad lines, which are poorly resolved in the bigger molecules, and which consist of the mutually overlapping resonances of the many magnetically slightly non-equivalent protons (Fig. III-12). The overall ^{1}H-NMR spectral features of a polypeptide chain can thus readily be used to distinguish between extended random coil forms and globular conformations (Figs. III-2 and III-12), e.g. for a rapid qualitative characterization of the solution conformation of a new protein, or for studies of the denaturation of a globular protein to a random coil form.

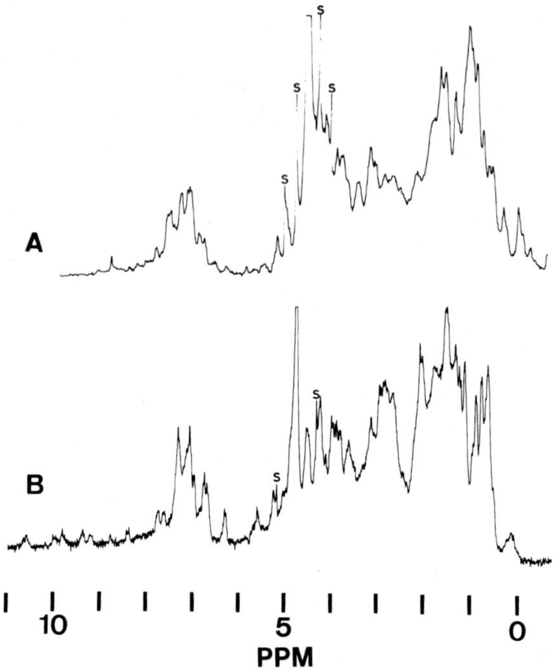

Fig. III-12. ^{1}H-NMR spectra at 220 MHz in D_2O-solutions. A. Hen egg white lysozyme (129 amino acid residues), pD = 4.8, $T = 50°$. B. The basic pancreatic trypsin inhibitor, "BPTI" (58 amino acid residues), pD = 7.7, $T = 20°$. The resonance of the residual solvent protons is at approx. 4.5 ppm in A, and 4.8 ppm in B. S identifies the spinning side bands of the water resonance.

More detailed information on various protein molecules has come from observations of a very limited number of resonance lines which could from their outstanding chemical shifts, or from other specific properties be assigned to certain regions of the protein molecule, or even to distinct amino acid residues in the polypeptide chain. It is a fortunate coincidence that some of these amino acids, which have often been found to be accessible to NMR observation even in rather big proteins, have also been reported to form an integral part of the active sites in numerous proteins. Thus even with the analysis of only a small fraction of the ^{1}H-NMR lines in a protein, it is often possible to obtain information on the active centers in these molecules.

As is to be expected from the ^{1}H-NMR of the individual amino acid residues (Table II-7), the regions between 0.5 and 5.0 ppm, and from 6.5 to 8.5 ppm in the spectra of proteins are crowded with resonance lines which have been only little or not at all displaced from their positions in the random coil polypeptide chain (Figs. III-2 and III-12). Therefore, natural NMR probes which produce resolved resonances of individual groups of protons outside of these crowded portions of the spectrum, will in principle be suitable for detailed studies. Typical examples are the unusual chemical shifts which arise in protons exposed to the local magnetic ring current fields in aromatic rings (Figs. II-2 and II-3). When an aliphatic methyl group is located near the ring plane of an aromatic amino acid side chain, its resonance can be shifted upfield by as much as 2 ppm from its position in the random coil polypeptide spectrum, and can thus appear as a well-resolved line between ca. 0.5 and − 1.5 ppm. The high field resonances of lysozyme (Fig. III-12) are a typical example. Since for a given ring size the extent of the ring current shifts depends only on the relative spatial arrangement of the ring and the observed protons (Fig. II-2), conformational changes can be clearly manifested in these resolved ring current shifted lines. Outstanding ^{1}H-NMR chemical shifts in a protein can also be produced by intramolecular hydrogen bonds involving labile protons of the polypeptide backbone, or of certain amino acid side chains. A typical example are the resonance lines between 8 and 11 ppm in the basic pancreatic trypsin inhibitor (Fig. III-12). In principle ^{1}H-NMR lines in the crowded central region of the spectrum are also potential natural probes, provided they can be resolved and identified. A classical example are the imidazole ring protons of histidine which have in a number of proteins been identified on the basis of their pH-dependence (Fig. II-13). With the modern high field spectrometers, individual resonances of other amino acids are also accessible for detailed studies (Section III.4). The NMR parameters, and possibly other conformation dependent physicochemical properties manifested in the NMR spectra, e.g. the pK-values of titratable groups, may then be used to monitor the protein conformation.

All the proteins molecules are not endowed with suitable natural NMR probes for a detailed NMR investigation. In some instances, this problem has been overcome through biosynthesis of partially deuterated protein analogues, where all but one or two kinds of amino acid residues would be fully deuterated[29,30]. The ^{1}H-NMR of the remaining protonated amino acid residues are then readily identified. Irrespective of their chemical shifts, they will in general yield quite well resolved lines, so that they can be used as ^{1}H-NMR probes in the selectively deuterated proteins.

3. CONFORMATIONAL STUDIES OF LINEAR AND CYCLIC OLIGOPEPTIDES

Over the last few years, ^{1}H-NMR investigations have been reported for a variety of natural and synthetic peptides. The motivation for these studies can be traced on the one hand to the interest in these compounds as models for protein structure and conformation, and on the other hand to the wide spectrum of biological functions of low molecular weight peptide molecules. It is reasonable to suppose, and has been experimentally substantiated in different examples, that there is a close relationship between the molecular conformations of the peptides and their biological roles as, for example, hormones, antibiotics, ionophores, or toxins. In this section, we shall first discuss the general strategies for ^{1}H-NMR studies of oligopeptides, which will be illustrated with two examples, i.e. cyclo-glycyl-L-prolyl-glycyl-glycyl-L-prolyl-glycyl, and cyclo-glycyl-L-alanyl-glycyl-glycyl-L-prolyl. This will then be followed by a brief introduction to the recent literature.

Peptide molecules with up to approximately ten amino acid residues will in general yield ^{1}H-NMR spectra with reasonably well resolved lines (Fig. III-13) from which the chemical shifts and the spin–spin coupling constants for the individual groups of protons can be determined. As was discussed in Section III.2, the ^{1}H-NMR studies can thus provide information on the "permissible ranges" for the torsion angles ϕ_i of all the amino acid residues (Fig. III-10), on intramolecular hydrogen bonding, and possibly also on the torsion angles χ_i^j (Fig. III-10) or on specific intramolecular interactions between different amino acid side chains. These data cannot in general in themselves be sufficient for a complete description of the molecular conformation. However, by judiciously combining the NMR data with theoretical investigations of the molecular geometries and the conformational energies in the peptide molecules, a very limited number of molecular conformations consistent with both sets of data can often be obtained[31]. The theoretical model studies[5] taken separately would in general result in a considerably larger number of favorable conformations. Combination with the NMR data will allow to eliminate all those among the energetically favored species, having one or several residues in a conformation which is not consistent with the observed $J_{HN\alpha}$ coupling, or with the observed formation of intramolecular hydrogen bonds.

Cyclo-Glycyl-L-Prolyl-Glycyl-Glycyl-L-Prolyl-Glycyl

This synthetic cyclohexapeptide[32] presents a relatively simple example in which two groups of three amino acid residues are related by the C_2-symmetry of the covalent structure. The C_2-symmetry is also clearly manifested in the ^{1}H-NMR spectrum (Fig. III-13A), which consists of the two ABX spin systems corresponding to the glycyl residues 1 and 4, and 3 and 6, respectively, and the resonances of the prolyl residues 2 and 5. From comparison with the individual amino acid residues (Table II-9), and on the basis of double resonance experiments, it could readily be shown that the resonances marked a, b, and e correspond to the two equivalent prolyl

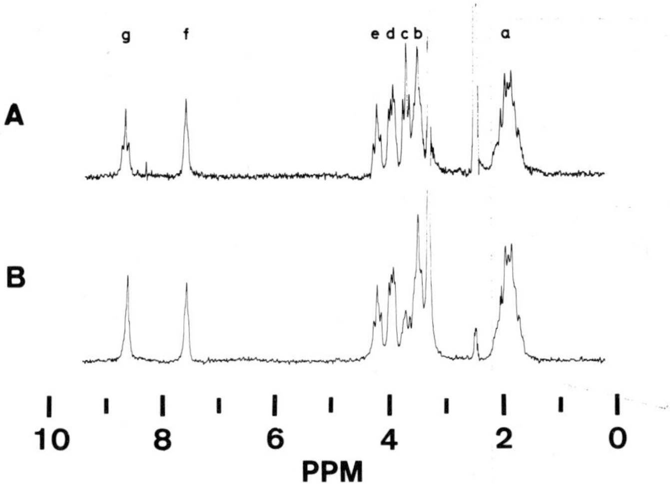

Fig. III-13. ^{1}H-NMR spectra at 100 MHz of A. *c*-[-Gly(1)-L-Pro(2)-Gly(3)-Gly(4)-L-Pro(5)-Gly(6)-]. B. *c*-[-Gly(1)-L-Pro(2)-Gly(d_2)(3)-Gly(4)-L-Pro(5)-Gly(d_2)(6)-] in DMSO-d_6. T = 27°. The residual protons of the solvent are observed at 2.5 ppm, and a trace of H_2O gives rise to the line at 3.3 ppm (from Schwyzer et al.[34]).

residues, the lines c and g to the one, and the lines d and f to the alternate ABX spin system of glycine. The temperature dependence of the amide proton line at 8.67 ppm corresponds very nearly to that observed in simple amides, which is e.g. 0.006 ppm/deg. in a DMSO solution of N-methylacetamide[33], whereas the resonance at 7.61 ppm is essentially independent of temperature (Fig. III-14). This shows that the amide protons observed at 7.61 ppm are shielded from the solvent, most probably by transannular hydrogen bonding in a β-type structure (Fig. III-15)[32]. At this point one was then left with the choice between the two different β-conformations shown in Fig. III-15. This ambiguity could be resolved with the partially deuterated peptide analogue *c*-[-Gly-L-Pro-Gly(d_2)-]$_2$, where the C^α-protons of Gly(3) and Gly(6) are replaced by deuterium [34, 35]. Comparison of the spectra A and B in Fig. III-13 shows that the resonances c and g correspond to the glycines 3 and 6. The resonance c is identified from its almost complete disappearance upon deuteration, the resonance g from the absence of the spin–spin coupling with the C^α-protons in spectrum B. Hence only the conformation A of Fig. III-15 would be compatible with the NMR data.

Some additional features of the conformation of *c*-[-Gly-L-Pro-Gly-]$_2$ can be derived from the observed chemical shifts and spin–spin coupling constants. Thus the chemical shift of the prolyl C^α-proton (Fig. 13e) is quite typical for the trans form of the Gly-L-Pro peptide group[36] (see also the

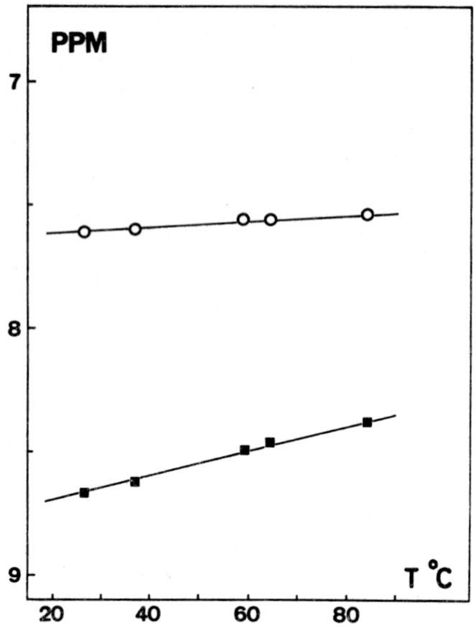

Fig. III-14. Dependence on temperature of the amide proton resonances of *c*-[Gly-L-Pro-Gly-]$_2$ in Fig. 13 (from Schwyzer et al.[34]).

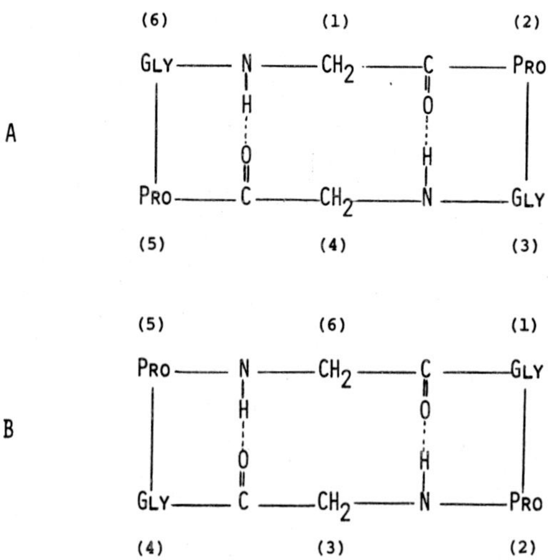

Fig. III-15. Two β-structures of *c*-[-Gly-L-Pro-Gly-]$_2$ which both contain two amide protons in transannular hydrogen bonds, and two solvent accessible amide protons.

TABLE III-3

^{1}H-NMR parameters for the glycyl residues in *c*-[Gly-L-Pro-Gly-]$_2$ obtained from spectrum simulation with the formalism of Section II.3*.

Residue	Spin system	Chemical shifts (ppm) and spin–spin coupling constants (Hz ± 1 Hz)	
Gly(1,4)	ABX	$\delta_A = 3.87$	$J_{AB} = (-)18.5$
		$\delta_B = 4.06$	$J_{AX} = 2.5$
		$\delta_X = 7.61$	$J_{BX} = 4.5$
Gly(3,6)	ABX	$\delta_A = 3.64$	$J_{AB} = (-)17.5$
		$\delta_B = 3.79$	$J_{AX} = 5.0$
		$\delta_X = 8.67$	$J_{BX} = 6.0$

*A good fit with the experimental spectra (Fig. 13) was obtained when these parameters were used in combination with a half-width of the lines at half-height of 1.5 Hz (from Schwyzer et al.[34]).

following example, and Chapter V), and approximate values for the torsion angles ϕ could be derived from the spin–spin coupling constants $J_{HN\alpha}$ in Table III-3, i.e. $\phi_1 = \phi_4 \approx -165°$, and $\phi_3 = \phi_6 \approx 100°$. Since in addition $\phi_2 = \phi_5$ is fixed at ca. 110° by the covalent structure of proline[5], the ^{1}H-NMR data for this symmetric peptide are almost sufficient in themselves for a complete description of the molecular conformation[34, 35].

As was mentioned in the foregoing section, the two types of amide protons can also be distinguished from their rate of exchange with hydrogen atoms of protic solvents. In a typical experiment, one adds a trace of D_2O to the peptide solution in DMSO-d_6, and measures the rate of the disappearance of the amide proton resonances. The data for *c*-[-Gly-L-Pro-Gly-]$_2$ in Table III-4 show on the one hand that the studies of the exchange rates confirmed the above interpretation of the temperature dependence (Fig. III-14) of the amide proton resonances. On the other hand, they also show that much care has to be taken in interpreting these measurements since little is known regarding the effect of DMSO on the amide-water hydrogen exchange. Furthermore absolute rates of exchange depend very sensitively on the pH of the water added, and in the reaction with 0.1-n DCl in D_2O (Table III-4) the "shielded" protons in this peptide appear to exchange even more rapidly than the solvent accessible protons. In this context it might be added that special care is also called for to exclude traces of metal ions, which can greatly affect the proton exchange rates.

TABLE III-4

Amide proton exchange rates in a solution of *c*-[Gly-L-Pro-Gly-]$_2$ in DMSO-d$_6$ after addition of D_2O. 20 μl of D_2O were added to 0.5 ml of a 0.05-M solution of the peptide in DMSO-d$_6$*.

pD of D_2O added (DCl/NaOD)	$\tau_{1/2}$ (minutes)	
	NH-3 and -6 (8.67 ppm)	NH-1 and -4 (7.61 ppm)
1.1	23	6
2.2	80	200
3.9	35	65
"neutral"	~ 180	~ 700
9.5	4	8

*The NMR parameters (Fig. 13, Table III-3) were not affected by this change in the solvent medium.

Cyclo-Glycyl-L-Alanyl-Glycyl-Glycyl-L-Prolyl

This second example was chosen to complement the description of the highly symmetrical c-[-Gly-L-Pro-Gly-]$_2$ with a more complex peptide illustrating some additional quite frequently encountered aspects. In the covalent structure of the cyclopentapeptide c-[-Gly(1)-L-Ala(2)-Gly(3)-Gly(4)-L-Pro(5)-], no amino acid residues are related by a symmetry operation and because two different conformations are simultaneously present in the solution, every one of the five amino acid residues gives rise to two distinct sets of resonances[37]. As is seen in Fig. III-16, the relative intensities of the two spectra are approximately 2:1. The resulting ^{1}H-NMR spectrum consists thus of two A_3PX spin systems of alanine(2), six ABX systems of Gly(1), Gly(3), and Gly(4), and of two seven-spin systems of proline(5). Following the procedures outlined in the previous example, i.e. partial deuteration of the α-methylene groups of glycine (Fig. III-17)[38], double resonance experiments[37], and spectrum simulation, this relatively complex spectrum was unravelled, and the resonances of all the individual amino acid residues identified in the two conformations M and m. The NMR parameters are collected in Table III-5. The temperature dependence of the amide proton resonances indicates that one proton in the species m, and two protons in the conformation M are shielded from the solvent by their environment in the peptide molecule (Fig. III-18).

In the species m and M, the chemical shifts of the C^α-protons of proline

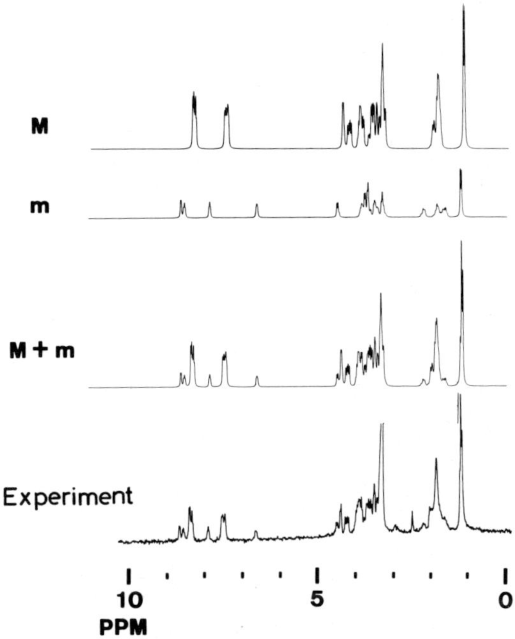

Fig. III-16. ^{1}H-NMR spectra at 220 MHz of *c*-[-Gly(1)-L-Ala(2)-Gly(3)-Gly(4)-L-Pro(5)-] in DMSO-d_6. In the experimental spectrum, two lines at 2.5 and 3.3 ppm correspond to the residual protons in DMSO-d_6, and to H_2O. The spectra M and m for the two species present in the solution have been simulated with the parameters of Table III-5 using a half-width of the lines at half-height of 2.2 Hz. The sum of the computed spectra, M + m, is to be compared with the experimental spectrum (from Meraldi et al.[37]).

(Table III-5) correspond quite closely to those which had previously been found for the cis- and trans-forms of the Gly-L-Pro peptide bond in DMSO-d_6[36]. Figs. III-16 and III-17 show that the cis and trans forms of proline can quite readily be distinguished in the ^{1}H-NMR spectra at 220 MHz or higher frequency. The high energy of activation for the cis-trans isomerization of proline can also explain the relatively slow rate of interconversion between the conformations M and m, which was studied with NMR double resonance techniques[39]. INDOR experiments showed that double resonance effects were mediated between corresponding resonances of the two species M and m by the exchange of peptide molecules between the two conformations, and it was found that τ_M, the life time in the conformation M, has to be within the limits $3 \cdot 10^{-1}$ sec $\geqslant \tau_M \geqslant 2 \cdot 10^{-2}$ sec[37]. This experiment further led to the identification of the corresponding resonances in the two molecular conformations (Table III-5) before the partially deuterated peptides became available. On the other hand, just as in the foregoing example,

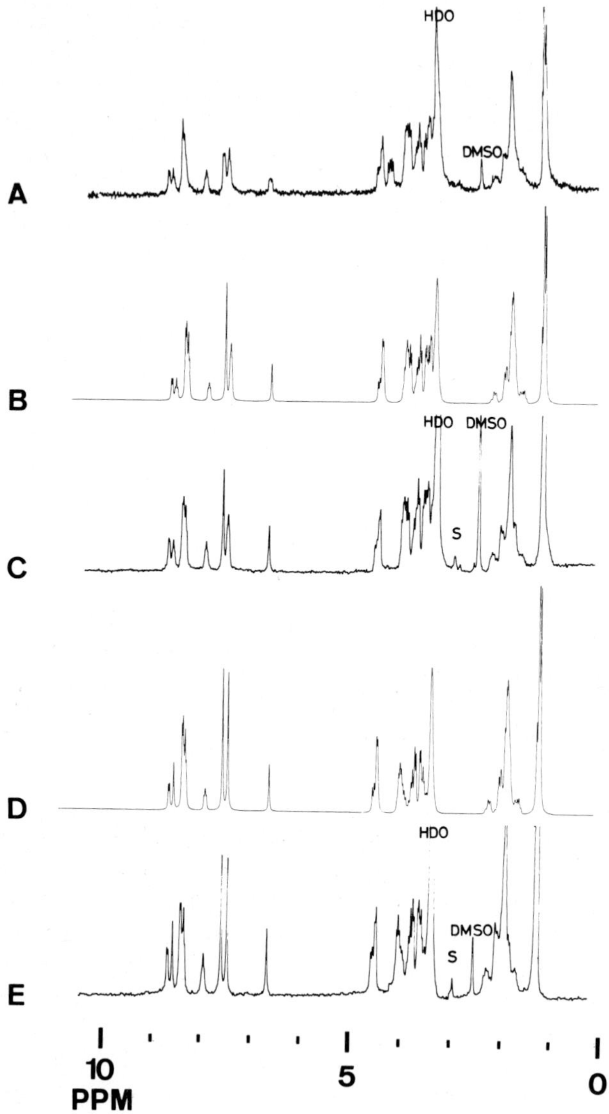

Fig. III-17. ^{1}H-NMR spectra at 220 MHz and 22° of solutions in DMSO-d_6 of A. *c*-[-Gly-L-Ala-Gly-Gly-L-Pro-] and the following partially deuterated analogues: B and C: *c*-[-Gly(d_2)-L-Ala-Gly-Gly-L-Pro-]; D and E: *c*-[-Gly(d_2)-L-Ala-Gly-Gly(d_2)-L-Pro-]. The spectra B and D had been computed with the parameters of Table III-5, using a half-width at half-height of the lines of 2.2 Hz. C and E are the observed spectra (from Meraldi[14]).

TABLE III-5

^{1}H-NMR parameters of the two molecular conformations M (65%) and m (35%) in solutions of *c*-[-Gly-L-Ala-Gly-Gly-L-Pro-] in DMSO-d$_6$ at 22°*.

Residue	Spin system	Conformation M		Conformation m	
Gly(1)	ABX	$\delta_A^M = 3.39$	$J_{AB}^M = 14.0$	$\delta_A^m = 3.49$	$J_{AB}^m = 14.0$
		$\delta_B^M = 4.32$	$J_{AX}^M = 3.1$	$\delta_B^m = 3.98$	$J_{AX}^m = 3.0$
		$\delta_X^M = 7.63$	$J_{BX}^M = 7.0$	$\delta_X^m = 6.72$	$J_{BX}^m = 7.7$
Ala(2)	A_3PX	$\delta_A^M = 1.27$	$J_{AP}^M = 7.0$	$\delta_A^m = 1.31$	$J_{AP}^m = 7.0$
		$\delta_P^M = 4.05$	$J_{AX}^M = 0.0$	$\delta_P^m = 3.96$	$J_{AX}^m = 0.0$
		$\delta_X^M = 8.42$	$J_{PX}^M = 4.6$	$\delta_X^m = 8.70$	$J_{PX}^m = 5.7$
Gly(3)	ABX	$\delta_A^M = 3.62$	$J_{AB}^M = 15.2$	$\delta_A^m = 3.57$	$J_{AB}^m = 15.7$
		$\delta_B^M = 3.77$	$J_{AX}^M = 4.8$	$\delta_B^m = 3.81$	$J_{AX}^m = 4.6$
		$\delta_X^M = 8.39$	$J_{BX}^M = 5.9$	$\delta_X^m = 7.96$	$J_{BX}^m = 6.8$
Gly(4)	ABX	$\delta_A^M = 3.52$	$J_{AB}^M = 14.0$	$\delta_A^m = 3.57$	$J_{AB}^m = 14.0$
		$\delta_B^M = 3.98$	$J_{AX}^M = 3.8$	$\delta_B^m = 3.80$	$J_{AX}^m = 5.8$
		$\delta_X^M = 7.53$	$J_{BX}^M = 6.0$	$\delta_X^m = 8.62$	$J_{BX}^m = 6.2$
Pro(5)	ABCDPQX	$\delta_A^M = 1.90$		$\delta_A^m = 3.57$	
		$\delta_B^M = 1.91$		$\delta_B^m = 1.91$	
		$\delta_C^M = 1.87$		$\delta_C^m = 1.68$	
		$\delta_D^M = 2.05$		$\delta_D^m = 2.27$	
		$\delta_P^M = 3.40$		$\delta_P^m = 3.42$	
		$\delta_Q^M = 3.43$		$\delta_Q^m = 3.45$	
		$\delta_X^M = 4.49$		$\delta_X^m = 4.55$	

*The chemical shifts δ are in ppm, the spin–spin coupling constants in Hz $\pm$ 1.0 Hz (from Meraldi[14]).

the different ABX spin systems could only on the basis of the partially deuterated molecules (Fig. III-17) be assigned to specific glycyl residues.

Preliminary studies indicate that even in this rather complex peptide the combination of the ^{1}H-NMR data with rigid model conformational energy calculations leads to a small number of NMR compatible low energy conformations[14]. However, it appears that further investigations will be needed in this case to decide whether the rigid model, in which all the peptide groups are assumed to be in the standard planar cis or trans form, is a reasonable approach, or if on the contrary the cyclisation conditions imposed by the rigid model are too restrictive in that they eliminate certain NMR compatible conformations which might be energetically quite favorable despite the inclusion of one or several distorted peptide groups.

Selected Studies of Linear Peptides

A large number of small linear peptides have been investigated in connection with NMR-studies of polypeptides and proteins, e.g. to obtain

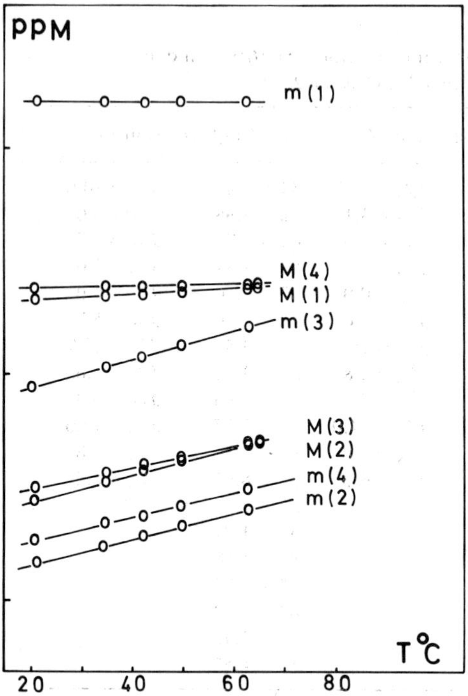

Fig. III-18. Temperature dependence of the amide proton resonance chemical shifts of *c*-[-Gly-L-Ala-Gly-Gly-L-Pro-] in DMSO-d_6 (Fig. III-16). M: more abundant conformation (65%), m: less abundant conformation (35%) (from Meraldi[14]).

the ^{1}H-NMR parameters of the individual amino acid residues (Tables II-7 and II-9), in the attempts to use NMR for amino acid sequence analysis[3,4], when using comparison with small peptide fragments for resonance identification in synthetic polypeptides (Figs. III-5 and III-6)[1], etc. Here, the discussion will be limited to a few selected synthetic and natural peptides which have recently attracted particular interest.

The general strategy for ^{1}H-NMR investigations of linear peptides follows that outlined at the outset of this section, and illustrated with the foregoing description of two cyclopeptides. However, in short linear peptides the combination of the spectroscopic data with theoretical studies of the molecular geometry and the conformational energy has so far not generally resulted in interpretations in terms of a precise description of the molecular conformations. The implication is that very often many energetically nearly equivalent conformations of linear oligopeptides are simultaneously present in a rapid equilibrium situation, so that the observed NMR parameters correspond to an average taken over a manifold of rotational states (e.g. Equation III-4). A detailed discussion of the interpretation of the ^{1}H-NMR parameters in terms of the populations of the different rotational states in small flexible peptides, in particular alanine dipeptides, was presented by Bystrov et al[7].

In linear peptides containing one or several imino acid residues, ^{1}H-NMR has been applied for investigations of the equilibria between cis and trans peptide bonds. In proline containing peptides in DMSO-solution, the cis-trans isomerization is manifested in much the same way as in the cyclopentapeptide shown in Figs. III-16 and III-17. In different solvents the relative chemical shifts of the cis- and trans-forms may be reversed, e. g. in $CDCl_3$ solution of t-Boc-Gly-L-Pro-OH, the C^{α}-proton resonance of cis-proline is at higher field than the corresponding trans proline resonance[36]. The hypothalamic thyrotropin-releasing factor, L-pyroglutamyl-L-histidyl-L-proline amide, is an example of a natural proline-containing tripeptide which has been extensively investigated by ^{1}H-NMR[40,41]. ^{1}H-NMR studies and conformational energy calculations have been combined in an investigation of the cis-trans equilibrium in aqueous solutions of sarcosylsarcosine, where again both forms of the peptide group are simultaneously present at ambient temperature[42]. Further examples will be discussed in Chapter V, which will show that ^{13}C-NMR is an even more powerful tool than ^{1}H-NMR for investigations of the cis-trans isomerization of peptide groups.

Extensive efforts have recently been directed to the determination of the solution conformations in the peptide hormone angiotensin II[43–46], and in neurohypophyseal peptides[47–61], which all consist of one peptide chain with less than 10 amino acid residues. As in many other NMR experiments with peptides, much depends on the choice of the solvent for the investigations of these peptide hormones (see also Table II-4). Thus, DMSO-d_6 is quite generally a good solvent in which the amide protons can readily be observed. However, even though much NMR information can be lost either because of the exchange of the labile protons in D_2O, or the overlap of the solvent resonance with the peptide spectrum in H_2O, aqueous solution data would often seem to be more relevant from the biologists point of view. In connection with studies of oxytocin and lysine vasopressin, a way out of this dilemma has been sought in terms of correlating ^{1}H-NMR assignments in different solvents[57]: Starting with the resonance assignments in DMSO-d_6, the course of each of the resonances was monitored as the proportion of H_2O to DMSO-d_6 in mixtures of the two solvents was gradually increased. With this procedure, certain inferences on the molecular conformations in the aqueous medium could thus be drawn from the NMR data in DMSO-d_6.

Selected Studies of Cyclopeptides

Because of the cyclization conditions imposed on the molecular geometry, conformational energy calculations for cyclopeptides result in general in a much smaller selection of low energy conformations than for the sterically less restricted linear peptides. Therefore, the combination of the computed low conformational energy species with the ^{1}H-NMR data has in many cases led to a rather detailed description of the backbone conformations in the cyclic peptide molecules. The method has been applied to a wide variety of synthetic and natural cyclopeptides. Reviews of some of the earlier work are available[63, 64].

Among the synthetic cyclopeptides, one finds molecules with ring sizes between two and approx. ten amino acid residues. In cyclic dipeptides, a rather rigid ring conformation is imposed by the covalent structure, and therefore the ^{1}H-NMR studies have also concentrated on the conformations of the amino acid side chains[11, 27, 65, 66]. For example, a preference for the folded conformation (Fig. III-19A) over the open forms (Fig. III-19B) of

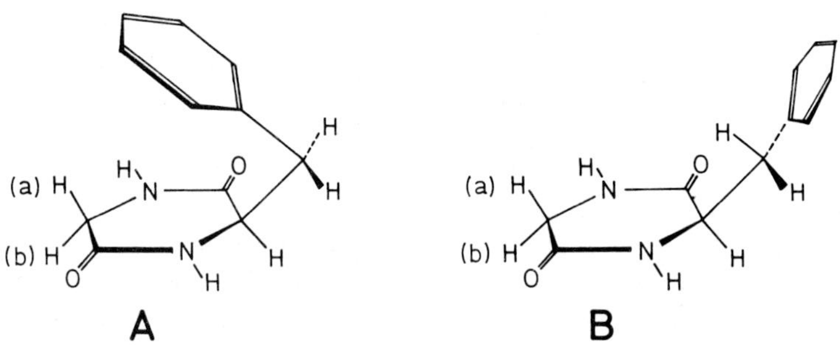

Fig. III-19. Cyclo-D-Phenylalanyl-Glycyl. A. Folded Conformation. B. Open Conformation.

cyclodipeptides containing an aromatic side chain was found to be manifested in an upfield shift of the glycyl C^{α}-proton a (Fig. III-19), which is due to the ring current field of the aromatic ring (Figs. II-2 and II-3)[65]. In cyclo-trisarcosyl[67] and cyclo-tri-L-prolyl[68], the ^{1}H-NMR imply that all the peptide groups are in the cis conformation, whereas in cyclo-tetrasar-cosyl the sequence of peptide group conformations was found to be cis-trans-cis-trans[67]. In a series of proline containing cyclopentapeptides, one of which was presented in more detail above (Figs. III-16 to III-18), two molecular species including respectively cis- and trans-proline could be distinguished[14, 37, 38]. ^{1}H-NMR has been applied for conformational studies of a large variety of cyclohexapeptides[34, 69–76]. β-type backbone conformations (Fig. III-15) are manifested in the ^{1}H-NMR of most of these molecules in much the same way as shown in Figs. III-13 and III-14. In some of the proline containing hexapeptides, however, the occurrence of multiple conformations was observed in similar spectral features as those shown for the cyclopentapeptide in Fig. III-16[74–76].

Representative examples for ^{1}H-NMR investigations of the molecular conformations of natural cyclopeptides are found among the ferrichromes. Evidence for unusually rigid solution conformations including a β-type backbone structure (Fig. III-15) was found for these cyclohexapeptides, which form extremely stable iron complexes in their natural environments[77–79]. ^{1}H-NMR studies include also some biological cycloheptapeptides, e.g. evolidine[80], ilamycin B_1[81], and phalloidin[82], and much effort has been invested in conformational studies of cyclodecapeptides, e.g. gramicidin S[33, 83–86], and antamanide[87]. With its ability to form metal ion complexes[88], antamanide can also be classified with the cation-specific cyclic peptides and depsipeptides which have been extensively investigated both in their metal free forms, and in complexes with different metal ions. Illustrative examples of this group of peptide molecules, which appear to have an important role in the transport of ions across membranes, are valinomycin[33, 89–92], and enniatin B[91, 93].

4. GLOBULAR PROTEINS

The strategy for a NMR investigation of the molecular conformations in a protein molecule will depend in each individual case on the amount of information already available from studies with other techniques. Thus the molecular weight, the amino acid composition, the amino acid sequence, single crystal X-ray data, or the occurrence of suitable mutations in proteins from different species, which could be used for resonance assignments, can all be important factors for the NMR studies. A common trait, however, of most of the conformational studies in globular proteins is that much of the experimental data come from observations of natural NMR probes of the type mentioned in Section III.2. In the following, this will be illustrated with selected experiments with diamagnetic proteins. In subsequent chapters, we shall see that the prosthetic groups in conjugated proteins represent additional very useful natural NMR probes.

Lysozyme and the Pancreatic Trypsin Inhibitor: Protein Denaturation

Hen egg white lysozyme consists of one polypeptide chain of 129 amino acid residues, and there are four disulfide bonds in the native form of the molecule. Six tryptophanyl residues are in the positions 28, 62, 63, 108, 111, and 123 of the polypeptide chain, and the molecule contains also three phenylalanyl and three tyrosyl residues. Lysozyme has been extensively investigated by ^{1}H-NMR techniques[94–96], and some typical NMR features are shown in Figs. III-12A and III-20. The ^{1}H-NMR spectrum of a lysozyme solution in D_2O (Fig. III-12A) contains a considerable number of resonance lines with chemical shifts which are not observed in the individual amino acid residues, e.g. in the spectral regions between −1 and 0.8 ppm, and 5.0 to 6.8 ppm. From their intensities, the resonances between 0.8 and −1.0 ppm correspond to approximately 45 protons, and these have been assigned to aliphatic amino acid side chains located near the aromatic amino acids, in particular the tryptophanyl residues[94]. Fig. III-20 shows the behavior of these high field resonances during the thermal denaturation of lysozyme[96]. During the early stages of the denaturation, the intensity of the ring current shifted lines decreases markedly, and the spectrum at 73° corresponds to

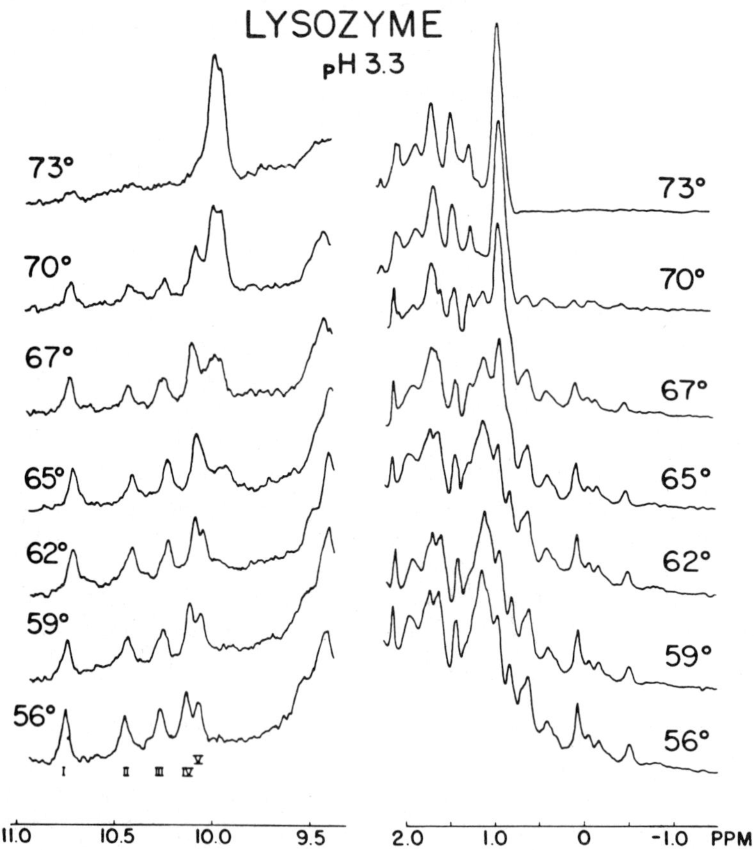

Fig. III-20. ¹H-NMR spectra of a 10% solution of hen egg white lysozyme in H_2O, pH = 3.3, through thermal denaturation. Different vertical scales are used in the representation of the low field and high field regions of the spectrum. 56°: native globular conformation (compare with the spectrum in D_2O, Fig. III-12). 73°: denatured "random coil" conformation with intact disulfide bonds (from McDonald et al.[96]).

that of the "random coil" form of lysozyme which still contains the intact disulfide bridges[95]. These and complimentary observations in different spectral regions have been used to differentiate between different possible mechanisms for the denaturation process[96]. In the lysozyme solution in H_2O (Fig. III-20), additional resolved lines occur at low field. These, and the unusual resonance positions between 5.0 and 6.8 ppm will be further discussed below.

The basic pancreatic trypsin inhibitor (BPTI) consists of one polypeptide chain with 58 amino acid residues, and there are three disulfide bonds in the native form of the molecule. The aromatic amino acid residues include four tyrosines, and four phenylalanines. This "miniprotein" forms a very stable globular conformation, and can be denatured only under rather extreme conditions of temperature and denaturing chemical agents. The unusual chemical shifts observed in a D_2O solution of the native BPTI (Fig. III-12B) for the resonances at 0 ppm, between 5 and 6.8 ppm, and between 8 and 11 ppm disappear in the heat-denatured protein (Fig. III-21B), which has a typical random coil spectrum[97]. The lines between 5 and 6.8 ppm, and the high field resonance at 0 ppm reappear when the solution is cooled again (Fig. III-21D), whereas the labile protons observed at low field had been replaced by deuterium atoms of the solvent. A random coil form of BPTI is also obtained when the protein is dissolved in DMSO[97]. Fig. III-22 shows that the ^{1}H-NMR of the labile protons of BPTI in DMSO give rise to a broad line centered at around 8.0 ppm. The rapid exchange of these protons after addition of some D_2O confirms the random coil character of the molecular conformation of BPTI in DMSO. The native form of BPTI is obtained when the solvent consisting primarily of DMSO is replaced by an aqueous medium (Fig. III-22).

In addition to their interest for investigations of protein denaturation, the spectral differences between the native and random coil forms of globular proteins (Figs. III-20 and III-22) are the most direct demonstration of the manifestations of the spatial molecular structures in the ^{1}H-NMR spectra. In the following, the discussion will concentrate on the relations of some of these conformation dependent spectral phenomena with specific structural features of the protein molecules.

Protein Conformations in Single Crystals and in Solution

When the single crystal atomic coordinates of a protein are available, and the local magnetic fields of the individual components are known, the conformation-dependent proton NMR chemical shifts which would arise if the molecular conformations in the crystal and in solution were identical, can be estimated. Comparison of the calculated chemical shifts with the

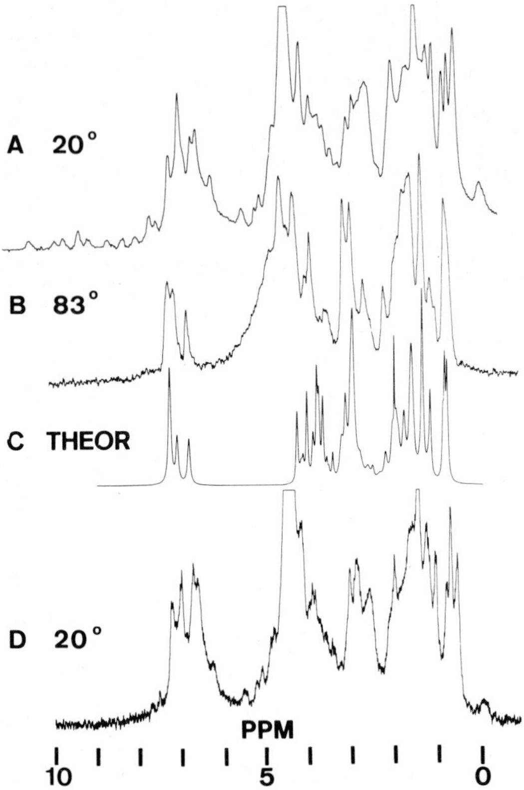

Fig. III-21. ^{1}H-NMR spectra of a D_2O solution containing 0.01 M of the basic pancreatic trypsin inhibitor (BPTI) and 6-M guanidium chloride, pD = 7.1. A. Native globular conformation of BPTI at 20°. B. Denatured "random coil" conformation with intact disulfide bonds. C. Hypothetical spectrum for the random coil polypeptide chain computed with the data of Table II-7. D. Spectrum of the solution B after cooling to 20° (from Masson and Wüthrich[97]).

observed spectra then provides a criterion for the investigation of the relations between the single crystal and solution conformations.

In practice such an analysis will depend most heavily on those resonance lines which are, as a consequence of the folding of the protein, shifted to extreme positions at high or low field. Therefore, ring current shifted high field lines have been extensively investigated. With d and θ (Fig. II-2) determined by the atomic coordinates, and using approximate local magnetic fields for the aromatic rings (Fig. III-3), the ring current shifts can be computed for all the atoms in the molecule. On the basis of such an analysis, Sternlicht and Wilson[98] found that the high field region of the spectrum of lysozyme between 1 and − 1 ppm, could be accounted for by the

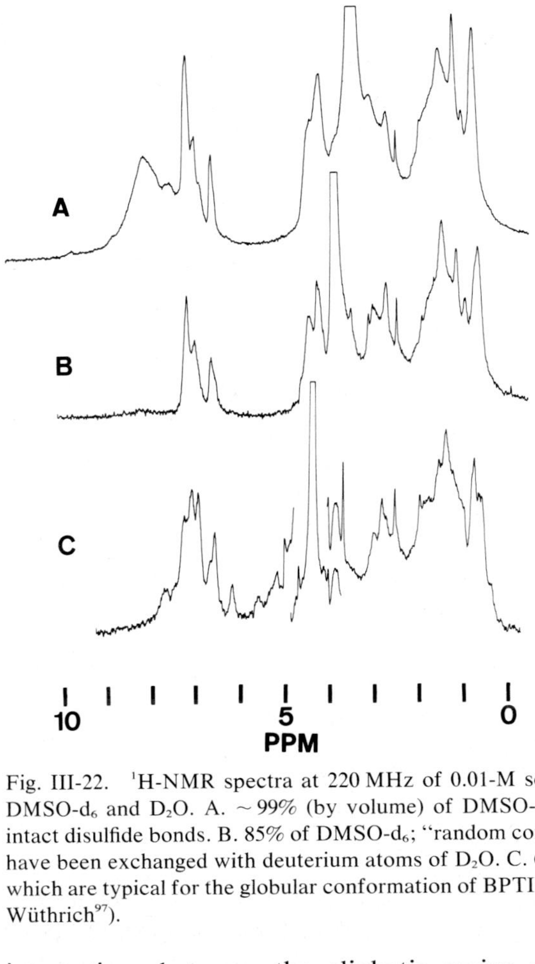

Fig. III-22. ^{1}H-NMR spectra at 220 MHz of 0.01-M solutions of BPTI in mixed solvents of DMSO-d_6 and D_2O. A. ~99% (by volume) of DMSO-d_6; "random coil" form of BPTI with intact disulfide bonds. B. 85% of DMSO-d_6; "random coil" spectrum, the labile hydrogen atoms have been exchanged with deuterium atoms of D_2O. C. 65% of DMSO-d_6; some of the features which are typical for the globular conformation of BPTI in D_2O are observed (from Masson and Wüthrich[97]).

interactions between the aliphatic amino acid side chains and the ring current fields of the aromatic rings.

When a close correspondence between the protein conformations in the crystal and in solution is evidenced by the above procedures, some of the largely shifted and well resolved resonance lines (Fig. III-20) can in general be assigned to specific amino acid residues[98]. These can then be used as probes for conformational changes in well defined regions of the molecule, e.g. during interactions of an enzyme with substrates, inhibitors, or other effector molecules.

While it appears that the consideration of the aromatic rings alone yields a reasonably good approximative explanation of the spectral features at high fields from 1.0 ppm (Figs. III-20 and III-22), it appears that observations of unusual chemical shifts in other spectral regions cannot in all cases satisfactorily be accounted for solely by aromatic ring current shifts. Typical examples are the resonances between 5.0 and 6.8 ppm in BPTI, the origin of which has not yet been ascertained.[1,99] Lysozyme (Fig. III-12) and ribonuclease A contain also a number of resonance lines between 5.0 and 6.8 ppm which appear not to be readily explicable by the aromatic ring currents[99]. This indicates that a thorough comparison of the experimental ^{1}H-NMR spectra with those computed on the basis of the single crystal atomic coordinates would have to include the local fields of other structural elements in addition to the aromatic rings. Thus for example, model computations of the influence of the helicity vs. random coil forms of the polypeptide chains have been described[98]. However, quite generally the extent of the effective local magnetic and electric fields of the individual fragments, and of combinations of different groups in the interior of a globular protein is difficult to assess with the present theoretical background, and much work is still to be done.

Additional qualitative correlations between single crystal and solution conformations of proteins can be obtained from observations of the labile protons. For example one may try to correlate the number of slowly exchanging amide protons in solutions of a protein with the number of amide protons that are buried in the interior of the molecule or involved in intramolecular hydrogen bonds in the single crystal conformation. The previously mentioned NMR studies of ferrichrome[79] have clearly shown the usefulness of such hydrogen exchange studies even for investigations of small peptides. Further illustrations will be presented below.

Studies of the Histidyl Residues

The pK-value for the imidazole ring of histidine is in the physiological pH region, and protonation-deprotonation reactions of histidine appear to be an important factor in many enzymatic processes. The pK value of histidyl residues located in interior regions of a protein molecule, or near other charged groups on the protein surface, can differ sizeably from that of free histidine, and in this way the reactivity of the histidines can to a certain extent be monitored by the protein conformation. On the other hand, the pK-value, and the chemical shifts of the histidyl protons can be used as natural probes for studies of the protein conformations.

We have already seen that a single histidyl residue in a protein can readily be identified in the ^{1}H-NMR spectrum, and its pK-value determined from the pH-dependence of the chemical shift (Fig. II-13). Thus, in calcitonin M the NMR data indicate that the single histidyl residue is freely exposed to the solvent[1]. In lysozyme, which is another example of a protein containing a single histidine, NMR observation of the C2 imidazole ring proton, which gives rise to the sharp line at approximately 8.8 ppm in Fig. III-12, provides evidence that the histidine side chain is appreciably affected by the neighboring groups in the native protein[100]. In proteins containing more than one histidyl residue, one encounters a considerably more complex system. Yet, in several cases so far, the histidine resonances could be identified from their pH dependence, and the individual pK_a values could then be determined. As an illustrative example, Figs. III-23 to III-25 show some

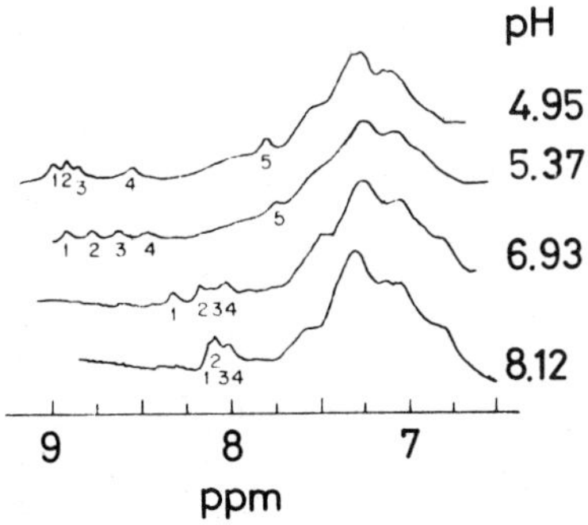

Fig. III-23. Aromatic region of the ^{1}H-NMR spectra at 100 MHz of 0.01-M solutions of ribonuclease A in deuteroacetate buffer at four different "pH-values" (measured with a glass electrode, without correction for the presence of D_2O). The signal: noise ratio had been improved by spectrum accumulation in a computer of average transients (CAT). The lines marked 1 to 4 correspond to the C-2 imidazole protons of the four histidyl residues, the peak 5 to one of the C-4 imidazole protons. Between 6.5 and 7.7 ppm one observes the mutually overlapping resonances of the three other histidine C-4 imidazole protons, and the aromatic protons of the six tyrosyl and three phenylalanyl residues. The chemical shift scale is referred to external HMDS (from Meadows et al.[101]).

spectra of bovine ribonuclease A. This enzyme consists of one polypeptide chain with 124 amino acid residues, including four histidines in the positions 12, 48, 105 and 119. Fig. III-23 shows the low field region of the ^{1}H-NMR spectrum of ribonuclease after the labile protons had been exchanged by deuterium[101]. The four C2-protons are clearly non-equivalent in the spectrum. On the basis of a long series of experiments, which included various chemical modifications of ribonuclease A, and studies of enzyme-inhibitor complexes[101–106], the individual resonances were assigned to specific ones of the four histidines. The resonance assignments, the titration curves, and the pK_a-values determined by NMR are summarized in Fig. III-24[105].

Once the histidine resonances have been identified and characterized in a protein, information can be obtained on how the histidyl residues are affected through interactions with other molecules. Thus NMR evidence

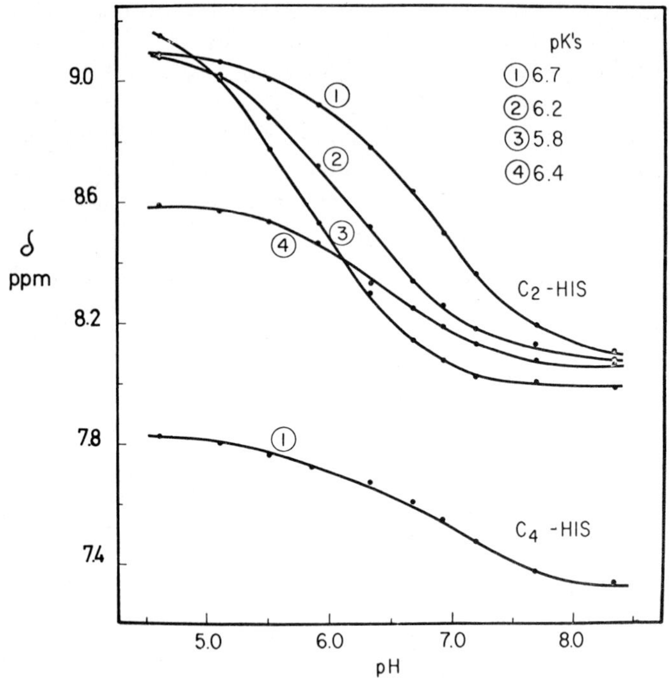

Fig. III-24. Titration curves of the four histidyl residues of ribonuclease. The chemical shifts δ are plotted vs. the "pH-value" (pH-meter reading in D_2O). The curves 1 to 4 for the C2 imidazole protons, from which the corresponding pK_a-values were obtained, correspond to the resonances 1 to 4 in Fig. III-23, and have been assigned as follows: 1 = His(105), 2 = His(12), 3 = His(119), 4 = His(48). The only C4 proton which could be resolved (Fig. III-23) comes from His(105) (from Meadows et al.[105]).

indicates that the single histidyl residue in lysozyme is not involved in the binding of the substrate[100]. In ribonuclease A, His(12) and His(119) had been implicated on the basis of X-ray studies to be part of the active site. In Fig. III-25 it is seen that these two histidines are indeed strongly affected when the inhibitor cytidine-3′-monophosphate is bound, whereas the chemical shifts and the pK_a's of His(48) and His(105) are essentially identical in the free enzyme, and the enzyme-inhibitor complex[107].

Histidine titrations by ^{1}H-NMR have been described for numerous other proteins, including, to name only a few, carbonic anhydrase[108], soybean trypsin inhibitor[109], and pancreatic ribonucleases from different species[110]. The current literature includes also systematic studies of the effects on the histidine protonation-deprotonation equilibrium of interactions with different charged or ionizeable groups in protein molecules[111–113].

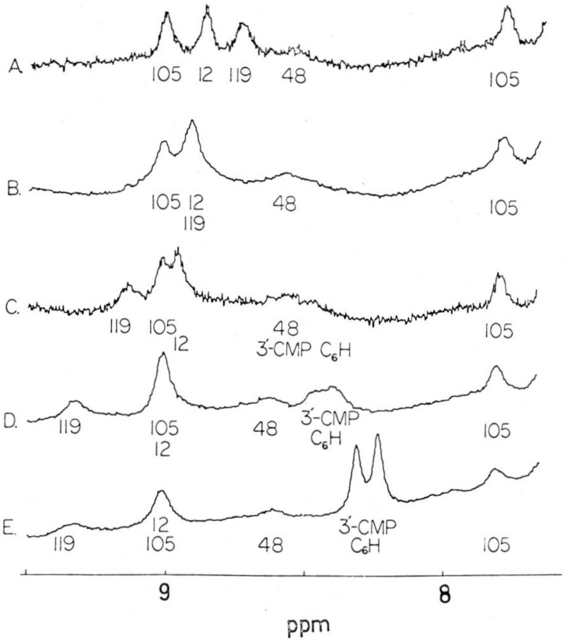

Fig. III-25. ^{1}H-NMR spectra between 7.5 and 9.5 ppm in D_2O solutions of 0.0065-M ribonuclease A, 0.2-M NaCl, and variable concentrations of the inhibitor cytidine-3′-monophosphate (3′-CMP). $T = 32°$, pH meter reading 5.5. The 3′-CMP concentrations were, A. no 3′-CMP, B. 0.002-M, C. 0.005-M, D. 0.01-M, E. 0.03-M. The resonance 105 at 7.8 ppm corresponds to the C4 imidazole proton of His(105). The resonances 48,119, 12, and 105 between 8.5 and 9.5 ppm, correspond to the C2 imidazole protons of these histidyl residues. The NMR of 3′-CMP are also identified. The signal-to-noise ratio had been improved by spectrum accumulation in a CAT (from Meadows and Jardetzky[107]).

Studies of Labile Protons

In the preceding section on peptides, the amide proton resonances observed in DMSO solutions had been used to study the formation of intramolecular hydrogen bonds, and in Fig. III-22 the amide protons of BPTI in the same solvent were observed as a broad featureless absorption band. Quite generally, similar phenomena are to be expected in other non-protic solvents (Table II-4). At this point, however, we want to consider the behavior of labile protons in proteins dissolved in protic solvents, i.e. in particular in H_2O and D_2O. On the basis of the introductory remarks on rate processes in Section II.5, three limiting cases can for practical purposes be distinguished. First, in the fast exchange limit the labile protons cannot be observed as separate resonances, either in D_2O or in H_2O. Second, in the slow exchange limit, e.g. with a life time with respect to chemical exchange of $\tau_e \gtrsim 5 \cdot 10^{-4}$ sec and a relative chemical shift between the water resonance and the labile protons of 1000 Hz, separate resonances are observed in H_2O solution, and more quantitative information on the exchange rate can then be obtained from a line shape analysis (Equation II-17). Third, separate resonances of the labile protons in D_2O solution will only be observed when the life time is sufficiently long for a ^{1}H-NMR spectrum to be recorded before extensive ^{1}H-D exchange has taken place. For CW experiments, this means in practice that τ_e must be of the order of one to several minutes.

When working with proteins, one may encounter all of these three different situations. The labile protons in a polypeptide chain exchange at fast rate when they are freely accessible to the solvent, and slower exchange rates result when the labile protons are shielded from the solvent, e.g. in the interior of a globular protein. Therefore, investigations of the proton exchange kinetics can yield information on the molecular conformation, and in particular on the dynamics of the spatial structures of proteins[114]. In the following this is illustrated with a few examples.

In the ^{1}H-NMR spectrum of BPTI (Figs. III-12B and III-21A), there are a series of resolved resonances of labile protons between 8 and 11 ppm. The time course of the disappearance of these lines in a D_2O solution has been followed in Fig. III-26. At neutral pH and ambient temperature, τ_e for these protons varies from less than one hour to several months and even years[97]. When the resonances of the individual protons are so well resolved, NMR has the big advantage over other methods for proton exchange studies, e.g. infrared spectroscopy or tritium tracer experiments[114], that the exchange of each individual proton can be studied separately. At temperatures above 80°, the proton exchange becomes too rapid for the resonances to be observed in an experiment in D_2O solution. However, for some of the protons one still has the "slow exchange" situation, so that separate resonances can be observed in H_2O (Fig. III-27). In D_2O solution[1,97,99,115], and in particular in the high temperature spectra in H_2O (Fig. III-27), the multiplet structures of these resonances are quite well resolved, and on the basis of the observed doublets these lines were assigned to backbone amide protons[97,115]. At this point one is then in principle in a position to determine some of the angles ϕ in this protein (Equation III-3).

When X-ray data on the protein are available, it can be of interest to relate the NMR observations on the labile protons with the molecular conformation in the single crystals. Since low field shifts of amide proton resonances in the extent observed in BPTI (Figs. III-26 and III-27) could hardly be caused by other interactions than intramolecular hydrogen bonding, a comparison of the number of slowly exchanging amide protons with the number of hydrogen bonds in the single crystal conformation provides a link

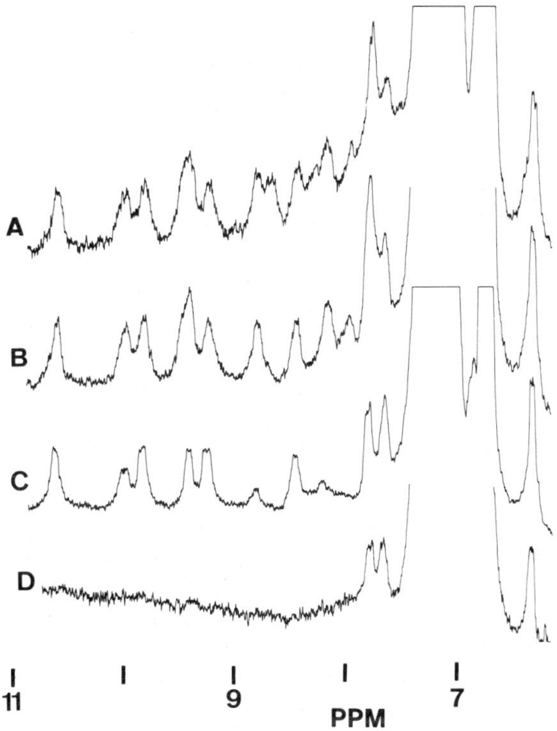

Fig. III-26. Spectral region from 7 to 11 ppm of the ^{1}H-NMR spectrum at 220 MHz of BPTI at different time intervals after preparation of a 0.01-M solution in D_2O, pD = 7.3, $T = 22°$. A. 25 min, B. 150 min, C. 660 hrs, D. after standing at 85° for a few minutes (from Masson and Wüthrich[97]).

between the two forms of the molecule[97]. Furthermore, since one expects $J_{HN\alpha}$ to be quite different for amino acid residues in α-helices and in β-sheets[31], it should in favorable cases be possible on the basis of the observed vicinal spin–spin coupling constants, to differentiate between amide protons in these two different types of hydrogen bonded secondary structures. Combining these resonance assignments with the studies of the proton exchange kinetics would then allow to differentiate between the dynamics of the molecular structures in different regions of the protein molecule[99].

In lysozyme the indole NH of the six tryptophanyl residues have been extensively investigated[94, 96, 116]. Fig. III-20 shows that the six tryptophans are magnetically equivalent in the random coil form of the protein, and give rise to a well resolved line at 10 ppm. In the native globular conformation, the individual indole rings are in different microenvironments, which results in

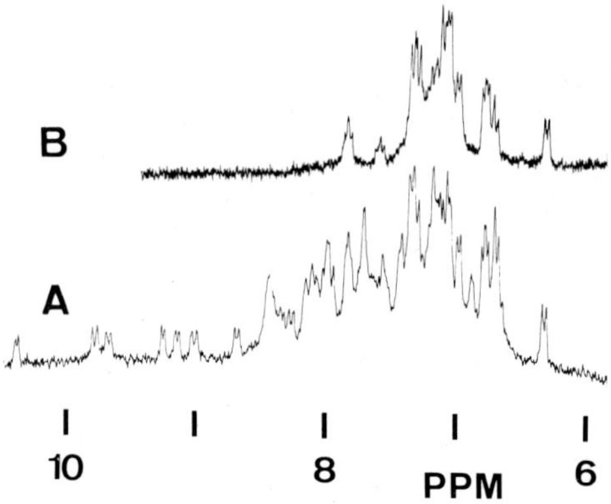

Fig. III-27. Spectral region from 6 to 11 ppm of the CW ^{1}H-NMR spectrum at 270 MHz of BPTI, $T = 87°$. A, in H_2O solution. B, in D_2O solution.

six separate ^{1}H-NMR lines. Five of these appear as well resolved resonances between 10 and 11 ppm (Fig. III-20), whereas the sixth line was apparently shifted to a position at higher field[96, 116]. Combining NMR studies of the exchange kinetics of these protons, of chemically modified lysozyme, and of lysozyme inhibitor complexes with the X-ray data on this protein, resulted in the assignment of the resonances I to V in Fig. III-20 to individual ones of the six tryptophans[116]. On this basis, the indole NH's can now be used to probe distinct areas of the lysozyme molecule, e.g. during reactions with substrates or inhibitors, or more generally under different conditions of solvent medium and other external factors.

^{1}H-NMR lines of the labile imidazole protons of histidine have been observed at extreme low field positions in H_2O solutions of different proteins. For example in ribonuclease A, a labile histidine proton is at approximately 13 ppm[117], and in chymotrypsin and chymotrypsinogen at around 17 ppm[118]. The assignment of these lines to the histidines was based on their pH dependence, which parallels that of the C2-protons of histidine (Figs. II-13 and III-24)[117–119]. These resonances promise to be of particular interest for studies of the formation of intramolecular hydrogen bonds or salt bridges involving the histidyl residues, both in the free proteins and the complexes formed with substrates, inhibitors, or other effector molecules.

Resonance Identification in the Central Spectral Region

The ^{1}H-NMR spectra of proteins presented so far in this chapter were recorded with CW spectrometers which are quite representative for the equipment available during the last decade. Improved spectral resolution can be obtained by the combination of high magnetic fields H_0 and suitable resolution enhancement techniques in modern Fourier transform NMR spectrometers*. This enables one to extend the detailed investigations to an increasing number of resonance lines in the crowded spectral region from 1 to 8 ppm (Fig. III-12).[120–125] In the following this is illustrated with the assignments of the aromatic protons and the ϵ-methylene protons of the lysyl residues in the 360 MHz convolution difference spectra* of BPTI.

When the labile protons in BPTI (Figs. III-21A, III-26, and III-27A) are replaced by D, the spectral region between 6 and 11 ppm contains only the resonances of the aromatic protons of the four tyrosyl residues and the four phenylalanyl residues in BPTI (Fig. III-27B), i.e. a total of 36 aromatic protons. First, the pH-dependence of this spectrum was investigated (Figs. III-28 and III-29). According to Equation II-8, the deprotonation of tyrosyl residues causes sizeable upfield shifts of the resonances of the aromatic protons. On the other hand, in the absence of pH-dependent changes of the protein conformation, the chemical shifts of the phenylalanines are independent of pH. Because the globular conformation of BPTI is essentially maintained at pH-values up to > 12.0, one can thus use pH titration studies to differentiate between Phe and Tyr proton resonances. At the same time the pK_a-values of the individual tyrosyl residues in the globular protein can be obtained[115, 123, 126].

Fig. III-28 shows the aromatic region of the CW ^{1}H-NMR spectrum at 360 MHz of BPTI at different pH values between 8.15 and 12.19†. It is readily apparent that variation of the pH causes sizeable spectral changes between 6 and 7 ppm. In addition, the one-proton line which is at 7.75 ppm at pH 8.15 titrates in the pH-range 10.25 to 12.19. The titration curves for the resonance lines which could be resolved in the spectra of Fig. III-28 and in additional spectra obtained with Fourier transform techniques (see e.g. Figs. III-30 and III-31) are represented in Fig. III-29. The pK_a-values (Equation II-8) given in the figure caption were obtained by least square fits of the experimental data to one-proton titration curves

$$\delta_{obs} = \delta_{HA} \frac{[H^+]}{[H^+] + K_a} + \delta_A \frac{K_a}{[H^+] + K_a} \qquad \text{(III-9)}$$

*An introduction to Fourier transform NMR spectroscopy and a description of resolution enhancement techniques will be presented in Chapter IV and in Section VI.5.

†pH titrations of the tyrosyl residues in BPTI had originally been described by Karplus, Snyder, and Sykes[115]. These authors have recently used a different approach from the one described here for the identification of the tyrosine resonances, which included the investigation of chemically modified BPTI molecules obtained by selective nitration of individual tyrosine rings. From these experiments, the assignments of the tyrosine resonances to specific residues (Fig. III-29) could also be obtained[126].

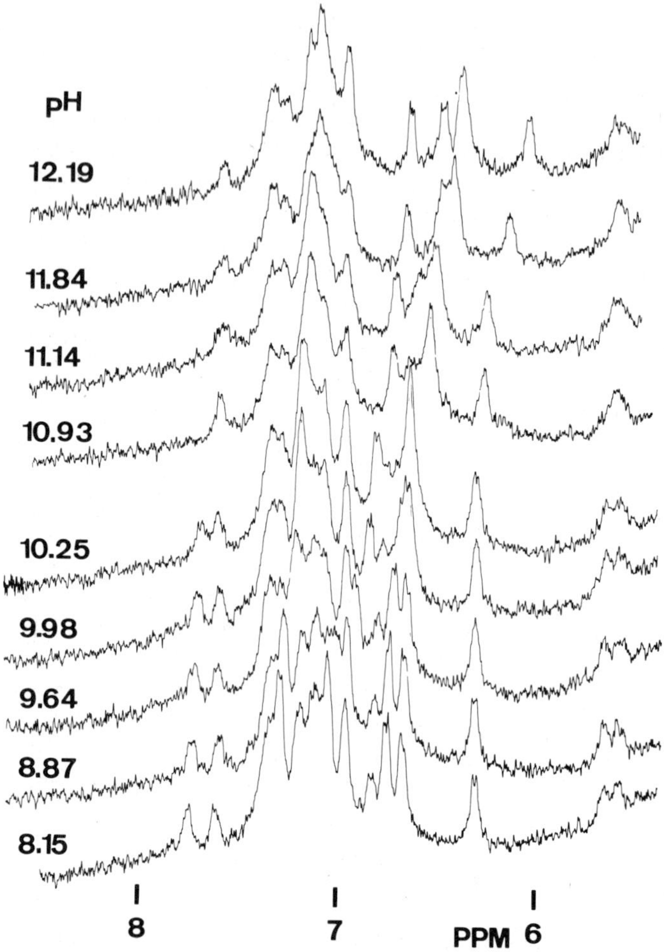

Fig. III-28. Aromatic region of the CW ^{1}H-NMR spectra at 360 MHz of a 0.02-M solution of BPTI in D_2O at different pH, $T = 22°$. Prior to these experiments, the labile protons had been exchanged with D of D_2O. The different pH-values were obtained by adding minute amounts of a KOD solution to the sample. The "pH"-values indicated in the figure correspond to the pH-meter readings with a combined glass electrode in the D_2O solution of BPTI (from Wagner and Wüthrich[123]).

where δ_{HA} and δ_A are the chemical shifts for the protonated and deprotonated tyrosines, respectively. The assignments of the resonance lines and pK_a-values to specific ones of the four tyrosyl residues in the positions 10, 21, 23 and 35 of the polypeptide chain were obtained from studies of selectively nitrated BPTI[126].

Since four different pK_a-values were observed in the Figs. III-28 and III-29,

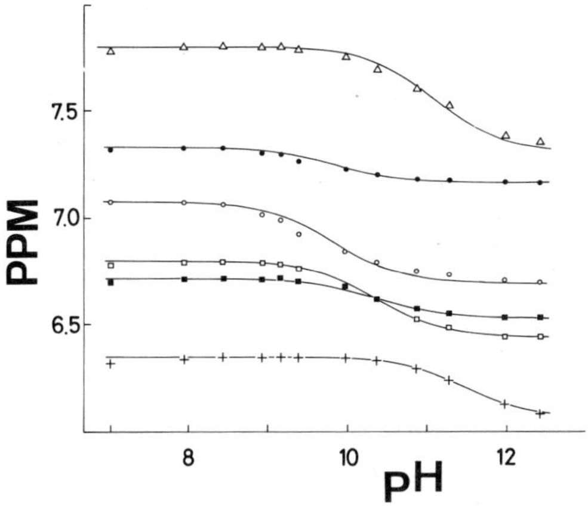

Fig. III-29. Plot vs. pH of the chemical shifts of six resonance lines in the aromatic region of the ^{1}H-NMR spectrum of BPTI which were assigned to tyrosine rings. $T = 22°$. The titration curves are least squares fits of the experimental points to one-proton titration curves (Equation III-9). The line at lowest field (△) corresponds in intensity to one proton, the other lines to two protons each. The resonances were also assigned to specific tyrosyl residues[126], and the following pK_a-values were obtained: +, Tyr(23), $pK_a = 11.5$; □■, Tyr(21), $pK_a = 10.4$; ○●, Tyr(10), $pK_a = 9.9$; △, Tyr(35), $pK_a = 11.1$. (Fig. and pK_a-values from Wagner and Wüthrich[123], resonance assignments from Snyder et al.[126]).

at least one resonance of each of the four tyrosines had thus been identified. The remaining tyrosine resonance lines were then assigned with double resonance experiments (Fig. II-9). As an illustration of the technique, Fig. III-30 shows the spin decoupling experiments which confirmed the assignments of the Tyr(10) resonances given in Fig. III-29. Once all the tyrosine lines had thus been identified[123, 125], it was obvious that the remaining resonances had to come from the four phenylalanines. Double resonance experiments were then also applied to differentiate between the resonances of the individual phenylalanine rings. Fig. III-31 illustrates the identification the AA′MXX′ type spectrum of one of the phenylalanines in BPTI.

The detailed analysis of the aromatic resonances in BPTi revealed a number of conformation dependent NMR features. Thus, comparison with the random coil ^{1}H-NMR spectra of BPTI (Figs. III-21B, C and III-22B) shows that there are sizeable conformation dependent chemical shifts, the extent of which is up to approximately 1 ppm for individual lines. Furthermore, for some of the aromatic rings, e.g. Tyr(35), the spectral type does not correspond to that expected from the symmetry of the covalent ring structure. This will be discussed in detail below. It is also worth noting that the sign of the relative chemical shift between the two two-proton lines of Tyr(21) changes at ca. pH 10.4 (Fig. III-29).

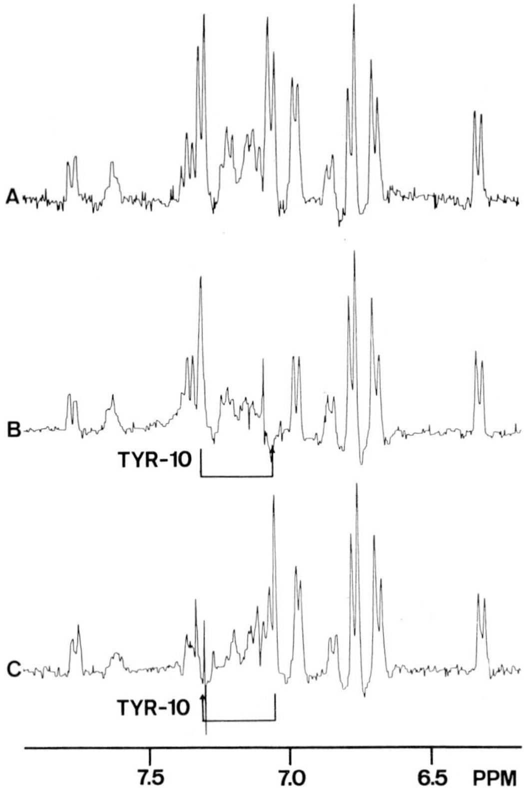

Fig. III-30. A. Aromatic region of the convolution difference (see Section VI.5) ^{1}H-NMR spectrum at 360 MHz of a BPTI solution in D_2O, pD = 7.8, $T = 26°$. B. Double resonance irradiation at 7.07 ppm (arrow) causes the collapse of a doublet at 7.33 ppm. C. Irradiation at 7.33 ppm causes the collapse of the doublet at 7.07 ppm. The two doublets correspond to the AA′BB′ spin system of Tyr(10) (from Wagner et al.[125]).

The Figs. III-30 and III-31 show that with the modern equipment of 1975, the spectral resolution obtained in the ^{1}H-NMR spectra of small globular proteins is at least in certain spectral regions reminiscent of the well resolved spectra of oligopeptides and other small molecules. Even the multiplet structures of the resonances can be partially resolved. Accordingly, many of the established techniques of high resolution NMR of organic molecules can now be used in studies of proteins. So far, individual aromatic resonances of Tyr, Phe and Trp have in addition to BPTI also been identified in the spectra of lysozyme[124] and cytochrome *c*[121, 122]. As is illustrated with the following example, well resolved spectra can also be obtained for the aliphatic amino acid residues in proteins.

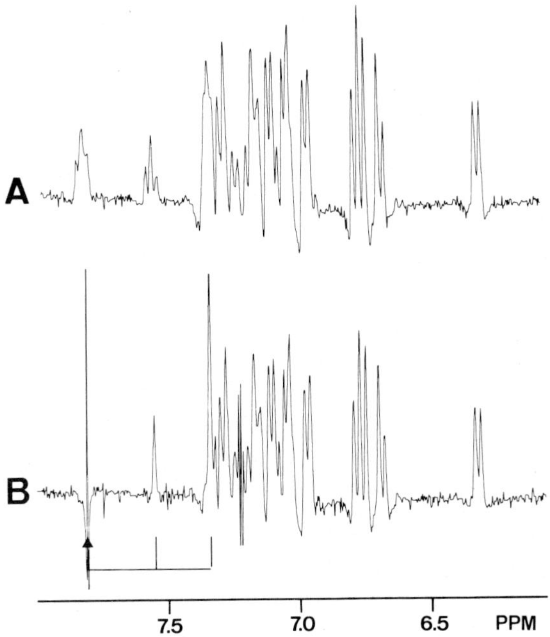

Fig. III-31. A. Aromatic region of the convolution difference (see Section VI.5) ^{1}H-NMR spectrum at 360 MHz of a BPTI solution in D_2O, pD = 7.8, T = 72°. B. Double resonance irradiation of the two-proton multiplet at 7.8 ppm causes the collapse of the one-proton multiplet at 7.5 ppm, and a two-proton multiplet at 7.3 ppm. These three resonances constitute an AA′MXX′ spectrum of a phenylalanine ring, where the 3,5-protons are at 7.8 ppm, the 4-proton at 7.5 ppm, and the 2,6-protons at 7.3 ppm (Wagner G., De Marco A. and Wüthrich K., unpublished).

Fig. III-32 shows the region from 0 to 4 ppm of the convolution difference ^{1}H-NMR spectra at 360 MHz of BPTI at different pH-values. The improved spectral resolution as compared to the CW ^{1}H-NMR spectra at 220 MHz (Figs. III-12B and III-21A) is readily apparent. The identification of individual lines by double resonance experiments and by comparison with homologous proteins[128], and studies of non-bonding interactions between different amino acid side chains[129] are still in progress. An interesting observation was made concerning the four lysyl residues in BPTI[127]. Since in the single crystal structure these four residues are on the protein surface, one might have expected that the four ε-methylene resonances should be essentially equivalent. However, from the pH-dependence two resonances with relative intensities 3 and 1 (A and B in Fig. III-32) were found to come from these protons. The chemical shift and the pK_a of 10.6 of the three essentially equivalent lysines in the resonance A appear to be representative for solvent accessible lysyl residues on the protein surface. The high field shift of the

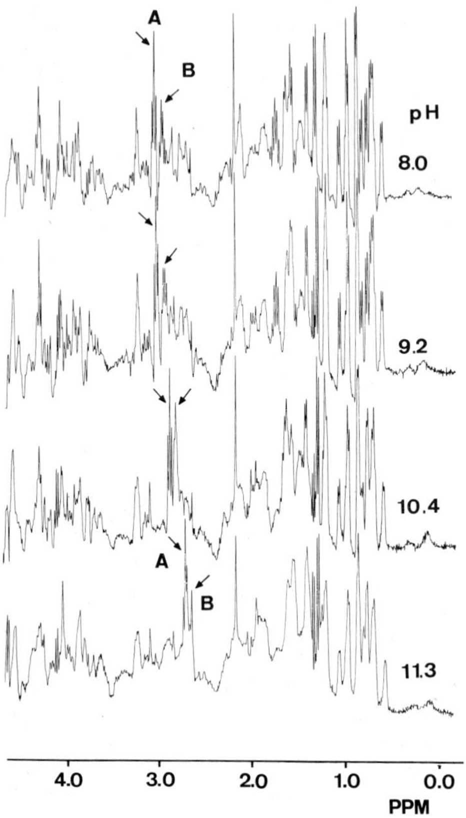

Fig. III-32. Spectral region from 0 to 4 ppm of the convolution difference (see Section VI.5) ^{1}H-NMR spectra at 360 MHz of a 0.01 M solution of BPTI in D_2O at the different pH meter readings indicated. A and B are the resonances assigned to ϵ-methylene protons of lysine (see text) (from Brown et al.[127]).

lysyl resonance B and its somewhat different pK_a-value of 10.8 were found to result from an interaction on the protein surface between Lys(41), to which this resonance was assigned, and Tyr(10). A confirmation for this structural feature of the protein surface was obtained from its manifestation in the titration behavior of Tyr(10) (Fig. III-29), and the effects of N^{ϵ}-dimethylation of the lysyl residues on the NMR parameters and the pK_a of Tyr(10)[127].

Dynamics of the Aromatic Amino Acids

Applications for studies of the dynamics of biopolymers in solution are one of the most exciting aspects of biological NMR studies. Throughout this book, different NMR experiments are described which can be used for this purpose. One of these techniques is based on detailed assignments of the aromatic resonances as described above. Its principles and the application to BPTI are discussed in the following.

In the covalent structure of the aromatic rings of tyrosine and phenylalanine, two pairs of protons in the positions 2 and 6, and 3 and 5, respectively, are related by a C_2-symmetry operation about the C^{β}—C^{γ} bond axis.

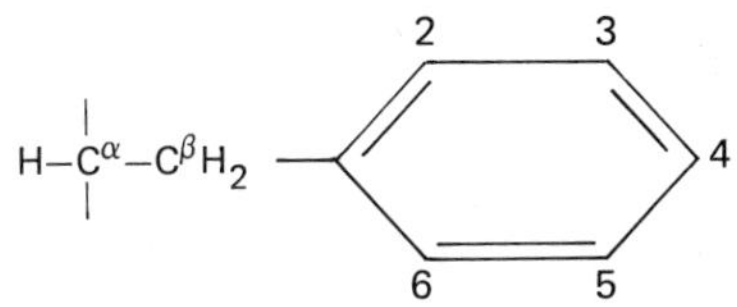

Therefore, when these rings are in an isotropic medium, AA′BB′ spectra for tyrosine and AA′BB′C spectra for phenylalanine are usually observed. In the interior of globular proteins, however, the environment of the aromatic rings is characterized by the non-periodic distribution of structural elements. As a consequence, the individual ring protons experience different microenvironments. In a static situation where the rings would be rigidly fixed in space in the interior of the protein, one thus has that $\Delta\delta_{2,6} = |\delta_2 - \delta_6|$ and $\Delta\delta_{3,5} = |\delta_3 - \delta_5|$ would in general be different from O. Therefore, if the aromatic rings in the interior of the protein are assumed to rotate about the C^{β}—C^{γ} bond axis*, the different spectral features of Fig. II-10 may result for each pair of symmetry-related protons. Three limiting cases have been observed in BPTI. (i) If the rotation frequency ν_e is small compared to both $|\nu_2 - \nu_6|$ and $|\nu_3 - \nu_5|$, ABCD type spectra for tyrosine and ABCDE type spectra for phenylalanine are observed. (ii) If ν_e is fast compared to the relative shifts of both pairs of protons, AA′BB′ spectra for tyrosine and AA′BB′C spectra for phenylalanine are observed. (iii) If the conformation dependent shifts $\Delta\delta_{2,6}$ and $\Delta\delta_{3,5}$ are sizeably different, one may at a given rotation rate ν_e have a rapid exchange situation for one of the two pairs of protons, and not for the other one. AA′BC type spectra for tyrosine and AA′BCD type spectra for phenylalanine may thus be encountered in proteins[125].

On the basis of these principal considerations, the above described assignments of the aromatic resonances at ambient temperature yield already some qualitative information on the dynamics of the aromatic rings. Thus the observation of an ABCD type spectrum for Tyr(35) (one of the four lines is included in Fig. III-29) shows that at 22° this ring has a static spatial

*In BPTI these rotational motions are actually 180° flips of the rings in and out of spatial orientations which correspond essentially to those observed in the single crystals.

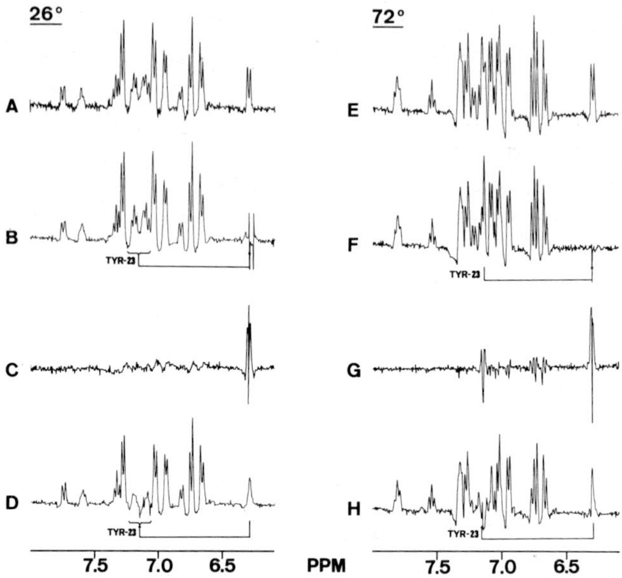

Fig. III-33. A. Aromatic region of the convolution difference (see Section VI.5) ^{1}H-NMR spectrum at 360 MHz of a BPTI solution in D_2O, pH = 7.8, T = 26°. B. Double resonance irradiation at 6.30 ppm causes no detectable changes of the resonances between 6.5 and 8.0 ppm, as revealed in the difference spectrum C=A—B. D. Broad band irradiation of the spectral region between approximately 7.0 and 7.3 ppm causes the collapse of the two-proton doublet at 6.30 ppm. E. Same as A except that T = 72°. F. At 72°, irradiation at 6.30 ppm causes spectral changes. The collapse of a sharp doublet at 7.14 ppm is revealed in the difference spectrum G=E—F. H. At 72°, selective irradiation at 7.14 ppm causes the collapse of the doublet at 6.30 ppm (from Wagner et al.[125]).

orientation on the NMR time scale. On the other hand, the AA′BB′ spectra of Tyr(10) (Fig. III-30) and Tyr(21), and the AA′BB′C spectrum for one of the phenylalanines show that at 22° these three rings are flipping rapidly about the C^{β}—C^{γ} bond axis. The forementioned third limiting situation giving rise to an AA′BC type tyrosine spectrum was observed for Tyr(23). The spectra A to D in Fig. III-33 show that at 26° the rotation rate of this ring is sufficiently rapid to produce a quite sharp two-proton doublet resonance of the 3,5-protons at 6.3 ppm, whereas the resonance intensity of the 2,6-protons is distributed over a frequency range of more than 100 Hz.

More quantitative data on the dynamics of the aromatic residues were obtained from studies of the temperature dependence of the ^{1}H-NMR spectra. As evidenced e.g. in the ^{1}H-NMR lines of the hydrogen bonded amide protons (Figs. III-26 and III-27A), the globular conformation of BPTI in aqueous solution is maintained over the temperature range 0° to 85°[97]. Spectra of the aromatic protons recorded over this temperature range are shown in the Figs.

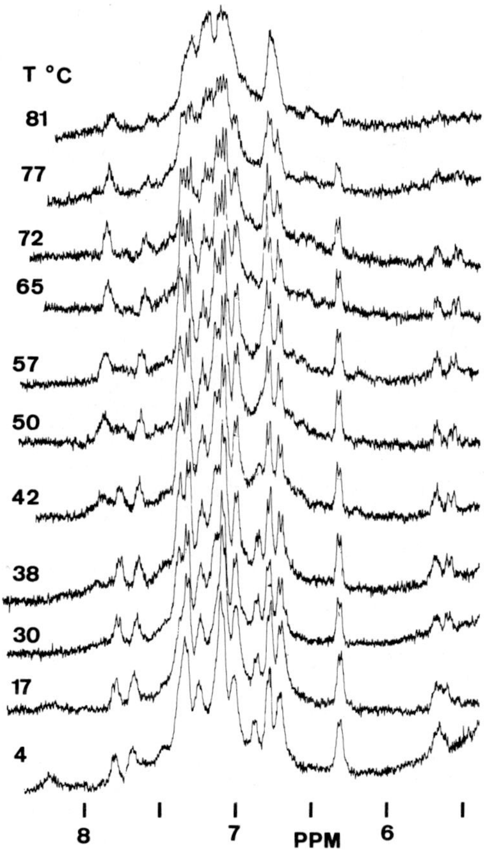

Fig. III-34. Temperature dependence between 4° and 81° of the spectral region from 5 to 9 ppm in the HCW ^1H NMR spectrum at 360 MHz of a 0.01-M solution of BPTI in D_2O, pD = 1.8. Prior to these experiments, the labile protons had been replaced by deuterium (from Wüthrich and Wagner[120]).

III-34 and III-35. Comparison of these two figures also presents a nice illustration of the potential of the digital resolution enhancement techniques employed in the experiments of Fig. III-35 (see Chapter IV and Section VI.5). The following observations are particularly relevant for the present discussion. First, certain resonances are essentially independent of temperature between 4° and 77°. These include e.g. the two-proton lines at 6.30, 6.69, 6.76, 6.98 and 7.07 ppm. All these resonances correspond to rings which had previously been found to rotate rapidly at ambient temperature. Second, the observation in the spectrum at 4° of five one-proton lines between 7.5 and 8.3 ppm is in itself conclusive evidence that at least one of the aromatic rings

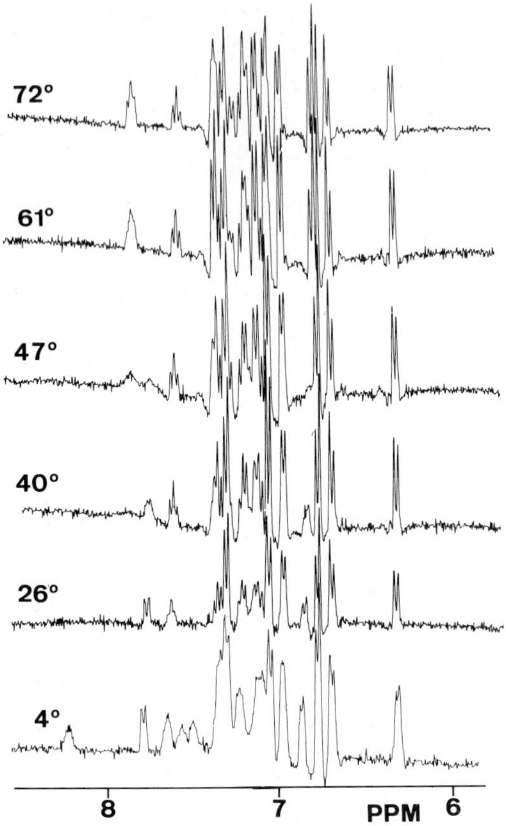

Fig. III-35. Temperature dependence of the spectral region from 6 to 9 ppm in the convolution difference (see Section VI.5) ^{1}H-NMR spectra at 360 MHz of a 0.01 M solution of BPTI in D_2O, pD = 7.8. Prior to these experiments the labile protons had been replaced by deuterium (Wagner G., De Marco A. and Wüthrich K., unpublished).

does not rotate rapidly about the C^{β}—C^{γ}-axis. Otherwise the entire spectrum of BPTI would contain only four aromatic one-proton lines, i.e. those of the C4-protons (Table II-6) of the phenylalanine rings. Third, the three one-proton lines at 8.2, 7.6 and 7.5 ppm in the spectrum at 4°, and an additional not resolved resonance at ca. 7.2 ppm disappear when the temperature is raised to approximately 30°. AT 40° two new broad resonances appear at 7.9 and 7.35 ppm which become narrower and move to slightly higher field positions at higher temperatures. These new lines correspond in intensity to two protons each. They are coupled with the one-proton line which moves from 7.67 ppm at 4° to 7.53 ppm at 72° (Fig. III-31). These temperature dependent features thus manifest the transition from a static situation at 4° to rapid rotation at 72° for one of the phenylalanine rings. The life time with respect to 180° flips about the C^{β}—C^{γ} axis for this phenylalanine can be estimated to be of the order $1 \cdot 10^{-2}$ sec at 4° and $8 \cdot 10^{-4}$ sec at 38°, and the free energy of activation for the rotational motions is $\Delta G^{\neq} = 13.7$ kcal · mole^{-1} at 27°. Fourth, the four one-proton lines of Tyr(35) at 7.75 (Fig. III-29), 6.80, 6.76 and 6.69 ppm[125, 126] disappear at approximately 50°. In the spectra at 61° and higher temperatures a new two-proton doublet resonance can be observed which corresponds to the average of the one-proton lines at 6.80 and 6.76 ppm. We thus have that over the temperature range from ca. 4° to 40°, Tyr(35) is static on the NMR time scale and gives rise to an ABCX type spectrum. Rapid rotation sets in at higher temperatures. From this the activation energy for 180° flips about the C^{β}—C^{γ} axis was found to be $\Delta G^{\neq} = 16.4$ kcal · mole^{-1} at 27°. It may be added that at 72° the rotation rate for Tyr(23) is sufficiently rapid to produce an AA'BB' type spectrum (Fig. III-33, E-H), that the transition from a static to a mobile situation for one of the two remaining phenylalanines is clearly manifested by the temperature dependence of the spectral region from 7.1 to 7.3 ppm (Fig. III-35), and that the fourth phenylalanine ring has an ABCDE spectrum over the entire temperature range studied.

With the possible exception of the measurements at very high temperatures, the experiments described here for BPTI should be generally applicable for small and medium sized proteins. The information obtained on the molecular dynamics represents a valuable complementation of the single crystal X-ray data. To mention only one of many attractive prospective uses of such results, the application for studies of the relations between temperature dependent structural properties and enzymatic activities of proteins from thermophilic organisms should be of much interest.

Selectively Deuterated Proteins

As was briefly mentioned in Section III-2, any of the twenty common amino acids can yield prominent ^{1}H-NMR lines in the spectrum of a selectively deuterated protein. This will now be illustrated by some experiments with staphylococcal nuclease[30]. In these studies a mixture of deuterated amino acids was obtained by hydrolysis of protein isolated from algae (*Scenedesmus obliquus*) grown in D_2O. *Staphylococcus aureus* was grown on a medium containing this mixture of deuterated amino acids, and an excess of the protonated form of those amino acids which should be observed in the protein. The selectivity deuterated nuclease was then isolated from the *Staphylococcus aureus* harvested from this culture. Fig. III-36 shows the aliphatic region of the ^{1}H-NMR spectra of the fully protonated enzyme, and of the analogue containing only tryptophan and methionine in the protonated form. It is seen that the proton resonances of

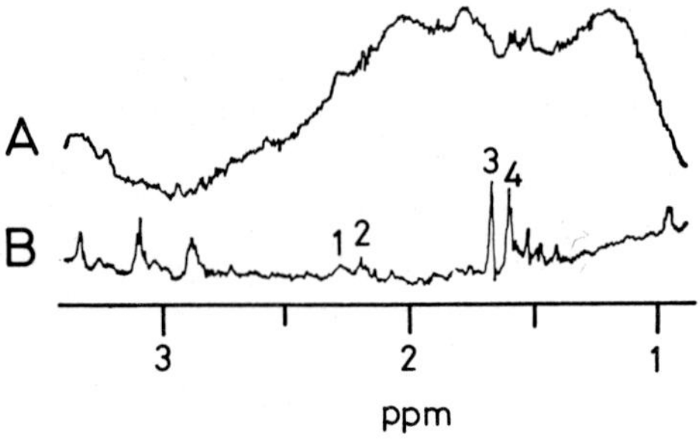

Fig. III-36. Spectral region between 1.0 and 3.5 ppm of the ^{1}H-NMR spectra at 100 MHz of staphylococcal nuclease, and a selectively deuterated nuclease analog. The solvent used was 0.3-M NaCl solution in D_2O, pH meter reading 8.0. $T = 32°$. A. Fully protonated staphylococcal nuclease. B. Selectively deuterated nuclease analog, where tryptophan and methionine were the only fully protonated amino acids. The peaks 1 to 4 had been tentatively assigned to the S-methyl resonances of the four methionyl residues. The protein concentrations were approximately 20% in A, and 6% in B. The signal: noise ratio was improved by time averaging of the spectra in a CAT (from Markley et al.[30]).

methionine are nicely resolved even at 100 MHz. Again, these lines can be used as NMR probes to detect conformational changes in the molecule when it is subjected to variable external conditions, or to interactions with other molecules, e.g. substrates or inhibitors.

Alternate possibilities for the combination of ^{1}H-NMR studies with deuterium labelling are given in certain conjugated proteins. The fully deuterated polypeptide chain may be combined with the protonated form of the prosthetic group, so that the ^{1}H-NMR of the latter can readily be observed in the protein molecule. In a typical example, the deuterated polypeptide chain of flavoprotein isolated from the blue-green thermophilic algae *Synechococcus lividus* was combined by exchange with the fully protonated flavin mononucleotide[130].

5. BIBLIOGRAPHY

1 Masson A., Konformationsstudien an Calcitonin M und am Trypsin-Inhibitor BPTI, Ph.D. Thesis, ETH, Zürich (1974)
2 Bemski G. and Hynes T., Nature *245*, 37 (1973)
3 Sheinblatt M., J. Amer. Chem. Soc. *88*, 2845 (1966)
4 Anteunis M. and Gelan J., J. Amer. Chem. Soc. *95*, 6502 (1973)
5 For a discussion of polypeptide conformation and the common nomenclature, see e.g.:
  (a) Ramachandran G. N. and Sasisekharan V., Adv. Protein Chem. *23*, 283 (1968)
  (b) Scheraga H. A., Adv. Phys. Org. Chem. *6*, 103 (1968)
  (c) Dickerson R. E. and Geis I., The Structure and Action of Proteins, Harper and Row, New York, 1969
  (d) IUPAC–IUB Commission on Biochemical Nomenclature, Biochemistry *9*, 3471 (1970)

6 Bothner-By A. A., Adv. Magn. Resonance *1*, 195 (1965)
7 Bystrov V. F., Portnova S. L., Tsetlin V. I., Ivanov V. T. and Ovchinnikov Y. A., Tetrahedron *25*, 493 (1969)
8 Ramachandran G. N., Chandrasekaran R. and Kopple K. D., Biopolymers *10*, 2113 (1971)
9 Barfield M. and Gearhart H. L., J. Amer. Chem. Soc. *95*, 641 (1973)
10 Bystrov V. F., Ivanov V. T., Portnova S. L., Balashova T. A., and Ovchinnikov Y. A., Tetrahedron *29*, 873 (1973)
11 Donzel B., Untersuchung der bevorzugten Konformationen einiger tryptophanhaltiger cyclo-Dipeptide in Lösung, Ph.D. Thesis, ETH, Zürich (1971)
12 Thong C. M., Canet D., Granger P., Marraud M. and Néel J., Comptes Rendus Acad. Sci. Ser C *269*, 580 (1969)
13 Solkan V. N. and Bystrov V. F., Tetrahedron Lett. *1973*, 2261
14 Meraldi J. P., Etude des Conformations Moléculaires de cinq Pentapeptides Cycliques par la Résonance Magnétique Nucléaire des Protons et par la Méthode des Potentiels Semi-Empiriques d'Energie, Ph.D. Thesis, ETH, Zürich (1974)
15 Kopple K. D., Go A., Logan R. H. and Savrda J., J. Amer. Chem. Soc. *94*, 973 (1972)
16 Karplus S. and Karplus M., Proc. Nat. Acad. Sci. *69*, 3204 (1972)
17 Sogn J. A., Gibbons W. A. and Randall E. W., Biochemistry *12*, 2100 (1973)
18 La Planche L. A. and Rogers M. T., J. Amer. Chem. Soc. *86*, 337 (1964)
19 Maia H. L., Orrell K. G. and Rydon H. N., Chem. Comm. *1971*, 1209
20 Neuman R. C., Jonas V., Anderson K. and Barry R., Biochem. Biophys. Res. Comm. *44*, 1156 (1971)
21 Drakenberg T. and Forsén S., Chem. Comm. *1971*, 1404
22 Love A. L., Alger T. D. and Olsen R. K., J. Phys. Chem. *76*, 853 (1972)
23 Davies D. B. and Khaled M. A., J. C. S. Perkins II, *1973*, 1651
24 Winkler F. K. and Dunitz J. D., J. Mol. Biol. *59*, 169 (1971)
25 Ramachandran G. N., Lakshminarayanan A. V. and Kolaskar A. S., Biochim. Biophys. Acta *303*, 8 (1973)
26 Kopple K. D., Wiley G. R. and Tauke R., Biopolymers *12*, 627 (1973)
27 Donzel B., Kamber B., Wüthrich K. and Schwyzer R., Helv. Chim. Acta *55*, 947 (1972)
28 Schwyzer R. and Ludescher U., Biochemistry *7*, 2514 (1968)
29 Crespi H. L., Rosenberg R. M. and Katz J. J., Science *161*, 795 (1968)
30 Markley J. L., Putter I. and Jardetzky O., Science *161*, 1249 (1968)
31 Gibbons W. A., Némethy G., Stern A. and Craig L. C., Proc. Nat. Acad. Sci. *67*, 239 (1970)
32 Schwyzer R., Carrión J. P., Gorup B., Nolting H. and Tun-Kyi A., Helv. Chim. Acta *47*, 441 (1964)
33 Ohnishi M. and Urry D. W., Biochem. Biophys. Res. Comm. *36*, 194 (1969)
34 Schwyzer R., Grathwohl Ch., Meraldi J. P., Tun-Kyi A., Vogel R. and Wüthrich K., Helv. Chim. Acta *55*, 2545 (1972)
35 Pease L. G., Deber C. M. and Blout E. R., J. Amer. Chem. Soc. *95*, 258 (1973)
36 Deber C. M., Bovey F. A., Carver J. P. and Blout E. R., J. Amer. Chem. Soc. *92*, 6191 (1970)
37 Meraldi J. P., Schwyzer R., Tun-Kyi A. and Wüthrich K., Helv. Chim. Acta *55*, 1962 (1972)
38 Meraldi J. P., Moeschler H., Schwyzer R., Tun-Kyi A. and Wüthrich K., J. Phys. Suppl. *34*, C8 41 (1973)
39 Hoffman R. A. and Forsén S., Progr. NMR Spectr. *1*, 15 (1966)
40 Fermandjian S., Pradelles P., Fromageot P. and Dunand J. J., FEBS Letters *28*, 156 (1972)
41 Blagdon D. E., Rivier J. and Goodmann M., Proc. Nat. Acad. Sci. USA, *70*, 1166 (1973)
42 Howard J. C., Momany F. A., Andreatta R. H. and Scheraga H. A., Macromolecules *6*, 535 (1973)

43 Weinkamm R. J. and Jorgensen E. C., J. Amer. Chem. Soc. *93*, 7038 (1971)
44 Glickson J. D., Cunningham W. D. and Marshall G. R., Biochemistry *12*, 3684 (1973)
45 Bleich H. E., Galardy R. E., Printz M. P. and Craig L. C., Biochemistry *12*, 4950 (1973)
46 Glickson J. D., Dadok J. and Marshall G. R., Biochemistry *13*, 11 (1974)
47 Johnson L. F., Schwartz I. L. and Walter R., Proc. Nat. Acad. Sci. *64*, 1269 (1969)
48 Deslauriers R. and Smith I. C. P., Biochem. Biophys. Res. Comm. *40*, 179 (1970)
49 Hruby V. J., Brewster A. I. and Glasel J. A., Proc. Nat. Acad. Sci. *68*, 450 (1971)
50 Von Dreele P. H., Brewster A. I., Scheraga H. A., Ferger M. F. and du Vigneaud V., Proc. Nat. Acad. Sci. *68*, 1028 (1971)
51 Feeney J., Roberts G. C. K., Rockey J. H. and Burgen A. S. V., Nature New Biol. *232*, 108 (1971)
52 Von Dreele P. H., Brewster A. I., Bovey F. A., Scheraga H. A., Ferger M. F., and du Vigneaud V., Proc. Nat. Acad. Sci. *68*, 3088 (1971)
53 Brewster A. I., Glasel J. A. and Hruby V. J., Proc. Nat. Acad. Sci. *69*, 1470 (1972)
54 Walter R., Glickson J. D., Schwartz I. L., Havran R. T., Meienhofer J. and Urry D. W., Proc. Nat. Acad. Sci. *69*, 1920 (1972)
55 Glickson J. D., Urry D. W., Havran R. T. and Walter R., Proc. Nat. Acad. Sci. *69*, 2136 (1972)
56 Von Dreele P. H., Brewster A. I., Dadok J., Scheraga H. A., Bovey F. A., Ferger M. F. and du Vigneaud V., Proc. Nat. Acad. Sci. *69*, 2169 (1972)
57 Glickson J. D., Urry D. W. and Walter R., Proc. Nat. Acad. Sci. *69*, 2566 (1972)
58 Von Dreele P. H., Scheraga H. A., Dyckes D. F., Ferger M. F. and du Vigneaud V., Proc. Nat. Acad. Sci. *69*, 3322 (1972)
59 Walter R. and Glickson J. D., Proc. Nat. Acad. Sci. *70*, 1199 (1973)
60 Brewster A. I., Hruby V. J., Glasel J. A. and Tonelli A. E., Biochemistry *12*, 5294 (1973)
61 Brewster A. I. and Hruby V. J., Proc. Nat. Acad. Sci. *70*, 3806 (1973)
62 Go N. and Scheraga H. A., Macromolecules *3*, 178 (1970)
63 Hassall C. H. and Thomas W. A., Chemistry in Britain *7*, 145 (1971)
64 Bovey F. A., Brewster A. I., Patel D. J., Tonelli A. E. and Torchia D. A., Acc. Chem. Res. *5*, 193 (1972)
65 Kopple K. D. and Marr D. H., J. Amer. Chem. Soc. *89*, 6193 (1967)
66 Kopple K. D. and Ohnishi M., J. Amer. Chem. Soc. *91*, 962 (1969)
67 Dale J. and Titlestad K., Chem. Comm. *1969*, 656
68 Deber C. M., Torchia D. A. and Blout E. R., J. Amer. Chem. Soc. *93*, 4893 (1971)
69 Kopple K. D., Ohnishi M. and Go A., Biochemistry *8*, 4087 (1969)
70 Kopple K. D., Ohnishi M. and Go A., J. Amer. Chem. Soc. *91*, 4264 (1969)
71 Portnova S. L., Shilin V. V., Balashova T. A., Biernat J., Bystrov V. F., Ivanov V. T. and Ovchinnikov Y. A., Tetrahedron Letters *33*, 3085 (1971)
72 Tonelli A. E. and Brewster A. I., Biopolymers *12*, 193 (1973)
73 Tonelli A. E. and Brewster A. I., J. Amer. Chem. Soc. *94*, 2851 (1972)
74 Deber C. M., Torchia D. A., Wong S. C. K. and Blout E. R., Proc. Nat. Acad. Sci. *69*, 1825 (1972)
75 Torchia D. A., di Corato A., Wong S. C. K., Deber C. M. and Blout E. R., J. Amer. Chem. Soc. *94*, 609 (1972)
76 Torchia D. A., Wong S. C. K., Deber C. M. and Blout E. R., J. Amer. Chem. Soc. *94*, 616 (1972)
77 Llinás M., Klein M. P. and Neilands J. B., J. Mol. Biol. *68*, 265 (1972)
78 Llinás M., Klein M. P. and Neilands J. B., Int. J. Peptide Protein Res. *4*, 157 (1972)
79 Llinás M., Klein M. P. and Neilands J. B., J. Biol. Chem. *248*, 915, 924 (1973)
80 Kopple K. D., Biopolymers *10*, 1139 (1971)

81 Cary L. W., Takita T. and Ohnishi M., FEBS Letters *17*, 145 (1971)
82 Patel D. J., Tonelli A. E., Pfaender P., Faulstich H. and Wieland Th., J. Mol. Biol. *79*, 185 (1973)
83 Liquori A. M. and Conti F., Nature *217*, 635 (1968)
84 Stern A., Gibbons W. A. and Craig L. C., Proc. Nat. Acad. Sci. *61*, 734 (1968)
85 Ovchinnikov Y. A., Ivanov V. T., Bystrov V. F., Miroshnikov A. I., Shepel E. N., Abdullaev N. D., Efremov E. S. and Senyavina L. B., Biochem. Biophys. Res. Comm. *39*, 217 (1970)
86 Ludescher U. and Schwyzer R., Helv. Chim. Acta *52*, 1637 (1971)
87 Patel D. J., Biochemistry *12*, 667 (1973)
88 Patel D. J., Biochemistry *12*, 677 (1973)
89 Haynes D. H., Kowalsky A. and Pressman B. C., J. Biol. Chem. *244*, 502 (1969)
90 Ivanov V. T., Laine I. A., Abdulaev N. D., Senyavina L. B., Popov E. M., Ovchinnikov Y. A. and Shemyakin M. M., Biochem. Biophys. Res. Comm. *6*, 803 (1969)
91 Shemyakin M. M., Ovchinnikov Y. A., Ivanov V. T., Antonov V. K., Vinogradova E. I., Shkrob A. M., Malenkov G. G., Evstratov A. V., Laine A. I., Melnik E. I., and Ryabova I. D., J. Membrane Biol. *1*, 402 (1969)
92 Patel D. J. and Tonelli A. E., Biochemistry *12*, 486 (1973)
93 Ovchinnikov Y. A., Ivanov V. T., Evstratov A. V., Bystrov V. F., Abdullaev N. D., Popov E. M., Lipkind G. M., Arkhipova S. F., Efremov E. S. and Shemyakin M. M., Biochem. Biophys. Res. Comm. *37*, 668 (1969)
94 McDonald C. C. and Phillips W. D., J. Amer. Chem. Soc. *89*, 6332 (1967)
95 McDonald C. C. and Phillips W. D., J. Amer. Chem. Soc. *91*, 1513 (1969)
96 McDonald C. C., Phillips W. D. and Glickson J. D., J. Amer. Chem. Soc. *93*, 235 (1971)
97 Masson A. and Wüthrich K., FEBS Lett. *31*, 114 (1973)
98 Sternlicht H. and Wilson D., Biochemistry *6*, 2881 (1967)
99 Wagner G. and Wüthrich K., to be published
100 Cohen J. S., Nature *223*, 43 (1969)
101 Meadows D. H., Markley J. L., Cohen J. S. and Jardetzky O., Proc. Nat. Acad. Sci. *58*, 1307 (1967)
102 Bradbury J. H. and Scheraga H. A., J. Amer. Chem. Soc. *88*, 4240 (1966)
103 Bradbury J. H. and Wilairat P., Biochem. Biophys. Res. Comm. *29*, 84 (1967)
104 Bradbury J. H. and Chapman B. E., Biochem. Biophys. Res. Comm. *49*, 891 (1972)
105 Meadows D. H., Jardetzky O., Epand R. M., Rüterjans H. H. and Scheraga H. A., Proc. Nat. Acad. Sci. *60*, 766 (1968)
106 Roberts G. C. K. and Jardetzky O., Adv. Protein Chem. *24*, 447 (1970)
107 Meadows D. H. and Jardetzky O., Proc. Nat. Acad. Sci. *61*, 406 (1968)
108 Cohen J. S., Yim C. T., Kandel M., Gornall A. G., Kandel S. I. and Freedman M. H., Biochemistry *11*, 327 (1972)
109 Markley J. L., Biochemistry *12*, 2245 (1973)
110 Migchelsen C. and Beintema J. J., J. Mol. Biol. *79*, 25 (1973)
111 Sachs D. H., Schechter A. N. and Cohen J. S., J. Biol. Chem. *246*, 6576 (1971)
112 Schechter A. N., Sachs D. H., Heller S. R., Shrager R. I. and Cohen J. S., J. Mol. Biol. *71*, 39 (1972)
113 Shrager R. I., Cohen J. S., Heller S. R., Sachs D. H. and Schechter A. N., Biochemistry *11*, 541 (1972)
114 Hvidt A. and Nielsen S. O., Adv. Protein Chem. *21*, 287 (1966)
115 Karplus S., Snyder G. H. and Sykes B. D., Biochemistry *12*, 1323 (1973)
116 Glickson J. D., Phillips W. D. and Rupley J. A., J. Amer. Chem. Soc. *93*, 4031 (1971)
117 Patel D. J., Woodward C. K. and Bovey F. A., Proc. Nat. Acad. Sci. *69*, 599 (1972)

118 Robillard G. and Shulman R. G., J. Mol. Biol. *71*, 507 (1972)
119 Griffin J. H., Cohen J. S. and Schechter A. N., Biochemistry *12*, 2096 (1973)
120 Wüthrich K. and Wagner G., FEBS Lett. *50*, 265 (1975)
121 Dobson C. M., Moore G. R. and Williams R. J. P., FEBS Letters *51*, 60 (1975)
122 Moore G. R. and Williams R. J. P., FEBS Letters *53*, 334 (1975)
123 Wagner G. and Wüthrich K., J. Magn. Resonance, in press
124 Campbell I. D., Dobson C. M. and Williams R. J. P., Proc. Royal Soc. London A, in press
125 Wagner G., De Marco A. and Wüthrich K., J. Magn. Resonance, submitted for publication
126 Snyder G. H., Karplus S., Rowan R. and Sykes B. D., Paper presented at the VIth International Conference on Magnetic Resonance in Biological Systems, Kandersteg, Switzerland, September 16–21, 1974
127 Brown L. R., De Marco A., Wagner G. and Wüthrich K., Eur. J. Biochem., submitted for publication
128 Wüthrich K., Wagner G. and Tschesche H., Proceedings XXIII Annual Colloquium on Protides of the Biological Fluids (Brugge, Belgium, 1975), in press
129 Campbell I. D., Dobson C. M. and Williams R. J. P., Chem. Comm. *1974*, 888
130 Crespi H. L., Norris J. R. and Katz J. J., Nature New Biology *236*, 178 (1972)

CHAPTER IV

SPIN RELAXATION AND MOLECULAR DYNAMICS

The high resolution ^{1}H-NMR spectra discussed in the foregoing Chapters II and III were obtained essentially exclusively by continuous wave techniques, where the radio frequency field is applied continuously while the spectrum is recorded. These experiments resulted in a description of the spin systems by the chemical shifts and the spin–spin coupling constants. However, as was briefly mentioned in Sections I.2 and I.3, additional information on the molecular structures can be obtained from measurements of the spin relaxation times T_1 and T_2, and for these studies different NMR techniques are usually employed. The sample, which is placed in a static polarizing magnetic field H_0, is irradiated with a short radio frequency pulse, and the observation of the spin system is made after the H_1-field is turned off. One therefore employs the terms "pulse methods", or "free precession techniques" for these experiments. Although pulse methods were developed almost as early as the CW studies, they have until recently found only limited application in chemical and biological studies. The recent introduction of Fourier transform techniques, however, greatly increased the potential of pulse methods for studies of complex molecules, so that they are now in many instances also much more efficient than CW experiments for studies of high resolution NMR spectra.

Many features of NMR spectra can only be understood by quantum mechanical considerations, which lead to the energy level schemes used in Sections I.2 and Appendix II to describe the high resolution ^{1}H-NMR experiments. On the other hand, numerous features of macroscopic samples of nuclear spin systems, i.e. in particular most of the pulse experiments, are more easily visualized through a classical treatment. Hence some selected features of the classical description of NMR will be presented at the outset of this chapter. This will then be applied to describe the principles of the pulse methods used for measurements of T_1 and T_2, and for Fourier transform NMR. Subsequent sections will present a survey of spin relaxation mechanisms, and the discussion in Chapters II and III of ^{1}H-NMR in peptides and proteins will be complimented with some remarks on proton spin relaxation in solutions of these molecules. The author hopes that the content of this chapter will also help the reader to gain access to the more thorough and more sophisticated treatments of nuclear spin relaxation in the classical textbooks by Abragam (Ref. 1 of Chapter I) and Slichter (Ref. 2 of Chapter I), or in the introductory textbook by Farrar and Becker[1].

1. THE CLASSICAL DESCRIPTION OF NMR BY THE BLOCH EQUATIONS

The Bloch equations are a set of differential equations which have been proposed on the basis of phenomenological arguments to describe the magnetic properties of ensembles of nuclear spins in external magnetic fields. As we shall see, the Bloch equations can account for the combined influence on the spin system of the static magnetic field H_0, the radio frequency field H_1, and the nuclear spin relaxation. While this description of a system of nuclear spins can in principle be developed in a fixed coordinate system of the laboratory, it is helpful for an easy visualization of most NMR phenomena to refeı the motion of the magnetization to a coordinate system that rotates about the static field H_0, and we shall therefore employ the rotating frame of reference in most of the following discussions.

Equations of Motion for the Macroscopic Magnetization—the Rotating Frame of Reference

The starting point of the classical description of NMR is the equation of motion of a magnetic moment $\boldsymbol{M}$ in a homogeneous magnetic field $\boldsymbol{H}$

$$\frac{\mathrm{d}\boldsymbol{M}}{\mathrm{d}t} = \gamma \boldsymbol{M} \times \boldsymbol{H} \qquad \text{(IV-1)}$$

The motion of $\boldsymbol{M}$ resulting from the torque $\boldsymbol{M} \times \boldsymbol{H}$ exerted on the magnetic moment by the applied field, is a precession about $\boldsymbol{H}$, as indicated in Fig. IV-1. For the nuclear magnetization of an ensemble of free spins, i.e. spins which are not subject to interactions with an environment that would induce spin relaxation, Equation IV-1 would provide a description of the time course of $\boldsymbol{M}$ in a coordinate system fixed in the laboratory (Fig. IV-1). Let us now transform the equation of motion into a coordinate system which rotates with respect to the laboratory frame with an angular velocity represented by a vector $\boldsymbol{\omega}$. The motion of the magnetic moment is then given by

$$\frac{\partial \boldsymbol{M}}{\partial t} = \gamma \boldsymbol{M} \times \left(\boldsymbol{H} + \frac{\boldsymbol{\omega}}{\gamma}\right) \qquad \text{(IV-2)}$$

It is seen that the form of Equation IV-1 is also valid in the rotating frame,

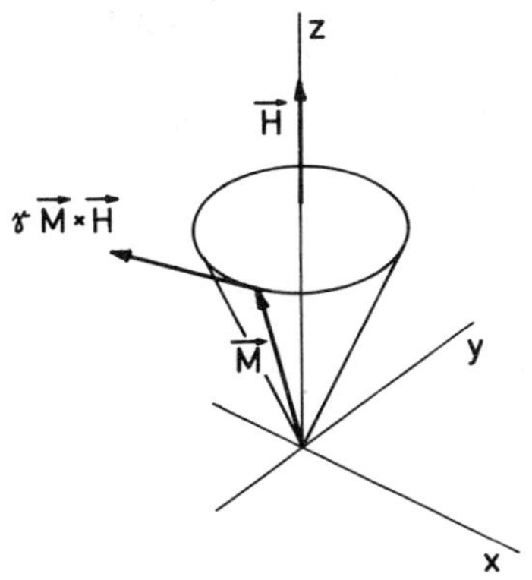

Fig. IV-1. Precession of the nuclear magnetization $\boldsymbol{M}$ about a static magnetic field $\boldsymbol{H}$. The carthesian coordinate system is chosen with the z-axis along $\boldsymbol{H}$.

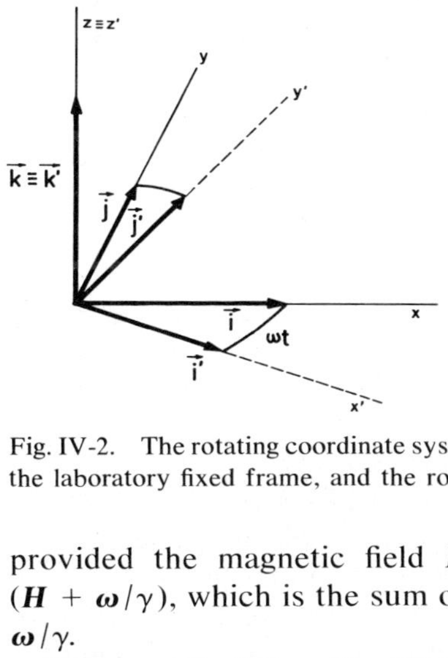

Fig. IV-2. The rotating coordinate system. $\boldsymbol{i}$, $\boldsymbol{j}$, $\boldsymbol{k}$, and $\boldsymbol{i}'$, $\boldsymbol{j}'$, $\boldsymbol{k}'$ are unit vectors along the axes of the laboratory fixed frame, and the rotating frame, respectively.

provided the magnetic field $\boldsymbol{H}$ is replaced by an effective field $\boldsymbol{H}_e = (\boldsymbol{H} + \boldsymbol{\omega}/\gamma)$, which is the sum of the laboratory field $\boldsymbol{H}$ and a fictitious field $\boldsymbol{\omega}/\gamma$.

If the result of Equation IV-2 is applied to a system where $\boldsymbol{H}$ is the static polarizing magnetic field $\boldsymbol{H}_0$, and one chooses a rotating frame with $\boldsymbol{\omega} = -\gamma \boldsymbol{H}_0$, the effective field $\boldsymbol{H}_e$ vanishes, and hence $\partial \boldsymbol{M}/\partial t = 0$. Thus, in this coordinate system which precesses with respect to the laboratory frame with an angular frequency* $\omega_0 = \gamma H_0$, which is called the Larmor frequency of the spins in the applied field H_0, the magnetic moment $\boldsymbol{M}$ is a fixed vector.

*Note the relation between the angular frequency ω in radians sec^{-1} and the frequency ν in hertz (Hz): $\omega = 2\pi\nu$.

In a typical NMR experiment, an alternating magnetic field $\boldsymbol{H}_1$ is applied perpendicular to the static field $\boldsymbol{H}_0$. This alternating field can be thought of as a field rotating in the xy plane at an angular frequency ω

$$\boldsymbol{H}_1 = H_1(\boldsymbol{i} \cos \omega t + \boldsymbol{j} \sin \omega t) \tag{IV-3}$$

where $\boldsymbol{i}$ and $\boldsymbol{j}$ are unit vectors along the x and y axes of the laboratory frame (Fig. IV-2). The effective field in a frame rotating with a frequency ω then becomes

$$\boldsymbol{H}_e = \boldsymbol{H}_0 + \frac{\boldsymbol{\omega}}{\gamma} + \boldsymbol{H}_1 \tag{IV-4}$$

At resonance the fictitious field exactly cancels $\boldsymbol{H}_0$, and leaves only $\boldsymbol{H}_1$ in the xy plane to interact with $\boldsymbol{M}$

$$\frac{\partial \boldsymbol{M}}{\partial t} = \gamma \boldsymbol{M} \times \boldsymbol{H}_1 \tag{IV-5}$$

Since $\boldsymbol{H}_1$ rotates at the same frequency as the frame, one can arbitrarily assign $\boldsymbol{H}_1$ along the x' axis of the rotating system (Fig. IV-2), and Equation IV-5 shows then that in the rotating frame, $\boldsymbol{M}$ precesses about the x' axis with the angular frequency γH_1. The angle θ through which $\boldsymbol{M}$ rotates during a time t_p is given in radians by

$$\theta = \gamma H_1 t_p \tag{IV-6}$$

This relation is fundamental for NMR pulse methods. By a proper choice of the pulse length t_p, precessions of $\boldsymbol{M}$ about definite angles θ can be obtained, e.g. a "90° pulse" or a "180° pulse" (Fig. IV-3).

Nuclear Spin Relaxation and Macroscopic Magnetization

In a real system of nuclear spins, the motion of the magnetization is further affected by relaxation processes. Different relaxation phenomena are in general observed for the different components of the magnetization $\boldsymbol{M}$. If the component M_z is chosen along the direction of the polarizing field $\boldsymbol{H}_0$, one has an equilibrium magnetization $M_0 > 0$ for $H_0 > 0$, which corresponds to the net macroscopic magnetization previously discussed with the use of Equation I-1. In the Bloch equations it is assumed that the trend of M_z towards its equilibrium value M_0 is a first order process with a characteristic time constant T_1

$$\frac{\mathrm{d}M_z}{\mathrm{d}t} = -\frac{M_z - M_0}{T_1} \tag{IV-7}$$

With the xy plane perpendicular to H_0, there is no way in an isotropic medium to distinguish between the x, $-x$, y, and $-y$ directions, and hence no reason for phase coherence in the xy plane. Therefore at equilibrium, the components M_x and M_y of the macroscopic magnetization will vanish. In the Bloch equations it is assumed that if the nuclear magnetization is given a component at right angles to H_0, e.g. by the application of a radio frequency pulse (Fig. IV-3), the decay of this "transverse magnetization" by spin

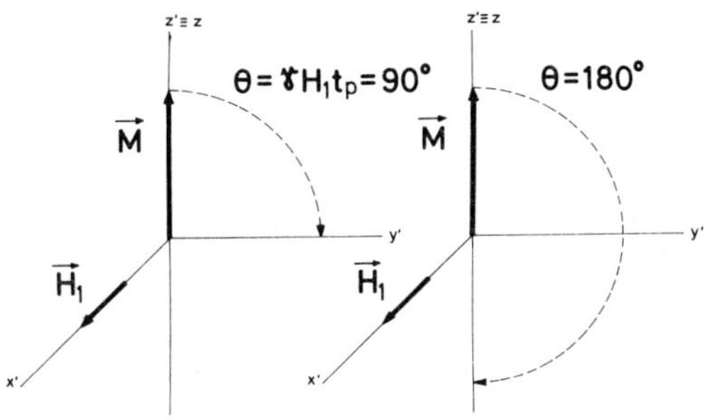

Fig. IV-3. Precession of $\boldsymbol{M}$ about H_1 in the frame rotating at the resonance frequency $\omega_0 = \gamma H_0$.

relaxation is a first order process with a characteristic time constant T_2

$$\frac{\mathrm{d}M_x}{\mathrm{d}t} = -\frac{M_x}{T_2}; \quad \frac{\mathrm{d}M_y}{\mathrm{d}t} = -\frac{M_y}{T_2} \tag{IV-8}$$

The relaxation times T_1 and T_2 are referred to as "longitudinal" or "spin-lattice" relaxation time, and "transverse" or "spin–spin" relaxation time, respectively. Experience has shown that the exponential decay of the magnetization used in the Bloch equations gives a quite accurate description of the relaxation phenomena observed in most nuclear spin systems.

The Bloch Equations

In writing the Bloch equations, the motion due to relaxation is superposed on the motion of the free spins, so that the equation of motion for the magnetization of a macroscopic ensemble of real spins becomes in the rotating frame

$$\frac{\mathrm{d}\boldsymbol{M}}{\mathrm{d}t} = \gamma \boldsymbol{M} \times \left[\left(H_0 - \frac{\omega}{\gamma}\right)\boldsymbol{k}' + H_1\boldsymbol{i}'\right] - \frac{M_{x'}\boldsymbol{i}' + M_{y'}\boldsymbol{j}'}{T_2} - \frac{M_z - M_0}{T_1}\boldsymbol{k}' \tag{IV-9}$$

where $\boldsymbol{i}'$, $\boldsymbol{j}'$ and $\boldsymbol{k}' \equiv \boldsymbol{k}$ are unit vectors along the x', y', and z' axes of the rotating coordinate system (Fig. IV-2), and $M_{x'}$ and $M_{y'}$ are the transverse components of $\boldsymbol{M}$ in that frame. For the three components of $\boldsymbol{M}$ in the rotating frame, one then has*

$$\frac{dM_{x'}}{dt} = -\frac{M_{x'}}{T_2} + (\gamma H_0 - \omega)M_{y'}$$

$$\frac{dM_{y'}}{dt} = -(\gamma H_0 - \omega)M_{x'} - \frac{M_{y'}}{T_2} + \gamma H_1 M_z \qquad \text{(IV-10)}$$

$$\frac{dM_z}{dt} = -\gamma H_1 M_{y'} - \frac{M_z - M_0}{T_1}$$

Steady State Solution of the Bloch Equations—Applications to High Resolution CW NMR

The Bloch Equations (IV-10) can be solved under various limiting conditions. We shall at this point briefly consider the steady state solutions obtained by setting $dM_{x'}/dt = dM_{y'}/dt = dM_z/dt = 0$, which is to a good approximation valid under the conditions of the slow passage CW experiments used to record high resolution ^{1}H-NMR spectra (Chapters II and III). If we substitute $\Delta\omega = (\gamma H_0 - \omega)$, the steady state solutions for the components of $\boldsymbol{M}$ in the Equations IV-10 can be written

$$M_{x'} = \frac{\Delta\omega\gamma H_1 T_2^2}{1 + (T_2\Delta\omega)^2 + \gamma^2 H_1^2 T_1 T_2} M_0 \qquad \text{(IV-11)}$$

$$M_{y'} = \frac{\gamma H_1 T_2}{1 + (T_2\Delta\omega)^2 + \gamma^2 H_1^2 T_1 T_2} M_0 \qquad \text{(IV-12)}$$

$$M_z = \frac{1 + (\Delta\omega T_2)^2}{1 + (T_2\Delta\omega)^2 + \gamma^2 H_1^2 T_1 T_2} M_0 \qquad \text{(IV-13)}$$

With the commonly employed experimental devices, $M_{x'}$ and $M_{y'}$ can be observed separately as the dispersion signal, and the absorption signal, respectively. In high resolution CW NMR measurements, the absorption signal $M_{y'}$ is usually observed.

Some of the general features of CW NMR spectra which were introduced in Section I.2, can readily be derived from the Equations IV-11 to IV-13. Thus it is seen that the three components of $\boldsymbol{M}$ are proportional to M_0, i.e. in the absence of initial polarization the NMR cannot be observed, and the sensitivity for the detection of the NMR signals will increase with increasing field strength of H_0. For low intensity of the radio frequency field H_1, i.e. for

*In a notation introduced by F. Bloch and widely used in the literature, the components of the magnetization in the rotating frame are denoted as u and v, where $u = M_{x'}$, and $v = -(\gamma/|\gamma|)M_{y'}$.

$\gamma^2 H_1^2 T_1 T_2 \ll 1$, M_y, increases proportional to H_1, or in other words "the saturation" of the NMR signal is negligible. As a consequence of the non-saturating experimental conditions employed in typical high resolution CW NMR experiments, the longitudinal relaxation time T_1 is not directly manifested in the absorption signal $M_{y'}$, which then has a Lorentz line shape with a half width at half height of $1/T_2$ (Equation IV-12; see also Equation I-9 and Fig. I-2). Equation IV-13 shows further that M_z is very little affected by H_1 in a CW experiment under non-saturating conditions. We thus conclude from Equations IV-11 to IV-13 that the relaxation time T_2 can in principle be obtained from a line shape analysis in the CW NMR spectra, and T_1 can be measured when the high resolution experiments are complemented with CW spectra obtained under conditions of appreciable saturation. However, there are numerous potential pitfalls and technical difficulties in CW studies of spin relaxation times, which makes pulse experiments by far the more reliable method for measurements of T_1 and T_2.

2. PULSE METHODS FOR MEASUREMENTS OF THE RELAXATION TIMES T_1 AND T_2

This section is mainly based on the Equation IV-6, which describes the angle θ through which a given pulse $(H_1 t_p)$ causes the magnetization in the rotating frame to precess. We shall consider the behavior of a spin system subjected to one or several subsequent 90° and/or 180° pulses (Fig. IV-3). General traits of the pulse methods to be discussed are further, that one chooses H_1 in Equation IV-6 sufficiently large for t_p to be short compared to T_1 and T_2, so that relaxation during the pulse is negligible, and that one considers only the NMR signal which arises after the radiofrequency pulse has been turned off. These techniques can readily be applied to spin systems with a single NMR line, e.g. for studies of the ^{1}H-NMR of water in dilute aqueous solutions of biopolymers. For investigations of more complex spin systems, the spectral information from a pulse experiment is first subjected to a Fourier analysis, as will be discussed in Section IV.3.

Free Induction Decay

If a 90° pulse is applied along the x' axis in the rotating frame, the magnetization $\boldsymbol{M}$ will lie entirely along the y' axis (Fig. IV-3). Because the apparatus is normally arranged to detect the magnetization $M_{y'}\boldsymbol{j}'$ (Fig. IV-2), the magnitude of $M_{y'}$ determines the intensity of the observed NMR signal. The latter is called a "free induction signal", since it is recorded while the nuclear magnetization precesses "freely" with respect to the laboratory frame, i.e. without applied H_1 field. With time elapsing after the pulse, the transverse relaxation causes the signal to decay (Fig. IV-4). In a perfectly homogeneous field H_0, this decay would be given by Equation IV-8, with the time constant T_2. The T_2 relaxation arises from molecular interactions which cause the individual nuclei to precess at slightly different frequencies, and thus to get out of phase with each other in the $x'y'$ plane. In practice, one will always have a finite field inhomogeneity ΔH_0 over the sample volume, which further contributes to the decay of the transverse magnetization. The overall free induction decay is then usually described by

$$\frac{dM_{y'}}{dt} = -\frac{M_{y'}}{T_2^*} \tag{IV-14}$$

where T_2^* includes contributions from both the transverse relaxation and the magnetic field inhomogeneity

$$\frac{1}{T_2^*} = \frac{1}{T_2} + \frac{|\gamma \Delta H_0|}{2} \tag{IV-15}$$

As we shall see in more detail in the following section, for a given spin system the free induction decay after a 90° pulse (Fig. IV-4) and the CW

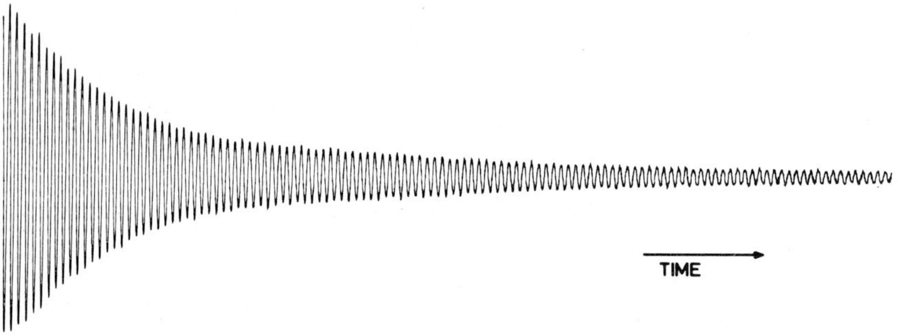

Fig. IV-4. Free induction decay of the ^{1}H-NMR signal of the residual protons in 99.8% D_2O following an off resonance 90° pulse.

NMR spectrum (Fig. I-2) are Fourier transforms of each other. As the corresponding phenomenon to the free induction decay by Equation IV-14, one has that the resonance line width observed in the CW experiment (Fig. I-2) is determined by

$$\delta\nu_{1/2} = \frac{1}{2\pi T_2^*} \tag{IV-16}$$

rather than by Equation I-9.

Measurements of the Longitudinal Relaxation Time T_1

The principles underlying pulse methods for measuring T_1 can best be illustrated by the so-called $180° - \tau - 90°$ sequence, which is a standard procedure. First, a 180° pulse inverts the magnetization along the z' axis (Fig. IV-3). During the period of time τ, longitudinal relaxation is allowed to occur, which causes M_z to go from the value $-M_0$ immediately after the 180° pulse through zero to its equilibrium value M_0. A 90° pulse applied after the time τ produces a free induction signal, the initial intensity of which is proportional to the value of M_z at the time τ (Fig. IV-3). From a combination of $180° - \tau - 90°$ experiments with different values of τ, the decay rate of M_z can be established (Fig. IV-5). Quantitatively, the decay of M_z is given

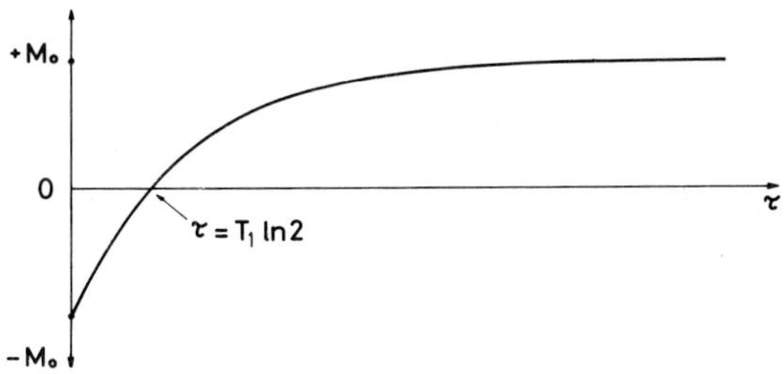

Fig. IV-5. Plot of the normalized values of $M_{z'}$ observed in a series of $180° - \tau - 90°$ experiments as a function of τ.

by Equation IV-7, of which the integration with $M_z = -M_0$ at $\tau = 0$ gives the function IV-17, which is plotted in Fig. IV-5.

$$M_z = M_0(1 - 2e^{-\tau/T_1}) \qquad \text{(IV-17)}$$

Other pulse experiments for measurements of T_1 include the $90° - \tau - 90°$ pulse sequence, and observation of the steady state value of M_z attained after a sequence $90° - \tau - 90° - \tau - 90°$. . . . Again, T_1 is obtained from the combination of measurements with different values of τ. The principal advantage of these experiments over the $180° - \tau - 90°$ method is that they are in general less time-consuming. On the other hand their application is restricted to spin systems with $T_2 \ll T_1$.[1]

Measurements of the Transverse Relaxation Time T_2

The foregoing discussion of the free induction decay has shown that the transverse relaxation time T_2^* observed in pulse experiments (Equation IV-14) or by CW NMR (Equation IV-16) can for any practical purposes only be used as an approximate measure of rather short T_2's, i.e. for $T_2 \ll |2/\gamma\Delta H_0|$ (Equation IV-15). However, pulse methods have been developed by which the field inhomogeneity problem can in principle be overcome. In Fig. IV-6, the basic idea underlying these methods is illustrated with the original "spin-echo" experiment by Hahn[2], which consists of a $90° - \tau - 180°$ pulse sequence followed by observation of $M_{y'}j'$ at the time 2τ. As Fig. IV-6 shows, in the absence of relaxation, the echo at the time 2τ (Fig. IV-6E) would have the same size as the initial signal after the 90° pulse (Fig. IV-6A). However, because the magnetization decays during the time period 2τ according to Equation IV-8, the echo amplitude of a real spin system will depend on T_2. The transverse relaxation time can thus in principle be determined from the combination of a number of spin-echo measurements with different values of τ, which results in a plot of the echo amplitude vs. τ

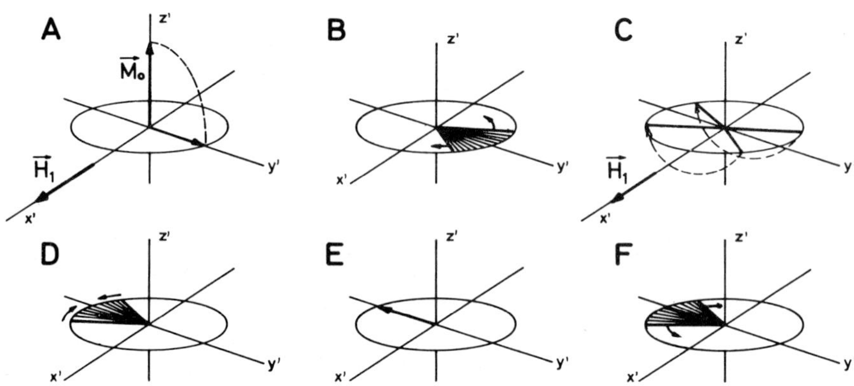

Fig. IV-6. The Hahn spin-echo experiment[2]. Note that it has been assumed in this scheme, that the dephasing of the magnetization comes entirely from field inhomogeneities, neglecting T_2-relaxation (Equation IV-8).
A. After a 90° pulse applied along the x'-axis of the rotating frame, the magnetization lies entirely along the y'-axis.
B. During the time period τ, the inhomogeneity of the field H_0 causes the contributions to the macroscopic magnetization from nuclei in different areas of the sample to dephase. (In a real system, the T_2 processes would cause additional dephasing.) Since the rotation rate of the frame has been chosen equal to the average precession rate of the nuclear spins, those nuclei precessing faster than the average appear to move clockwise with respect to the rotating frame, and v.v.
C. At the time τ, a 180° pulse along x' causes the magnetization to rotate by 180° about the x'-axis.
D. As a result of the 180° pulse, the magnetization is again in the $x'y'$ plane, but the faster components follow now behind those precessing at the average rate, and v.v. During another time period τ, the field inhomogeneity thus causes the magnetization to rephase.
E. At the time 2τ, the faster components have again reached those precessing at the average rate, and the slower ones have fallen back to the average position. At this time the spin echo is measured. (The dephasing from the T_2 processes in a real system would not be recovered.)
F. With more time elapsing, the magnetization dephases again.

in much the same way as shown for the $180° - \tau - 90°$ sequence in Fig. IV-5.

The argumentations of the Hahn spin-echo experiment are strictly valid only if any given nuclear spin in the sample experiences a constant local magnetic field during the entire period 2τ, i.e. if there were no diffusion of molecules in the NMR sample. Therefore, more sophisticated spin-echo pulse sequences have been developed to overcome the difficulties arising from diffusion, e.g. the Carr-Purcell[3] and the Meiboom-Gill[4] methods[1].

3. FOURIER TRANSFORM NMR

We have seen that an exponential decay is observed for the free induction signal of H_2O (Fig. IV-4), which gives a single Lorentzian line in a CW ^1H-NMR experiment (Fig. I-2). For a more complex spin system, the free induction decay consists of a correspondingly complex interferogram of the different resonance frequencies, which can no longer be analyzed by inspection (Fig. IV-7). After Fourier transformation (FT) of the free induction decay, however, a NMR spectrum corresponding to that obtained in a CW experiment is obtained. Among the principal advantages of FT NMR spectroscopy over the conventional CW mode, which have also greatly affected the biological applications of NMR, are (i) the improved sensitivity, (ii) the capacity for simultaneous measurements of T_1 for all the resonances in a complex spectrum, (iii) the inherent short observation time which allows investigations of short-lived species, and thus enlarges the potential of NMR for studies of reaction kinetics.

Fourier Transformation

The free induction decay after a radiofrequency pulse is often referred to as the "time domain response" or "response function" $f(t)$, and the corresponding NMR spectrum as the "frequency domain spectrum" or "spectral distribution function" $F(\omega)$. It has been shown that under certain conditions which can readily be met in a typical experiment, $f(t)$ following a 90° pulse (Fig. IV-3) and the corresponding spectrum $F(\omega)$ are Fourier transforms of each other[5]

$$F(\omega) = \frac{1}{\sqrt{2\pi}} \int_{-\infty}^{\infty} f(t)\, e^{-i\omega t}\, dt \qquad \text{(IV-18)}$$

$$f(t) = \frac{1}{\sqrt{2\pi}} \int_{-\infty}^{\infty} F(\omega)\, e^{i\omega t}\, d\omega \qquad \text{(IV-19)}$$

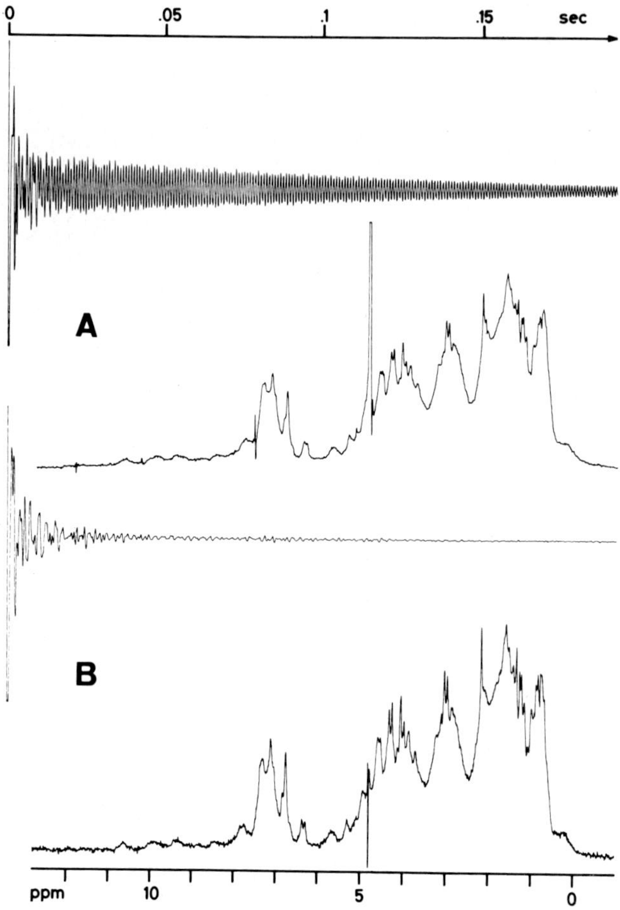

Fig. IV-7. FT ^{1}H-NMR spectra at 100 MHz of a 0.01-M solution of the basic pancreatic trypsin inhibitor in D_2O. 100 transients had been accumulated with an acquisition time of 1 sec. A. Free induction decay after a 90° pulse (only the decay during 0.2 sec after the pulse is shown), and the spectrum obtained through Fourier transformation. B. Same as A, except that the solvent resonance had been eliminated by a $180° - \tau - 90°$ pulse sequence.

The FT NMR method then consists basically in the application of a radio frequency pulse to a spin system, and Fourier transforming the response $f(t)$ of the system in order to obtain the spectral information in the more explicit form of the frequency domain function $F(\omega)$ (Fig. IV-7).

Instrumentation for Fourier Transform NMR

The basic equipment for FT NMR consists of a pulse NMR spectrometer, and an on-line digital computer properly interfaced to the spectrometer. In the spectrometer the sample is subjected to the desired pulse sequence, and the resulting response of the system is recorded. The computer is used to store the free induction decay in digital form, to accumulate repeated signals to improve the signal-to-noise ratio, to perform the Fourier transformation, and for miscellaneous types of data processing. For a more thorough discussion of the instrumental requirements for FT NMR, the reader is referred to the references cited in this section. Here we shall briefly dwell on the requirement for a large intensity of the radio frequency field H_1, which illustrates a basic aspect of the functioning of the method.

In Section IV.2 we discussed the situation which arises when a pulse of intensity H_1 and duration t_p is applied exactly at the resonance frequency of the spins (Equations IV-5 and IV-6, Fig. IV-3). In a complex spin system with n resonance lines, the frequency of the pulse will necessarily be off-resonance for $(n-1)$ lines. For practical reasons, one actually selects a pulse frequency which is outside of the range of chemical shifts occurring in the system (Fig. IV-8). In the rotating frame, the effective magnetic field for the i different spins in the system is therefore given by (see also Equation IV-4)

$$|\mathbf{H}_e| = \left[\left(\frac{\omega_i - \omega}{\gamma}\right)^2 + H_1^2\right]^{1/2} \qquad \text{(IV-20)}$$

where ω_i are the Larmor frequencies of the individual nuclei, ω is the "carrier frequency" at which the pulse is applied (Fig. IV-8), and H_1 is the

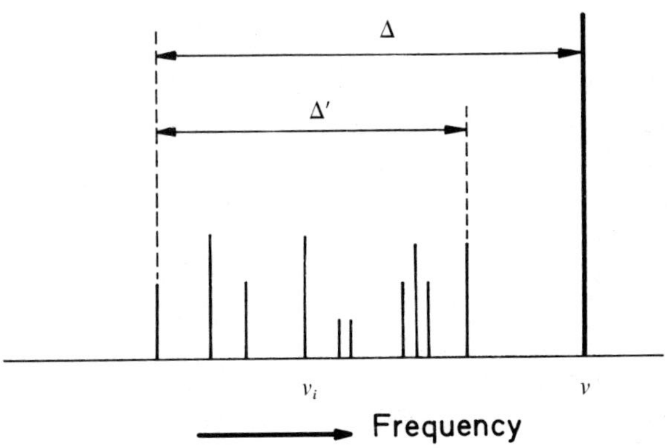

Fig. IV-8. Schematic spectrum extending over a range of chemical shifts Δ'Hz. If a pulse of radiofrequency ν is applied, the resonance lines extend as far as Δ Hz from the carrier frequency. As indicated in the scheme, the carrier frequency is in practice usually chosen outside the spectral region Δ'. (In this scheme, ν was placed near the high frequency end of the range Δ'. In an alternative and in general essentially equivalent experiment, it could be placed near the low frequency limit of Δ'.) Note that $\omega = 2\pi\nu$ and $\omega_i = 2\pi\nu_i$.

amplitude of the radio frequency field. With

$$\gamma H_1 \gg 2\pi\Delta \tag{IV-21}$$

where Δ is the entire range of chemical shifts in Hz relative to the carrier frequency (Fig. IV-8), the term $(\omega_i - \omega)$ in Equation IV-20 can be neglected, and hence one has that $\boldsymbol{H}_e \approx \boldsymbol{H}_1$ for all the frequencies ω_i in the spectrum. Introducing Equation IV-21 into Equation IV-6, we obtain for a $\pi/2$ ($\equiv 90°$) pulse

$$t_p \ll \frac{1}{4\Delta} \tag{IV-22}$$

Overall this discussion shows that a pulse of frequency ω can tip nuclear spin moments precessing at frequencies different from ω, and that the condition of Equation IV-5 is to a good approximation valid for all the components of a complex spin system, provided that a short intense pulse is applied which satisfies the conditions of Equations IV-21 and IV-22.

The Gain of Sensitivity in FT NMR over CW NMR

The increased sensitivity has been foremost in the motivations for the development of FT NMR, and it has indeed opened an entirely new era for studies of less abundant and less sensitive nuclei, in particular ^{13}C. The following argument, which was adapted from Ernst and Anderson's[5] original investigation of the conditions under which FT NMR could be more efficient than direct observation of the CW spectrum, can make the possible sensitivity enhancement intuitively clear. Let us assume that we deal with a quite typical ^1H-NMR spectrum of spectral width 1000 Hz, which consists of resonances with a line width of 0.1 Hz. In scanning this spectrum in the CW mode, one collects information on any given line only during 1/10000 of the time of duration of the experiment. It is obvious that the rate of accumulation of information could be greatly increased by observation of the entire spectrum during the full duration of the experiment, e.g. by using several transmitters and receivers with different frequencies, of which each one would record a given line at all times. The FT NMR experiment makes use of the potential of such a "multichannel spectrometer". As we have seen above, a short pulse is equivalent to the simultaneous application of a wide range of radio frequencies, encompassing in full the spectral region Δ (Fig. IV-8), and after the pulse the response of all the spins in the system can therefore be recorded simultaneously. In practice this results in an increase of the sensitivity by approximately one order of magnitude[1,5]. When sensitivity is a critical factor, one will for both FT and CW experiments improve the signal-to-noise ratio by spectrum accumulation in a computer.

Since time averaging improves the signal-to-noise with the square root of the accumulation time (Section II.4), the gain in time needed to obtain a given signal-to-noise ratio by FT NMR as compared to CW NMR thus becomes of the order of a factor 100.

More generally, the theoretical saving in time with the FT method as compared to CW NMR can be given approximately by $(\Delta'/\delta\nu_{1/2})^{1/2}$, where Δ' is the total range of chemical shifts in the spectrum (Fig. IV-8), and $\delta\nu_{1/2}$ is the width of a typical line in the spectrum (Fig. I-2). This shows that the method is particularly advantageous for nuclei with large chemical shift ranges giving high resolution spectra with much fine structure.

Practical Aspects: Selection of the Experimental Parameters and Data Processing in the Computer

As in the conventional CW mode, the sensitivity and spectral resolution attainable in a FT experiment are in principle limited by the hardware, i.e. in particular the magnetic field H_0, the specifications of the pulse generator for producing intense short pulses, the rate and the dynamic range† available for acquiring the digital data during the free induction decay, and the memory size of the computer. However, the quality of a spectrum recorded with a given set of equipment will also depend on the choice of a number of experimental parameters. The impact of some of these will now be briefly considered.

When working with the common FT spectrometers, the spectral width Δ is chosen so as to include all the resonance lines of the spectrum, with the carrier frequency outside of the spectral region Δ' (Fig. IV-8). A radiofrequency pulse of "pulse width" (PW $\equiv t_p$ in Equations IV-6 and IV-22) is then applied, and the resulting response digitally stored during a preselected "acquisition time" (AT), which is usually of the order 1 sec. In a typical experiment, this sequence of pulse and acquisition is repeated at regular time intervals, which are given by the sum of AT and a waiting period between pulses, the "pulse delay" (PD), and the consecutive signals are time-averaged in the computer. Let us assume that the PW corresponds to a 90° pulse (Fig. IV-3). The ensuing free induction decays with a time constant T_2^* (Equation IV-15), which limits the attainable spectral resolution (Equation IV-16). To actually achieve the optimal resolution given by T_2^*, one would tend to observe the signal for a very long time. In practical terms this means that one would select AT $\approx 3T_2^*$, because at the time $3T_2^*$ after the

†These are specifications of the digital computer system. The "rate" means the data aquisition rate of the computer. The "dynamic range" is limited by the size of the computer memory, and is in particular closely related to the length of the words used for computer storage.

pulse, the signal intensity has been reduced by a factor $e^3 = 20$, and there is not much to be gained by extending the AT beyond this value. If $T_1 \approx T_2^*$, M_Z would after the time $3T_2^*$ also have recovered nearly to its equilibrium value M_0, and the system would be ready immediately after data acquisition for the next pulse to be applied. In general, however, one has that $T_1 > T_2^*$, and for the initial signal intensity after subsequent pulses to be nearly equal to that after the first pulse (Fig. IV-3), the time interval between pulses has to be lengthened beyond AT by the inclusion of a pulse delay, so that $(\mathrm{AT}+\mathrm{PD}) \gtrsim 3T_1$. Note that only after a time $4.6T_1$, M_z has recovered to better than 99% of M_0.

If the signal : noise ratio after a given time is of prime importance, one will tend to sacrifice resolution for sensitivity by selecting somewhat different parameters, i.e. in particular by shortening AT, and possibly employing a short PD, or no PD at all. The resulting increased efficiency seems intuitively clear if one considers that data taken in the tail of a free induction decay (Fig. IV-4 and IV-7) have a poorer signal-to-noise ratio than those obtained shortly after the pulse, and that one would wish for maximum efficiency to repeat the pulse immediately after a period of data acquisition. On the other hand, in the generally prevailing situation that $T_1 > T_2^*$, the condition $(\mathrm{AT}+\mathrm{PD}) \gtrsim 3T_1$ may then no longer be fulfilled, so that little restoration of M_z towards M_0 can take place, and the signal after subsequent pulses is greatly attenuated (Fig. IV-3). Eventually a steady state is reached with the signal intensity immediately after the pulse determined by $(\mathrm{AT}+\mathrm{PD})/2T_1$.[6] In practice, a maximal signal-to-noise ratio after a fixed time is achieved by combining relatively small values for AT and PD with a shortened PW, so that $\theta < 90°$ (Equation IV-6). The optimum value of θ depends on T_1 and T_2^*, and is usually determined empirically to give the best signal : noise ratio[5].

A consequence of practical importance arises because the T_1's of the individual lines in a complex spectrum are in general different from each other, so that the signal intensities per spin observed under steady state conditions with $M_z \neq M_0$ will be different for the different lines. Therefore, if the integrated resonance intensities in a FT NMR spectrum are to be used for the determination of the number of spins (Fig. II-1), one has to be particularly careful in the selection of the parameters PW, AT, and PD, and should strictly adhere to the condition $(\mathrm{AT}+\mathrm{PD}) \gtrsim 5$ times the longest T_1 in the spectrum. Multiple pulse techniques have been proposed to shorten the waiting time between pulses[1]. It appears, however, that these have not as yet been widely applied in practice.

After the signals resulting from a specific set of values for PW, AT and PD have been accumulated for a given period of time, the noise can be reduced at the expense of some artificial line broadening through multiplication of the time averaged free induction decay by a simple exponential. Alternatively, one can by a similar digital filtering process enhance the resolution at the expense of some reduction in signal-to-noise (Fig. IV-9). The time constants for these sensitivity enhancement and resolution enhancement functions can be selected by the operator. Since the digital filter has to be applied to the free induction decay prior to Fourier transformation, and its

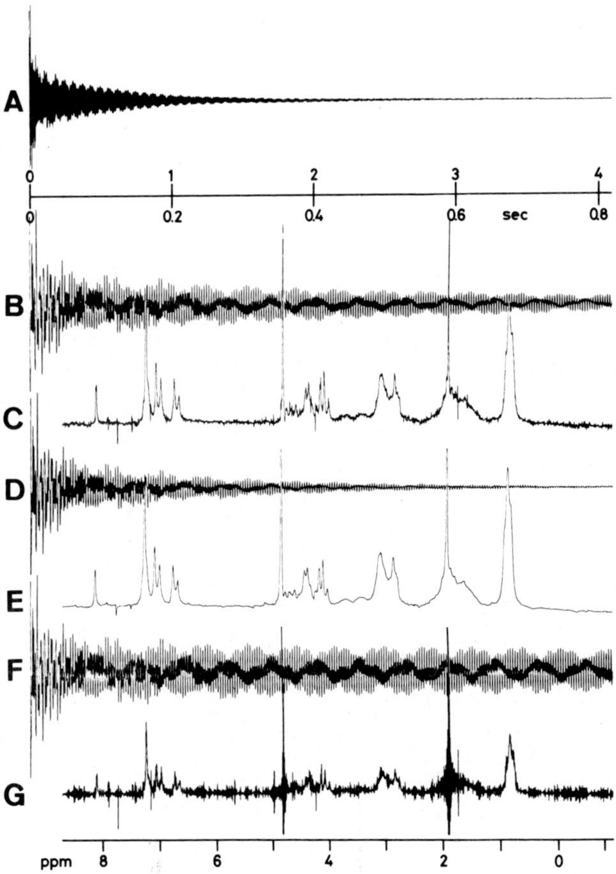

Fig. IV-9. Effects of sensitivity enhancement and resolution enhancement by digital filtering of the 100 MHz ^{1}H-NMR spectrum of a 0.1-M solution in D_2O of angiotensin II, which is a linear polypeptide hormone with the amino acid sequence H-Asp-Arg-Val-Tyr-Ile-His-Pro-Phe-OH. 6 transients were accumulated with an acquisition time of 4 sec. The two intense resonances at 1.8 and 4.8 ppm correspond to acetate, and the residual solvent protons, respectively. A. Free induction decay. B. Expanded plot of the region from 0 to 0.8 sec in A. C. Fourier transform of A. D and E. Free induction decay and spectrum after sensitivity enhancement with a time constant of 0.5 sec. F and G. Free induction decay and spectrum after resolution enhancement with a time constant of 1 sec.

effect becomes in general only clearly manifested in the frequency domain spectrum, it is very helpful to be able to store the free induction decay on a magnetic disc or a tape. From there it can be recalled for the application of digital filtering with various different time constants. Such capabilities for repeated manipulation of the stored free induction decay enable one also to go to more sophisticated data processing, as we will see in the following chapters. For biological studies they are of particular relevance, because repeated aquisition of the free induction decay would in most cases be prohibitively time consuming.

T_1 Measurements in Complex Spin Systems

Next to the increase in sensitivity, the relative ease with which T_1's can be measured in complex molecules is one of the big assets of FT NMR. It adds T_1 to the chemical shift and the spin–spin coupling constant as a third spectral parameter routinely available for resonance assignments and structure elucidation.

T_1's for the different nuclei in a spin system can readily be measured by the usual $180° - \tau - 90°$ pulse sequence (Fig. IV-5), with Fourier transformation of the free induction signal following the 90° pulse. In this way spectra are obtained in which the nuclei are partially relaxed, the degree of relaxation depending on the individual T_1's (Fig. IV-10)[7]. An alternate method for T_1 measurements, the so-called "progressive saturation technique", has the inherent advantage of being less time consuming than the $180° - \tau - 90°$ pulse sequence. It makes use of the above mentioned manifestation of T_1 in the steady state condition attained upon application of successive 90° pulses at intervals which are short compared to $5T_1$.[8]

In principle, spin–spin relaxation times T_2 may be determined by a similar approach based on Fourier transformation of the free induction signal at a suitable point during a spin echo experiment (Fig. IV-6). However, additional difficulties can arise in spin systems involving spin–spin coupling between the different nuclei[1,9], which have so far prevented extensive use of FT NMR for measurements of T_2.

4. RELAXATION MECHANISMS

There are two essential requirements for an efficient relaxation mechanism. First, there needs to be some interaction with the environment which acts directly on the spins, and second, this interaction must fluctuate with time. We can thus tentatively express the relaxation rates by

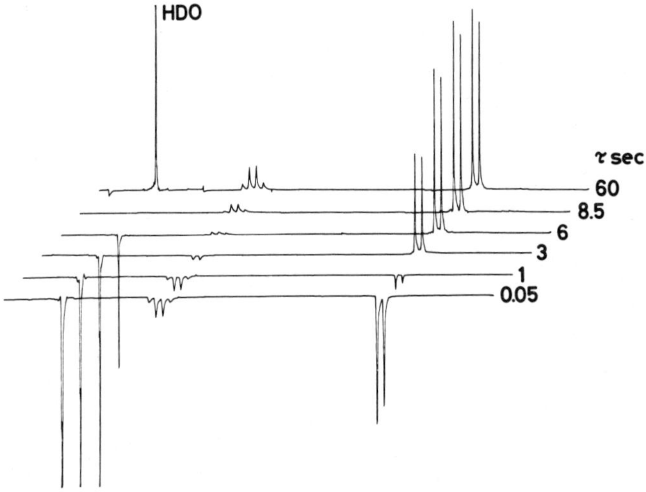

Fig. IV-10. Partially relaxed ^{1}H-NMR spectra at 100 MHz of a 0.3-M solution of L-alanine in D_2O obtained by the $180° - \tau - 90°$ pulse sequence and Fourier transformation. Spectra obtained with several different τ's are shown. For $\tau = 8.5$ sec, a solvent eliminated spectrum was obtained. It is seen that T_1 increases from the methyl protons ($T_1 = 2.0$ sec) to the C^α-proton ($T_1 = 5.9$ sec) to the solvent resonance ($T_1 \approx 12$ sec) (see Appendix V).

$$\frac{1}{T_1} = E_1 \cdot f_1(\tau); \qquad \frac{1}{T_2} = E_2 \cdot f_2(\tau) \tag{IV-23}$$

where E_1 and E_2 describe the interaction with the nuclear spin, and $f_1(\tau)$ and $f_2(\tau)$ account for the time fluctuations of these interactions. One can readily see that these basic conditions for relaxation are met for a solute in H_2O, or in any other solvent containing hydrogen atoms. The magnetic moments of the protons produce local magnetic fields which induce interaction with the nuclear spins of nearby solvent and solute molecules. The strength of this interaction depends on the relative coordinates of the interacting spins, and hence the Brownian motions of the molecules will cause the required time fluctuations of the interactions. In addition to these intermolecular processes, the coupling between spins located on the same molecule, e.g. the two protons of H_2O, can also be modulated by the rotational tumbling of the molecules, and thus induced relaxation. In most spin systems, intramolecular relaxation mechanisms are actually found to make the dominant contributions to the observed relaxation rate.

Five different relaxation mechanisms are quite typical for nuclear spin systems, i.e. (i) magnetic dipole–dipole interaction (DD), (ii) chemical shift anisotropy interaction (SA), (iii) scalar coupling interaction (SC), (iv) spin-rotation interaction (SR), and (v) electric quadrupole interaction (Q).

For any given system, the observed relaxation times will in general result from contributions by various different relaxation mechanisms, and can be expressed as

$$\frac{1}{T_1} = \frac{1}{T_1^{DD}} + \frac{1}{T_1^{SA}} + \frac{1}{T_1^{SC}} + \frac{1}{T_1^{SR}} + \frac{1}{T_1^{Q}} \qquad \text{(IV-24)}$$

and a corresponding expression for $1/T_2$. To relate the observed relaxation times with the molecular properties, one will therefore have to devise means by which the relative importance of the different terms on the right hand side of Equation IV-24 can be established.

To provide some intuitive understanding of the processes which induce spin relaxation, a qualitative discussion of the relations between relaxation rate and molecular motions will next be presented. This is then followed by the explicit quantitative expressions for the different relaxation times in Equation IV-24, which had been derived elsewhere (pp. 264–353, Ref. 1 of Chapter I), and which provide a basis for the experimental differentiation of the different relaxation mechanisms.

Spin Relaxation and Molecular Motions

Let us consider the effects of the oscillating microscopic magnetic fields $\boldsymbol{h}(t)$ produced by the Brownian tumbling of individual molecules on the macroscopic magnetization $\boldsymbol{M}$ after a 90° pulse. In the rotating frame (Fig. IV-2), $\boldsymbol{M}$ is in the $x'y'$ plane, with $\boldsymbol{M} = M_0\boldsymbol{j}'$ immediately after the pulse (Fig. IV-3). We want to investigate how the z-component of the magnetization returns to its equilibrium position $\boldsymbol{M}_0$ under the influence of the microscopic fields $\boldsymbol{h}(t)$, i.e. T_1 relaxation, and how the macroscopic magnetization $M_0\boldsymbol{j}'$ vanishes as a result of the interactions with the microscopic fields $\boldsymbol{h}(t)$, i.e. T_2 relaxation. Equations IV-1 and IV-6 show that $\boldsymbol{M}$ can be perturbed from any given position, if a torque is exerted on it over a certain period of time. This is the case when both $\boldsymbol{M}$ and a component of $\boldsymbol{h}(t)$ are fixed in the rotating frame.

To investigate the situation more thoroughly, we rewrite the microscopic fields $\boldsymbol{h}(t)$ in the rotating frame

$$\boldsymbol{h} = h_{x'}\boldsymbol{i}' + h_{y'}\boldsymbol{j}' + h_{z'}\boldsymbol{k}' \qquad \text{(IV-25)}$$

As was mentioned above, any of the three components can perturb $\boldsymbol{M}$, provided they are static in the rotating frame. From the choice of the rotation axis about $z' \equiv z$ (Fig. IV-2), a static field $h_{z'}\boldsymbol{k}'$ is also static in the laboratory frame, whereas a static field $h_{y'}\boldsymbol{j}'$ or $h_{x'}\boldsymbol{i}'$ rotates in the laboratory system at or near the Larmor frequency of the nucleus. Viewed from the laboratory system, relaxation can thus be induced by low frequency perturbations along the z axis, and by high frequency perturbations in the xy plane.

The next step of the argument is based on some elementary vector algebra, i.e. the relations governing the cross products of the unit vectors in the rotating frame (Fig. IV-2): $\boldsymbol{i}' \times \boldsymbol{j}' = \boldsymbol{k}'$, $\boldsymbol{j}' \times \boldsymbol{k}' = \boldsymbol{i}'$, $\boldsymbol{k}' \times \boldsymbol{i}' = \boldsymbol{j}'$, $\boldsymbol{j}' \times \boldsymbol{i}' = -\boldsymbol{k}'$, etc, and $\boldsymbol{i}' \times \boldsymbol{i}' = \boldsymbol{j}' \times \boldsymbol{j}' = \boldsymbol{k}' \times \boldsymbol{k}' = 0$. Applying these rules to compute the torque exerted on $\boldsymbol{M}$ by $\boldsymbol{h}$

$$\boldsymbol{h} \times \boldsymbol{M} = (h_{x'}\boldsymbol{i}' + h_{y'}\boldsymbol{j}' + h_{z'}\boldsymbol{k}') \times (M_{x'}\boldsymbol{i}' + M_{y'}\boldsymbol{j}' + M_{z'}\boldsymbol{k}') \qquad \text{(IV-26)}$$

we see that a field $h_{z'}\boldsymbol{k}'$ can only perturb the magnetization in the $x'y'$ plane. Because different nuclei in the macroscopic sample experience somewhat different perturbations, the phase coherence of the spins in the $x'y'$ plane will decay under the influence of the $h_{z'}\boldsymbol{k}'$ fields, which is a pure T_2-process. Equation IV-26 shows further that the microscopic fields $h_{x'}\boldsymbol{i}'$ and $h_{y'}\boldsymbol{j}'$ exert torques on the components of M both in the $x'y'$ plane and along the z' axis, and are thus effective in T_1 and T_2 relaxation. Viewed from the laboratory frame, T_1 can thus only be affected by rapid molecular motions with frequency components near the nuclear Larmor frequency ω_0, and, as we will see below, also near $2\omega_0$. Therefore T_1 increases steadily when the molecular motions become either considerably slower or considerably faster than what corresponds to the Larmor frequency of the nucleus studied. T_2, on the other hand, is affected by the same high frequency processes as T_1, and in addition also by low frequency processes. As will be shown in more detail below (Fig. IV-12), one thus has that $T_2 \leqslant T_1$, and that T_2 mimics the frequency dependence of T_1 only in the range $\omega \gg \omega_0$, where ω are the characteristic frequencies of the relaxation processes, whereas in contrast to T_1 it decreases further with decreasing ω in the range $\omega < \omega_0$.

Up to this point we have referred to "slow" or "rapid" molecular motions exclusively in terms of the corresponding spectral densities in the frequency domain, which are quantitatively described by the "spectral density function" $J(\omega)$. $J(\omega)$ is related by a Fourier transformation (Equations IV-18 and IV-19) to a time domain function $K(\tau)$

$$J(\omega) = \frac{1}{\sqrt{2\pi}} \int_{-\infty}^{\infty} K(\tau)\, \mathrm{e}^{-i\omega\tau}\, \mathrm{d}\tau \qquad \text{(IV-27)}$$

$K(\tau)$ is called the "autocorrelation function" of a set of expressions $g(t)$ which describe the time dependence of the location and orientation in space of the molecules, or of other molecular rate processes. As the name implies, $K(\tau)$ characterizes the correlation between the values which a function $g(t)$ assumes at different times t and $(t + \tau)$. To get an intuitive appreciation for the behaviour of $K(\tau)$, let us again consider the Brownian motions in a liquid. This tumbling of the molecules is mainly determined by the random molecular collisions in the liquid, which cause the direction and rate of the rotational and translational thermal molecular motions to vary in a random fashion with time. In other words, the molecular collisions during a certain period of time τ_c cause the molecules to loose the memory of their foregoing motional history, i.e. "there is no correlation between $g(t)$ and $g(t+\tau_c)$". From this it is intuitively clear that the autocorrelation function will decrease with time, and it could actually be shown that $K(\tau)$ has the form

$$K(\tau) = K(0)\,e^{-|\tau|/\tau_c} \tag{IV-28}$$

where $K(0)$ is the value of K at the time $\tau = 0$, and τ_c is the "correlation time" which characterizes the time course of $K(\tau)$. τ_c is a characteristic quantity for a given system, and as we shall see in the following, it is intimately related to the size, shape, and rigidity of the molecules, and to the viscosity of the solution. Whenever τ_c is such that it corresponds to a high spectral density $J(\omega)$ (Equation IV-27) at the nuclear Larmor frequency, the molecular motions make for an effective relaxation process, provided there is also a sizeable interaction E (Equation IV-23) with the nuclear spins.

Magnetic Dipole–Dipole Relaxation

Magnetic dipole–dipole relaxation is omnipresent in NMR. Whenever two nuclei I and S with magnetic moments μ_I and μ_S are placed in a static magnetic field $\boldsymbol{H}_0$, they experience a total magnetic field $\boldsymbol{H}_{tot} = \boldsymbol{H}_0 + \boldsymbol{H}_{loc}$, where $\boldsymbol{H}_{loc}$ is the local field of the neighboring nucleus. It has been shown that this "dipole–dipole coupling" through space between two nuclei depends on the magnetic moments μ_I and μ_S, on the distance $|\boldsymbol{r}_{IS}|$ between the nuclei, and on the angle ξ_{IS} between $\boldsymbol{r}_{IS}$ and the direction of the static field $\boldsymbol{H}_0$ (Fig. IV-11). Therefore, if the two spins are bound to the same molecule, e.g.

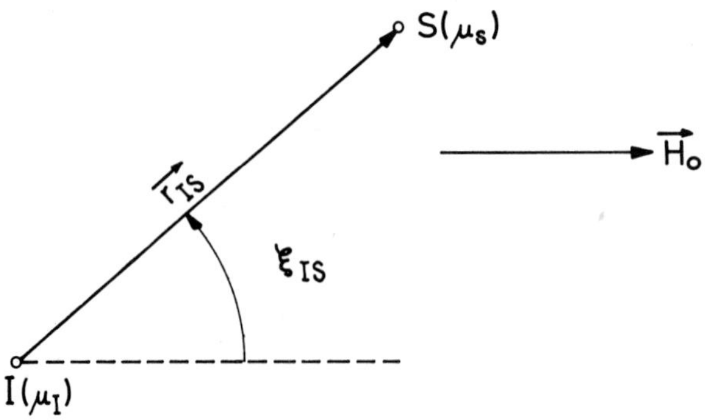

Fig. IV-11. Diagram showing the positions with respect to the magnetic field $\boldsymbol{H}_0$ of two nuclei I and S with magnetic moments μ_I and μ_S. $\boldsymbol{r}_{IS}$ is the "relaxation vector".

the two protons of H_2O, and thus rigidly fixed at a given distance $|\boldsymbol{r}_{IS}|$, $\boldsymbol{H}_{loc}$ will still vary with time because of its dependence on ξ_{IS}, which is obviously affected by the rotational Brownian motions of the molecules. If the two interacting spins are located in different molecules, the time dependence of the dipole–dipole coupling will arise from the effects of the Brownian motions on both the internuclear distance $|\boldsymbol{r}_{IS}|$, and the angle ξ_{IS}.

We present here the quantitative expressions for T_1^{DD} and T_2^{DD} for the case of two identical spins with spin quantum number I and magnetic moment $\mu = \gamma I\hbar$, which are bound to the same molecule, so that only the orientation of the "relaxation vector" $\boldsymbol{r}_{IS}$ relative to H_0 (Fig. IV-11, ξ_{IS}) but not $|\boldsymbol{r}_{IS}| \equiv r$ is modulated by the molecular motions[10]

$$\left.\frac{1}{T_1^{DD}}\right)_{rot} = \frac{2\gamma^4\hbar^2 I(I+1)}{5r^6}\left(\frac{\tau_c}{1+\omega_I^2\tau_c^2} + \frac{4\tau_c}{1+4\omega_I^2\tau_c^2}\right) \tag{IV-29}$$

$$\left.\frac{1}{T_2^{DD}}\right)_{rot} = \frac{\gamma^4\hbar^2 I(I+1)}{5r^6}\left(3\tau_c + \frac{5\tau_c}{1+\omega_I^2\tau_c^2} + \frac{2\tau_c}{1+4\omega_I^2\tau_c^2}\right) \tag{IV-30}$$

In these equations, ω_I is the nuclear Larmor frequency, and τ_c the effective correlation time for rotational reorientation of the relaxation vector $\boldsymbol{r}_{IS}$. In simple rigid molecules, e.g. H_2O, the modulation of the dipole–dipole coupling is essentially restricted to the Brownian rotational tumbling of the whole molecule, which we may characterize by the correlation time τ_R. Assuming that the molecule can be represented by a sphere of radius a, τ_R is usually related to the molecular dimensions, and the viscosity of the solution, η, by

$$\tau_R = \frac{4\pi\eta a^3}{3kT} \tag{IV-31}$$

The behavior of the functions in Equations IV-29 and IV-30 corresponds to the qualitative features expected from the foregoing discussion of the dependence of the relaxation rates on the rate of the molecular motions. Fig. IV-12 depicts the situation for two nuclei with fixed distance r at two different resonance frequencies ω_I. At both frequencies considered, $1/T_1^{DD}$ goes through a maximum. The maximal value of $1/T_1^{DD}$ is seen to be proportional to $1/\omega_I$, and its location on the τ_c scale is given approximately by $\omega_I\tau_c = 1$. For rapid molecular motions, i.e. in the region where $\omega_I\tau_c \ll 1$ ("extreme motional narrowing"), the relaxation rate is proportional to τ_c, and we have that the longitudinal and the transverse relaxation rates are identical

$$\frac{1}{T_1^{DD}} = \frac{1}{T_2^{DD}} = \frac{2\gamma^4\hbar^2 I(I+1)}{r^6}\tau_c\,;\ \omega_I\tau_c \ll 1 \tag{IV-32}$$

In contrast to the behavior of Equation IV-32 found under extreme motional

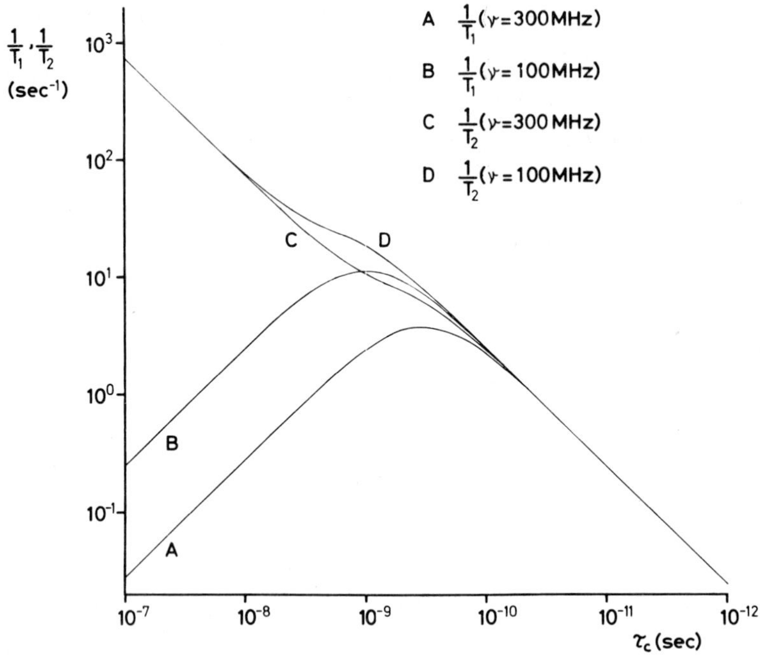

Fig. IV-12. Plot of $1/T_1^{DD}$ and $1/T_2^{DD}$ vs. τ_c (Equations IV-29 and IV-30) at two different frequencies, i.e. $\nu = 100$ MHz and $\nu = 300$ MHz. The curves have been computed for the two protons of a methylene group, with $r = 1.8$ Å. τ_c decreases from left to right.

narrowing conditions the relaxation rates are frequency dependent in the region of less rapid molecular motions, where T_2 is no longer equal to T_1. Typical correlation times for molecular rotations in liquids are $\tau_R \approx 3 \cdot 10^{-12}$ sec for H_2O, $\tau_R \approx 1 \cdot 10^{-11}$ to $1 \cdot 10^{-10}$ sec for amino acids and peptides in aqueous solution, and $\tau_R \approx 10^{-9}$ to 10^{-7} sec for globular proteins with molecular weights from 10000 to 100000, which is roughly what one would estimate from Equation IV-31.

Chemical Shift Anisotropy Relaxation

As was expressed in Equation I-4, a nucleus in a NMR sample experiences a polarizing magnetic field $\boldsymbol{H}_0(1-\sigma)$. The shielding factor σ is in general anisotropic, i.e. its value depends on the orientation of the molecule with respect to $\boldsymbol{H}_0$. As a consequence, the rotational tumbling of the molecules produces random fluctuations of the magnetic field experienced by the nucleus, and thus the basic requirements for a successful relaxation mechanism can in principle be met.

Experience so far has shown that in all but very few cases the chemical shift anisotropy is too small for $1/T_1^{SA}$ or $1/T_2^{SA}$ to contribute noticeably to the observed relaxation rates (Equation IV-24)[1]. However, since $1/T_1^{SA}$ and $1/T_2^{SA}$ increase proportional to H_0^2 (p. 315, Ref. 1 of Chapter I), chemical shift anisotropy interactions might play a more important role in future work at extremely high fields, e.g. quite likely in high field ^{31}P-NMR. The H_0^2 dependence is also the most direct evidence from which to evaluate separately the chemical shift anisotropy contributions to the overall relaxation rate (Equation IV-24).

Scalar Coupling Relaxation

Between nuclei linked together through one or several chemical bonds, one has in addition to the dipole–dipole interactions the so-called spin–spin coupling, which was discussed in Section II.2. As can be seen from Equation II-4, the J-coupling is an isotropic or "scalar" interaction, i.e. its strength is independent of the orientation of the molecules relative to $\boldsymbol{H}_0$. In contrast to the two previously discussed relaxation mechanisms, the scalar coupling is thus not affected by the Brownian movements of the molecules, and in the absence of alternate time modulation it is a static interaction which is exclusively manifested in the fine structure of the NMR spectra (Sections II.2 and Appendix II). On the other hand, the basic conditions for effective relaxation through scalar coupling $\hbar A\boldsymbol{I}\boldsymbol{S}$ between two different spins I and S, where $A = 2\pi J$ stands for the scalar coupling constant, can be met whenever A, or $\boldsymbol{I}$, or $\boldsymbol{S}$ varies randomly with time.

Let us first consider "scalar relaxation of the first kind", i.e. when $A = A(t)$ carries the time dependence. This comes about when the coupling between the two spins is intermittently interrupted by chemical exchange, i.e. by breakage of the chemical linkage between two spins I and S followed by formation of new bonds linking I with a different spin S. The effective correlation time for the resulting relaxation process is τ_e, the average life time of the linkage between a particular pair of spins I and S, and we then have for the relaxation rates of the spin I (p. 308, Ref. 1 of Chapter I)

$$\frac{1}{T_1^{SC}} = \frac{2S(S+1)A^2}{3} \frac{\tau_e}{1+(\omega_I - \omega_S)^2\tau_e^2} \qquad \text{(IV-33)}$$

$$\frac{1}{T_2^{SC}} = \frac{S(S+1)A^2}{3} \left(\tau_e + \frac{\tau_e}{1+(\omega_I - \omega_S)^2\tau_e^2}\right) \qquad \text{(IV-34)}$$

where ω_I and ω_S are the Larmor frequencies of the two nuclei.

"Scalar relaxation of the second kind" arises from rapid relaxation of the spin S, which causes the interaction field $\hbar A\boldsymbol{S}$ seen by the nucleus I to fluctuate randomly with time. Sufficiently rapid relaxation is usually encountered for nuclei S with a spin quantum number $I > 1/2$ (Table I-1), and for unpaired electronic spins in paramagnetic radicals. The effective correlation time is then determined by the relaxation times of the spins S. With $\tau_1 = T_{1S}$ and $\tau_2 = T_{2S}$, the resulting relaxation rates are

$$\frac{1}{T_1^{SC}} = \frac{2S(S+1)A^2}{3} \frac{\tau_2}{1+(\omega_I - \omega_S)^2\tau_2^2} \tag{IV-35}$$

$$\frac{1}{T_2^{SC}} = \frac{S(S+1)A^2}{3} \left(\tau_1 + \frac{\tau_2}{1+(\omega_I - \omega_S)^2\tau_2^2}\right) \tag{IV-36}$$

These formulae have exactly the same form as the Equations IV-33 and IV-34 if we make $\tau_1 = \tau_2 = \tau_e$. This is not very surprising if one considers that all the spin I sees, is the fluctuating interaction field $\hbar AS$, and that it cannot recognize whether the correlation time is determined by the time dependence of A or S.

From the molecular geometries of the systems considered here, it is obvious that any pair of spins interacting through scalar coupling, interact also through dipole–dipole coupling. It is then of importance to distinguish between the contributions to the observed relaxation rates arising from the two different relaxation mechanisms (Equation IV-24). For this, the following situation, which is quite typically encountered, is of practical importance. Assume that the correlation time τ_R for the rotational motions of the molecules is sufficiently short so that $T_1^{DD} = T_2^{DD}$ (Equation IV-32), and the effective correlation time for scalar relaxation, i.e. τ_e (or τ_2), sufficiently long so that $(\omega_I - \omega_S)\tau_e \gg 1$. We then have that T_1 is essentially unaffected by scalar coupling (Equations IV-33 and IV-35), whereas with the relatively long correlation time τ_e (or $\tau_1 \geq \tau_2$), scalar coupling can be an effective T_2 relaxation process (Equations IV-34 and IV-36). If this is the case, the observed relaxation times T_1 and T_2 are then no longer equal and we have, for example, for scalar relaxation of the first kind

$$\frac{1}{T_{2I}} - \frac{1}{T_{1I}} = \frac{1}{T_{2I}^{SC}} = \frac{1}{3} S(S+1)A^2\tau_e \tag{IV-37}$$

which is often readily accessible for experimental verification.

Spin Rotation Relaxation

Spin rotation relaxation arises from time dependent coupling of the nucleus with molecular magnetic moments produced by rapid rotation of the molecule, or a segment of a molecule, about a given axis, e.g. the rapid rotation of a methyl or a phenyl group about the C—C bond by which it is attached to a bigger molecule. The electrons of the molecules participating in this rotation can produce a local magnetic field at the resonant nucleus which is proportional to the molecular angular momentum operator. Because of the molecular collisions, a molecule will remain in any given rotational state only for a limited period of time. The spin rotation interaction thus varies with time, and hence the basic requirements for an effective relaxation process are in principle met.

In practice, relatively few examples have been encountered, where spin rotation relaxation plays an important role. These concern mostly small molecules with axial symmetry[1], and nuclei like ^{13}C and ^{15}N which occur at natural abundance (Table I-1), and are not linked to protons or other nuclei with large magnetic moments.

Electric Quadrupole Relaxation

Nuclei with a spin quantum number $I > \frac{1}{2}$ possess an electric quadrupole moment (Table I-1), and the nuclear spin I is then coupled to the electric field gradient at the nucleus. Provided the chemical environment of the nucleus is not spherically symmetric but has axial or lower symmetry, the extent of the electric quadrupole interaction depends on the orientation of the molecule with respect to H_0. The rotational Brownian motions of the molecules will therefore cause fluctuations with time of the strength of the quadrupole coupling, and thus make for a very efficient relaxation mechanism. ^{14}N and ^{35}Cl may serve to illustrate these basic notions. In the symmetrical environment of $NH_4^{\oplus}$ or $N(CH_3)_4^{\oplus}$, a relatively narrow ^{14}N NMR line is readily observed, whereas the amide ^{14}N NMR lines in peptides or proteins are broadened beyond detection by quadrupole relaxation. The symmetrical structure of the chloride ions in aqueous NaCl solutions allows again the observation of the ^{35}Cl NMR line, whereas extremely broad lines are characteristic for chlorine nuclei in organic molecules, and for chloride ions bound to metal ions.

We present here the quantitative expression for the electric quadrupole relaxation rate in the extreme narrowing case $\omega_I \tau_c \ll 1$ (p. 313, Ref. 1 of Chapter I)

$$\frac{1}{T_1^Q} = \frac{1}{T_2^Q} = \frac{3(2I+3)}{40I^2(2I-1)}\left(1+\frac{\eta^2}{3}\right)\left(\frac{eQq}{\hbar}\right)^2 \tau_c \qquad \text{(IV-38)}$$

e is the electron charge, Q the nuclear quadrupole moment (Table I-1), q the electric field gradient, τ_c the correlation time for rotational motions of the quadrupole nucleus which is in rigid molecules given by τ_R (Equation IV-31), and η the "asymmetry parameter" which describes deviations of the nuclear environment from axial symmetry. For a spherically symmetric environment one has that $q = 0$, and as stated above, the quadrupole relaxation term in Equation IV-24 then vanishes.

Electric quadrupole relaxation is a very effective process which causes considerable broadening of the NMR lines even for nuclei in small molecules. Nuclei with $I > \frac{1}{2}$ (Table I-1) are therefore quite generally not suitable for "high resolution" NMR studies. On the other hand, investigations of the relaxation times in quadrupole nuclei can provide an unambiguous measurement of the molecular correlation time τ_c, and hence possibly τ_R (Equation IV-31), if the parameters η, Q, and q (Equation IV-38) can independently be determined.

5. PROTON SPIN RELAXATION IN PEPTIDES AND PROTEINS

In dilute solutions of a peptide or a protein in D_2O, or in some other non-protonated solvent, the proton relaxation comes essentially entirely from intramolecular dipole–dipole interactions (Equations IV-29 and IV-30). In protonated solvents, intramolecular dipolar interactions still account for the dominant contribution to the relaxation rates, but the interactions of the solvent accessible nuclei with the protons of the solvent must then also be taken into consideration. For the amide protons, scalar coupling with the amide nitrogen atom is an additional possible relaxation process, which does not in general, however, produce a dramatic effect. This can be seen for example in Figs. III-5, III-6, III-13, III-17, III-26, and III-27, where the amide proton resonances are not appreciably broader than those of the protons bound to carbon atoms.

As compared to the discussion of intramolecular dipole–dipole relaxation for a system of two spins in a rigid molecule (Section IV.4), more delicate problems may be encountered in peptides and proteins. Thus, intramolecular interactions between more than two protons may contribute to the relaxation, and the rotational motions of the relaxation vectors r_{IS} (Fig. IV-11) for the individual pairs of nuclei need not be independent of each other. A typical example are the three protons in a methyl group. Furthermore, as a consequence of the inherent flexibility of polypeptide chains, the rotational motions of the relaxation vectors linking pairs of protons are often not restricted to the overall tumbling of the molecule. Experimental investigations have actually shown that for given peptide chain segments, very similar T_1's can occur in molecules of largely different size. This implies that in these cases the intramolecular motions make the dominant contribution to the effective correlation time manifested in the relaxation rate[11]. In addition to the influence on the effective correlation time τ_c, intramolecular segmental motions may also affect $|r_{IS}|$ for dipole–dipole interactions between nuclei in different segments of the polypeptide chain, which may, as a consequence of the molecular conformation, be located in close proximity of each other. The following discussion will first concentrate on the methyl proton relaxation in proteins, which is of considerable importance in practice, and then include an illustration of the dependence of the dipolar interactions on the segmental flexibility.

^{1}H Relaxation in Methyl Groups

Relaxation studies of the protons of methyl groups are of much interest because of the practical advantages in measurements of T_1 and T_2 for methyl groups as compared to the other types of protons in polypeptide chains, i.e. the higher sensitivity for detection, and in certain cases the absence of fine structure from spin–spin coupling[11,12]. However, the interpretation of the data is by no means straightforward. As indicated above, the motions of the relaxations vectors $\mathbf{r}_{IS}$ for the individual pairs of protons within the methyl group are obviously not independent of each other, and furthermore, non-isotropic motion must also be taken into account for a methyl group spinning about the bond linking it to the adjacent carbon in the molecule.

In the general case of dipolar coupling between more than two spins, which includes the cross correlation between the motions of the different relaxation vectors, the relaxation will be non-exponential and not describable by one relaxation time[12]. However, the cross correlation effects are usually small[12], and the relaxation time for a proton being relaxed by interactions with i protons at distances r_i may to a good approximation be described by a summation in Equation IV-29 over the individual pairs of protons[11].

$$\frac{1}{T_1^{\mathrm{DD}}} = \frac{3\gamma_{\mathrm{H}}^4\hbar^2}{10}\sum_i \frac{1}{r_i^6}\left(\frac{\tau_c}{1+\omega_{\mathrm{H}}^2\tau_c^2} + \frac{4\tau_c}{1+4\omega_{\mathrm{H}}^2\tau_c^2}\right) \quad \text{(IV-39)}$$

In a treatment which allows for the spinning of the methyl group with a correlation time $\tau_{\mathrm{M}e}$, and rotational reorientation of the spinning axis with a correlation time τ_c, T_1 becomes[11]

$$\frac{1}{T_1} = \frac{3\gamma_{\mathrm{H}}^4\hbar^2}{10}\sum_i \frac{1}{r_i^6}\left[\frac{1}{4}\left\{\frac{\tau_c}{1+\omega_{\mathrm{H}}^2\tau_c^2} + \frac{3\tau_{c\mathrm{M}}}{1+\omega_{\mathrm{H}}^2\tau_{c\mathrm{M}}^2}\right\} + \right.$$

$$\left. + \left\{\frac{\tau_c}{1+4\omega_{\mathrm{H}}^2\tau_c^2} + \frac{3\tau_{c\mathrm{M}}}{1+4\omega_{\mathrm{H}}^2\tau_{c\mathrm{M}}^2}\right\}\right] \quad \text{(IV-40)}$$

where

$$\frac{1}{\tau_{c\mathrm{M}}} = \frac{1}{\tau_c} + \frac{1}{\tau_{\mathrm{M}e}}$$

The summation is over the three pairs of spins within the methyl group, since coupling with other protons is comparatively inefficient due to the r_i^{-6} factor. The term in square brackets, $f(\tau_c, \tau_{Me})$, is to be compared to the term $f(\tau_c)$ given in the brackets in Equation IV-39. It is seen that $f(\tau_c, \tau_{Me})$ reduces to $f(\tau_c)/4$ when $\tau_{Me} \ll \tau_c$, i.e. for rapid spinning of the methyl group, and to $f(\tau_c)$ for $\tau_{Me} \gg \tau_c$, i.e. when the spinning of the methyl group is hindered. Since the distance r_i between pairs of methyl protons may readily be evaluated, τ_c and τ_{Me} can separately be obtained from the T_1-values measured at different frequencies ω_H (Equation IV-40).[11]

Two additional comments on systems including methyl groups attached to macromolecules are called for. First, rapid intramolecular spinning of the methyl groups could in principle cause spin rotation interactions to affect the overall relaxation rate. In practice, this contribution seems, however, not to be noticeable.[12] Second, rapid internal rotation of a methyl group attached to a slowly tumbling molecule may give rise to a ^{1}H-NMR signal consisting of two components, a well known phenomenon in the limiting case of a rotating methyl group attached to a static particle.[13, 14]

T_1 Relaxation of the Amide Protons in Ferrichrome

As is quite generally the case for protons in diamagnetic proteins, the amide proton T_1 relaxation comes mainly from dipolar coupling with neighboring protons. It is then readily apparent from the covalent structure of the peptide group, that the length of the relaxation vectors linking the amide proton with the surrounding protons depends on the molecular conformation, i.e. on the dihedral angles ϕ_i, ψ_i, and χ_i^j (Figs. III-9 and III-10). This is e.g. clearly manifested in the relaxation times T_1 of the amide protons in the ferrichrome peptides.[15]

Ferrichrome is a cyclic hexapeptide which has an extremely high affinity for iron(III), and other trivalent metal ions (Fig. IV-13). Metal coordination locks the molecule in a very rigid conformation, which was shown to be essentially identical in solution and in single crystals (Ref. 77–79 of Chapter III). In this rigid species, the effective correlation times for all the intramolecular relaxation vectors between pairs of protons are essentially identical to the overall rotational correlation time τ_R (Equation IV-31). Therefore, the differences between the relaxation times for the three glycyl residues, and the three ornithyl residues, respectively, have to come mainly from the different lengths of the relaxation vectors, $|r_{IS}|$, in the three dimensional molecular structure. This was also substantiated by calculations based on the geometry of the single crystal structure.[15] It is seen that between Orn(2) and Orn(3), T_1 differs by a factor 2 (Fig. IV-14).

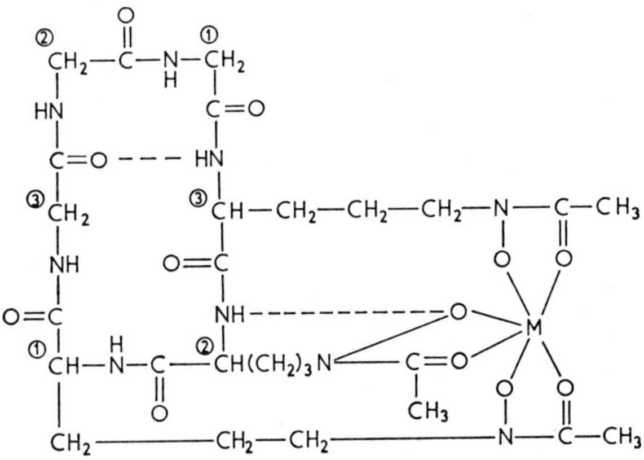

Fig. IV-13. Structure of ferrichrome: cyclo-triglycyl-tri(δ-N-acetyl-δ-N-hydroxy-L-ornithyl). The numbers in circles label the glycyl residues from 1 to 3, and the ornithyl residues from 1 to 3. The metal ion is denoted by *M*, and is coordinated by the three hydroxamate groups of the ornithyl side chains. For the NMR studies, Fe^{3+} was replaced by the diamagnetic Al^{3+}. The dotted lines indicate the intramolecular hydrogen bonds found in crystalline ferrichrome A.

Metal-free ferrichrome has a quite flexible solution conformation, which is not unusual for peptides of this size. The relaxation times for the three glycyl residues are then nearly identical, with $T_1 = 0.31$, 0.31, and 0.30 sec, respectively, and markedly different from those of the ornithyl residues, for which $T_1 = 0.23$, 0.24, and 0.17 sec. The different T_1's in the metal-bound and metal-free ferrichrome arise from the combined influence of the different effective correlation times for corresponding relaxation vectors, and from the different effective values of $|r_{IS}|$. The importance of the latter is particularly evident because the T_1's for some of the amide protons are shorter in the flexible form of the molecule.

In metal free ferrichrome, and in many other peptides and proteins, it is of much interest to investigate the segmental mobility[11] of the peptide backbone via NMR relaxation measurements. However, as we have seen in the foregoing, the influence of variations of the effective correlation time τ_c, where $1/\tau_c = 1/\tau_R + 1/\tau_g$, and τ_g characterizes the intramolecular segmental rotational motions, cannot readily be evaluated separately when dipolar coupling between flexibly linked protons is an effective relaxation mechanism. Therefore, the ^{13}C relaxation times T_1 of the α-carbon atoms, which are essentially entirely determined by dipolar coupling with the covalently attached protons, are a more reliable probe for studies of peptide backbone segmental motions than the amide proton relaxation rates. Details of ^{13}C relaxation studies will be given in the next chapter.

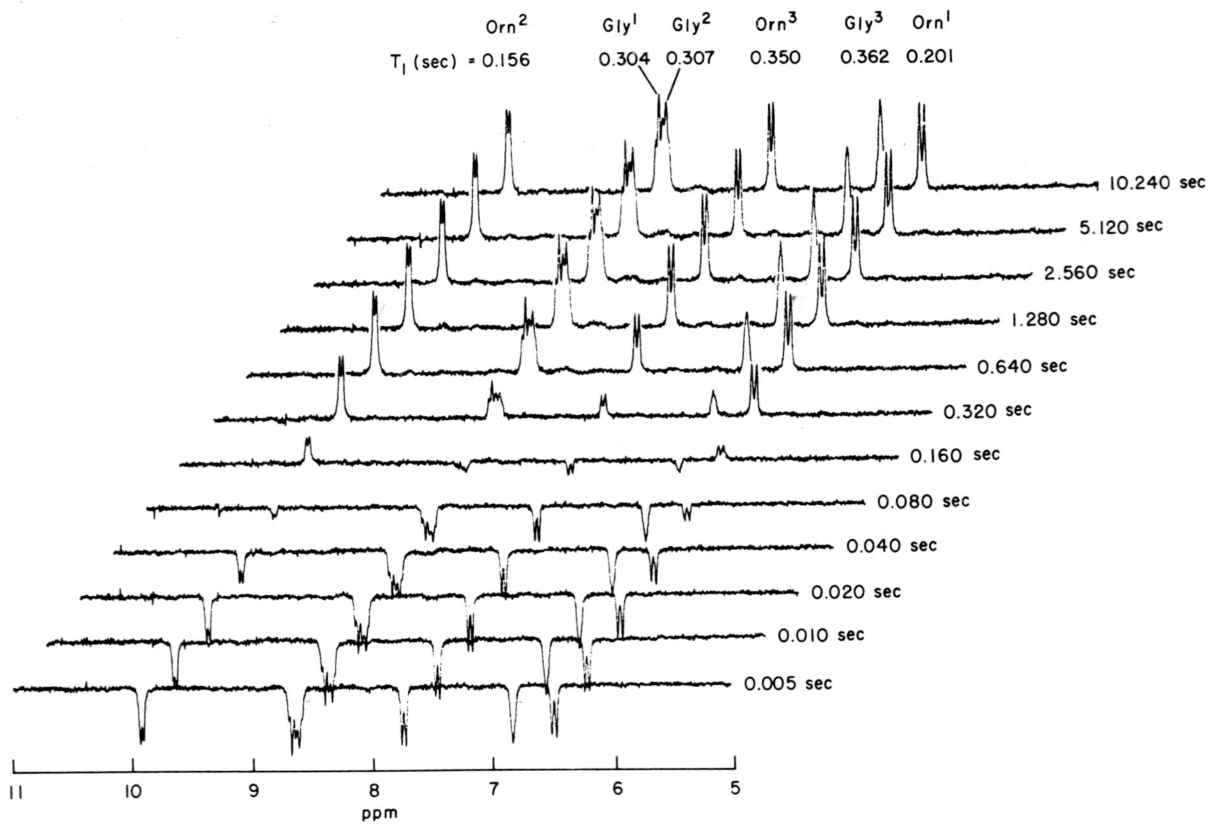

Fig. IV-14. Partially relaxed FT ^{1}H-NMR spectra of the amide protons in the Al^{3+} complex of ferrichrome. Solvent d_6-DMSO, $T = 81°$, the delay times τ in the $180° - \tau - 90°$ pulse sequence are indicated on the right. The resonance assignments (see Fig. IV-13 for the molecular structure) and the values for T_1 are indicated in the top trace (from Llinás and Klein[15]).

6. ELIMINATION OF THE SOLVENT NMR LINES IN DILUTE SOLUTIONS OF MACROMOLECULES

When working with dilute solutions in D_2O, one usually faces problems which arise from the presence at approximately the center of the spectrum, of the comparatively strong resonance of the residual protons of the solvent. The principal difficulty in CW NMR experiments is that the HDO line and its spinning side bands cover a sizeable range of the spectrum, so that observations of the C^α protons are in most cases not readily practicable (Figs. III-2 and III-12). These problems are of course greatly accentuated if one works in H_2O solution.

FT NMR combined with the use of larger size sample tubes enables one in principle to extend high resolution ^{1}H-NMR studies to very small solute concentrations of less than 10^{-4} M. The intensity of the residual HDO resonance will then be more than 3 orders of magnitude greater than those of the solute molecules. In addition to the above mentioned obscuring of a region of interest, the large solvent line will often cause difficulties in phasing the spectrum, loss of accuracy due to the limited dynamic range available for the analog-to-digital conversion of the signal, and the appearance of spurious resonances in the FT NMR spectrum. With the advent of FT NMR, the need for procedures which would eliminate the solvent resonance has thus become more urgent, and several experiments have been designed to this end. Since these are mainly based on the different T_1 relaxation rates of the solute and the solvent, their discussion at this point presents also an illustrative example of a practical application of NMR relaxation data.

One of these experiments uses a $180° - \tau - 90° - T$ pulse sequence, the application of which is based on the large differences between the T_1's of HDO (5–15 sec) and the solute (< 1 sec) in D_2O solutions of macromolecules[16,17]. If $\tau = \ln 2 \cdot T_1(\mathrm{HDO})$, the HDO resonance will pass through $M_Z = 0$ at the time the 90° pulse is applied (Fig. IV-5). The magnetization from the macromolecule, on the other hand, has at this same time essentially fully recovered to its equilibrium value. Hence the signal after the 90° pulse corresponds to a solvent eliminated spectrum (Figs. IV-7 and IV-10).

An alternate experiment makes use of the T_1 dependence of the steady state attained in a pulse sequence $(90° - T)_n$, where $T = (\mathrm{AT} + \mathrm{PD}) < 5T_1$.[17] Because the steady state signal intensity is proportional to $1/T_1$, the HDO resonance is in this experiment largely attenuated as compared to the NMR of the high molecular weight solute molecules.

Other experiments suggested for peak elimination include the saturation of a given resonance by a relatively long single frequency pulse applied prior to the nonselective 90° pulse of a typical FT NMR experiment[18], or careful application of a relatively long weak pulse instead of the typical 90° pulse, which would be in resonance with the lines of interest and at the same time off resonance for H_2O.[19]

Different relaxation times can also be the basis for recording only the narrow lines in a spectrum where these are superimposed upon broader resonances. This can be of practical importance for studies of systems which contain both largely immobilized and mobile components, e.g. membranes. In a "delayed Fourier Transform NMR experiment", a delay time τ_d is introduced between the end of the radio frequency pulse and the start of data acquisition[20]. If $I(0)$ is the intensity immediately after the pulse, the intensity of each component of the transient signal at the start of the data acquisition is then $I = I(0)\,e^{-\tau_d}/T_2$, where T_2 is the transverse relaxation time of the particular component. It is readily seen that the broad components of the spectrum can be effectively attenuated by a proper choice of τ_d.

7. BIBLIOGRAPHY

1 Farrar T. C. and Becker E. D., Pulse and Fourier Transform NMR, Academic Press, New York 1971

2 Hahn E. L., Phys. Rev. *80*, 580 (1950)

3 Carr H. Y. and Purcell E. M., Phys. Rev. *94*, 630 (1954)

4 Meiboom S. and Gill D., Rev. Sci. Instr. *29*, 688 (1958)

5 Ernst R. R. and Anderson W. A., Rev. Sci. Instr. *37*, 93 (1966)

6 Waugh J. S., J. Mol. Spectroscopy *35*, 298 (1970)

7 Vold R. L., Waugh J. S., Klein M. P. and Phelps D. E., J. Chem. Phys. *48*, 3831 (1968)

8 Freeman R. and Hill H. D. W., J. Chem. Phys. *54*, 3367 (1971)

9 Freeman R. and Hill H. D. W., J. Chem. Phys. *55*, 1985 (1971)

10 Solomon I., Phys. Rev. *99*, 559 (1955)

11 Coates H. B., McLauchlan K. A., Campbell I. D., and McColl C. E., Biochim. Biophys. Acta *310*, 1 (1973)

12 Werbelow L. G. and Marshall A. G., J. Amer. Chem. Soc. *95*, 5132 (1973)

13 Andrew E. R. and Bersohn R., J. Chem. Phys. *18*, 159 (1950)

14 Wüthrich K., Proc. Nat. Acad. Sci. *63*, 1071 (1969)

15 Llinás M. and Klein M. P., Paper presented at the VIth International Conference on Magnetic Resonance in Biological Systems, Kandersteg 1974

16 Patt S. L. and Sykes B. D., J. Chem. Phys. *56*, 3182 (1972)

17 Benz F. W., Feeney J., and Roberts G. C. K., J. Magn. Res. *8*, 114 (1972)

18 Schaefer J., J. Magn. Res. *6*, 670 (1972)

19 Redfield A. G. and Gupta R. K., J. Chem. Phys. *54*, 1418 (1971)

20 Seiter C. H. A., Feigenson G. W., Chan S. I., and Hsu M. C., J. Amer. Chem. Soc. *94*, 2535 (1972)

CHAPTER V

CARBON-13 NMR OF AMINO ACIDS, PEPTIDES AND PROTEINS

For quite some time ^{13}C NMR has been recognized as a valuable complementation of ^{1}H NMR for studies of the molecular structures in organic materials. For example it had long been known that a better spectral resolution could be obtained with ^{13}C than with ^{1}H, that ^{13}C NMR could provide a more direct observation of the molecular backbones, and that certain carbon-containing functional groups with no protons attached would be accessible to ^{13}C-NMR observation. On the other hand the sensitivity for NMR observation of ^{13}C at natural abundance is down by a factor 6000 as compared to ^{1}H (Table I-2), and the resulting experimental difficulties have until after 1970 prevented practical applications of ^{13}C-NMR on a broad scale. Advances in instrument design then opened new avenues for the method, which can now also be applied for direct observation of macromolecules. This chapter starts with a brief introduction to some basic features of ^{13}C NMR, and will then concentrate on applications for studies of structure and conformation in peptides and proteins.

1. ^{13}C-NMR PARAMETERS

The chemical shift and the longitudinal relaxation time T_1 are, at the present stage of development, the parameters of greatest practical interest in ^{13}C-NMR. Spin–spin coupling constants, even though omnipresent in the generally encountered spin systems, are for technical reasons often not directly manifested in the spectra, and measurements of T_2 have so far been too difficult to be of much practical importance (see Section IV.3). This section presents a brief introduction to the ^{13}C NMR parameters of organic materials. For a thorough survey, the reader is referred to two recent monographs on the subject[1,2].

Chemical Shifts

In a very similar way as for protons (Section II.1), the ^{13}C chemical shifts are primarily determined by the covalent molecular structures, and can be further influenced by secondary bonds and by non bonding intra- and intermolecular interactions, e.g. with the solvent. The theoretical basis of these different contributions to the ^{13}C chemical shifts is discussed in the above mentioned textbooks[1,2]. Here the interpretation of the chemical shifts in amino acids and random coil peptide chains will be based on comparison with observations in a variety of molecular structures, and on a set of empirical rules relating ^{13}C chemical shifts with the nature of the substituents in structures deriving from aliphatic hydrocarbons[3,4].

Table V-1 indicates chemical shift ranges for a selection of functional groups which occur in the common amino acids. It is seen that the ^{13}C chemical shifts in these structures extend over approximately 200 ppm, as compared to 10 ppm for the corresponding ^{1}H chemical shifts (Table II-1). Qualitatively, the trends of the ^{1}H and ^{13}C chemical shifts in different structures are quite similar. Thus, aliphatic methyl groups are observed at high field near TMS, which is the standard for ^{13}C chemical shifts used

TABLE V-1

Carbon-13 chemical shifts in some diamagnetic organic structures*.

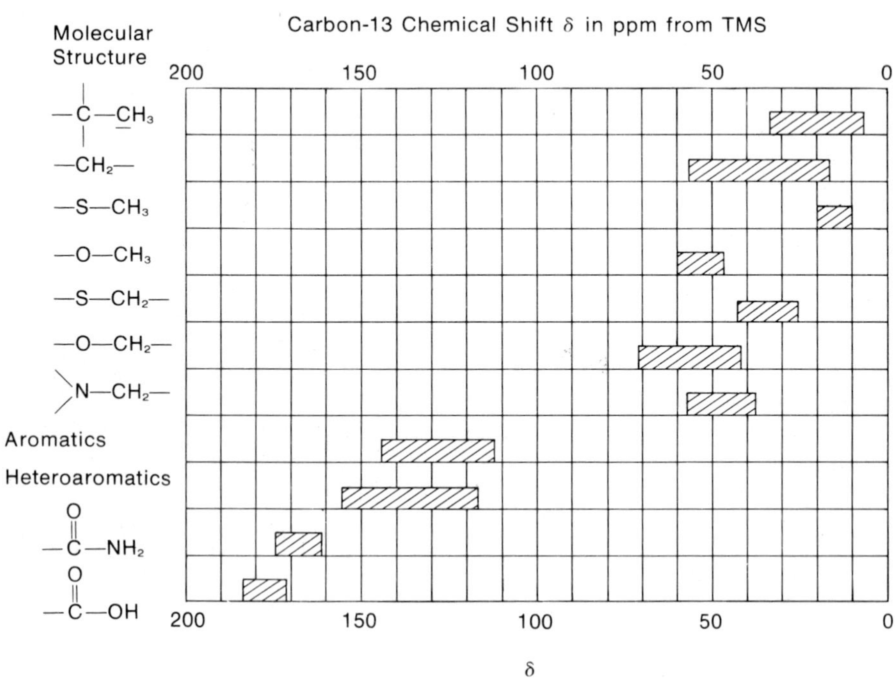

*The bars indicate the spectral regions where the resonances may occur under different conditions of solvent and temperature.

throughout in this book, methylene carbons bound to an electronegative atom at around 50 ppm, aromatic ring carbons at 110 to 150 ppm, and carbonyl carbons at around 175 ppm. Fig. V-1A shows that the ^{13}C chemical shifts of L-alanine correspond quite well to what one might have predicted from Table V-1.

Grant and coworkers[3] have shown that with the use of a suitable set of empirical parameters, the chemical shifts of many substituted alkanes can be predicted from the parent hydrocarbons, which can in turn be treated as substituted methane. For an illustration of these procedures, let us consider the situation in linear alkanes, where the chemical shift for the *i*-th carbon atom can be predicted by[3]

$$\delta^i = B + \sum_j n_{ij} \cdot A_j^{CH_3} \qquad \text{(V-1)}$$

where $B = -2.6$ ppm represents the chemical shift of methane, $A_j^{CH_3}$ is the chemical shift parameter for replacement by a methyl group of a proton bound to the carbon atom in position *j* relative to the *i*-th carbon atom, and n_{ij} is the number of substitutions in position *j*. The parameters for the "α, β, γ, δ, and ϵ effects", respectively, are

$$\begin{aligned} &A_\alpha^{CH_3}\ (\underline{C}\text{—H} \rightarrow \underline{C}\text{—CH}_3) && 9.1\text{ ppm} \\ &A_\beta^{CH_3}\ (\underline{C}\text{—C—H} \rightarrow \underline{C}\text{—C—CH}_3) && 9.4\text{ ppm} \\ &A_\gamma^{CH_3} && -2.5\text{ ppm} \qquad \text{(V-2)} \\ &A_\delta^{CH_3} && 0.3\text{ ppm} \\ &A_\epsilon^{CH_3} && 0.1\text{ ppm} \end{aligned}$$

where negative numbers indicate shifts to higher field at constant radio frequency. Application of the expressions V-1 and V-2 to *n*-pentane gives

$CH_3—CH_2—CH_2—CH_2—CH_3$

observed δ	34.5	22.6	13.7
predicted δ	34.4	22.5	13.7

For example, the predicted value for the methyl carbon atom is $\delta = -2.6 + 9.1 + 9.4 - 2.5 + 0.3$. Extension of this approach to branched or substituted alkanes requires the inclusion of additional parameters to account for specific structural features[1,2], as will be further illustrated with the spectra of the amino acids in Section V.3.

Spin–Spin Coupling Constants

A selection of representative spin–spin coupling constants of ^{13}C is presented in Table V-2. It is seen that couplings of 100 to 300 Hz occur

TABLE V-2

Typical values for homonuclear and heteronuclear spin–spin coupling constants of ^{13}C (in Hz).

Molecular structure	$^{1}J_{^{13}CH}$	$^{1}J_{^{13}C^{13}C}$	Others	
CH_3—CH_3	126	35	$^{2}J_{^{13}CCH}$	−4.8
CH_3—O—CH_3	140			
H_2N—CO—CH_3	129			
C_6H_6	159	57.0	$^{2}J_{^{13}CCH}$	1.0
			$^{3}J_{^{13}CCCH}$	7.4
			$^{4}J_{^{13}CCCCH}$	−1.1
Imidazole C_2H	208			
C_4H	190			
$(CD_3)_2SO$			$^{1}J_{^{13}CD}$	21
F_3C—CF_3			$^{1}J_{^{13}CF}$	260
			$^{2}J_{^{13}CCF}$	46
$^{14}N(CH_3)_4^+$			$^{1}J_{^{13}C^{14}N}$	4.2
CH_3—CO—$^{15}NH_2$			$^{1}J_{^{13}C^{15}N}$	≈ 13
CH_3—$C^{17}O$—CH_3			$^{1}J_{^{13}C^{17}O}$	≈ 20

between directly bonded ^{13}C and ^{1}H, or ^{19}F, respectively. Long range couplings with these two nuclei, and couplings with other nuclei directly bonded to ^{13}C, are mostly in the range 0 to 50 Hz[1,2].

The manifestation of the proton-carbon-13 spin–spin couplings in amino acids is illustrated with the ^{13}C NMR spectrum of L-alanine in Fig. V-1D. It shows that the resonances of C^{β} and C^{α} are split into a quartet, and a doublet, respectively, by the directly bonded protons. Additional much smaller splittings result from the long range couplings, which are also manifested in the C = 0 resonance. For practical reasons to be discussed in the following section, biological applications of ^{13}C-NMR have so far almost exclusively used ^{1}H decoupled spectra (Fig. V-1A), where all the ^{1}H—^{13}C couplings are eliminated.

In as far as the spin–spin couplings with the naturally occurring isotopes other than ^{1}H are concerned, the natural abundance of D, ^{13}C, ^{15}N, and ^{17}O (Table I-1) is too small for the interactions with ^{13}C to be detectable in the ^{13}C-NMR spectra. The quadrupole nuclei, e.g. ^{14}N and ^{17}O (Table I-1), will often also relax too rapidly for their spin–spin couplings to be manifested in the fine structure of the ^{13}C NMR spectra.

Overall, structural information from spin–spin coupling with ^{13}C has so far been obtained in rather few studies of biological molecules. With further improvements of instrument sensitivity, and in particular with the increased use of isotope enriched molecules, it is to be expected that spin–spin couplings with ^{13}C will become an important source of structural information in much the same way as ^{1}H—^{1}H couplings (Section III.2)[1,2,5]. Some

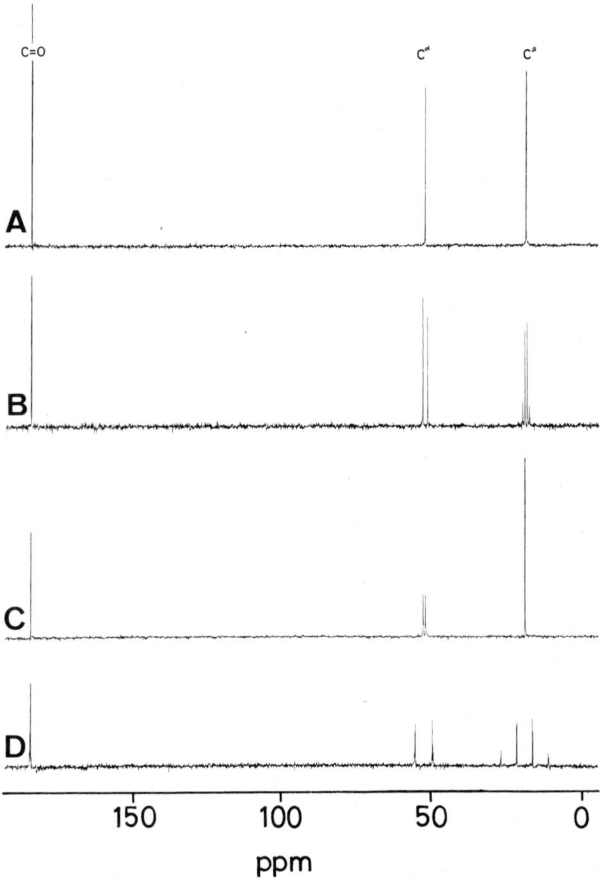

Fig. V-1. FT ^{13}C-NMR spectra at 25.16 MHz of a ca. 2-M solution of L-alanine in D_2O solution, pD = 10.5. A. Proton noise decoupled. B. Proton off resonance irradiation at −1.0 ppm. C. Specific double resonance irradiation of the methyl protons at 1.2 ppm (Fig. I-11). D. Without proton double resonance irradiation. For the spectra A, B, and C, 50 transients were accumulated, for spectrum D 200 transients. The spectral width was 5000 Hz, the pulse width 14 μsec (which corresponded to a 90° pulse), the acquisition time 0.8 sec, and a delay time of 60 sec was chosen.

potential applications are mentioned in different sections of this book (e.g. Section III.2).

Longitudinal Relaxation Times

Four of the five spin relaxation mechanisms introduced in Section IV.4 can in principle be operative for ^{13}C, i.e. magnetic dipole–dipole coupling, chemical shift anisotropy, scalar coupling, and spin rotation interaction. In practice, intramolecular ^{13}C—^{1}H dipolar coupling is with very few exceptions the dominant relaxation mechanism in diamagnetic molecules of biological interest. Possible exceptions might include sizeable contributions from spin rotation to the relaxation rate of certain carbon atoms, and scalar coupling with quadrupole nuclei. In many cases the latter affects mainly the spin–spin relaxation of ^{13}C (Equation IV-37), and thus results in broadening of the resonance lines, e.g. in carbon atoms bonded to ^{14}N. Relaxation by anisotropy of the chemical shift does not appear to be important for ^{13}C, not even at the high fields H_0 obtained with superconducting solenoids[6].

As a consequence of the r^{-6} dependence of $1/T_1^{DD}$, largely different ^{13}C relaxation times will be encountered for carbons with and without attached protons. Typical values for small molecules are $T_1 \approx 2-20$ sec for carbons with attached protons, and $T_1 \gtrsim 30$ sec for carbons with no protons attached. Because of its inverse proportionality to the correlation time for rotational tumbling (Equations IV-29 and IV-32), T_1 will be appreciably shorter in larger molecules. T_1's of 1 second or shorter for methyl and methylene carbons, and 1 to 10 sec for carbonyl carbons are quite typically encountered in peptides and proteins.

T_1 measurements of ^{13}C are of particular interest for studies of the dynamic properties in larger molecules, because the relaxation rates of carbon atoms with attached protons are essentially entirely determined by the intramolecular ^{1}H—^{13}C dipolar coupling. Differences between the relaxation times of a given structural segment in different environments, e.g. a given amino acid residue in different peptide chains, can therefore directly be related to the different effective rotational correlation times (Equations IV-29 to IV-32). This makes T_1's of ^{13}C a more reliable parameter for investigations of segmental motions in chain molecules than e.g. T_1's of ^{1}H. As an illustration, let us consider the data observed in *n*-decanol[7]. The overall rotational tumbling of this molecule is restricted by intermolecular hydrogen bonding. The short T_1's expected as a consequence of the restricted mobility of the molecules are observed only for the carbon atoms nearest to the hydroxyl group, whereas considerably longer T_1's occur along the chain towards the methyl end.

	CH_3—	CH_2—	CH_2—	CH_2—	CH_2—CH_2—CH_2—	CH_2—CH_2—	CH_2—OH
T_1 (sec)	3.1	2.2	1.6	1.1	0.84	0.77	0.65
τ_c (10^{-12} sec)	5	11	15	21	28	30	36

From the effective correlation times τ_c (Equation V-3) one sees that the contributions from the overall rotational motions τ_R and the intramolecular segmental motions τ_g, which are both assumed to be isotopic rotations,

$$\frac{1}{\tau_c} \approx \frac{1}{\tau_R} + \frac{1}{\tau_g} \qquad \text{(V-3)}$$

are on the average comparable in magnitude, and that τ_g decreases markedly along the hydrocarbon chain towards the methyl end.

2. EXPERIMENTAL ASPECTS: OBSERVATION AND IDENTIFICATION OF ^{13}C NMR-LINES

The experimental difficulties arising from the inherent low sensitivity of ^{13}C-NMR at natural abundance are by now routinely overcome through the combined use of Fourier transform NMR, and certain phenomena in double resonance experiments. The modern technology includes also larger sample volumes, i.e. sample tubes with 10, 12, 20 and 25 mm diameter[8], and the investigation of ^{13}C isotope enriched compounds. Because of the somewhat different methods used for recording ^{1}H and ^{13}C-NMR spectra, the techniques for the identification of the individual resonance lines are also somewhat different. This section deals with some of those practical aspects of ^{13}C-NMR which have not been encountered in the discussion of ^{1}H-NMR.

Sample Preparation for ^{13}C-NMR

For a typical ^{13}C-NMR experiment using a 10 or 12 mm sample tube, a sample volume of approximately 1.5 ml is required. For practical purposes the lower limit on the solute concentration is 0.01 M. For this lower concentration limit, which corresponds to using approximately 300 mg of a protein with molecular weight 20000, an observation time of ca. 10 to 20 hours is required to obtain a signal-to-noise ratio of 10:1. The large diameter sample tubes are routinely fitted with a plug to prevent vortexing (Fig. II-8).

The same solvents are commonly used for ^{1}H and ^{13}C-NMR. With the exception of water, these compounds give rise to ^{13}C-NMR lines. The chemical shifts of a selection of solvents are given in Table V-3. Because the modern commercial spectrometers operate with heteronuclear deuterium locks, deuterated solvents are generally used. Table V-3 shows that the ^{1}H—D isotope effects on the ^{13}C chemical shifts have to be taken into account. Also, the resonances of the deuterated solvents are usually split into multiplets by the spin–spin coupling with D (Table V-2) even in proton noise-decoupled spectra. In principle, similar procedures to those described in Section IV.6 for ^{1}H-NMR can be used to eliminate or reduce the intensity of the solvent peaks. For an alternative approach to eliminate the solvent resonances, ^{13}C depleted solvents are commercially available.

TABLE V-3

Chemical shifts of some common solvents and standards for ^{13}C-NMR of biological molecules*.

Solvent	Chemical shift (ppm from TMS)
DMSO	40.5[(a)]
d_6-DMSO	39.6[(a)]
p-Dioxane[(b]	(67.8)[(a, b)]
$CHCl_3$	77.2[(a)]
$CDCl_3$	76.9[(a)]
CH_3OH	49.3
CD_3OD	49.0
TFA	121, 166
CS_2	192.8[(a)]
CS_2 ext. capillary	193.7[(a)]

*Note that multiplet resonances will appear for the deuterated and fluorinated molecules even in ^{1}H noise-decoupled spectra. (a) from Ref. 10 (b) often used as a secondary reference to convert data taken in aqueous solution to the TMS scale. The chemical shift depends somewhat on the concentration.

^{13}C Isotope Enriched Compounds

Because the signal-to-noise ratio improves proportional to the square root of the number of transients accumulated, the observation time required to arrive at a predetermined quality of the spectrum is drastically reduced when the concentration of ^{13}C nuclei is increased. For macromolecules the solute concentration cannot in general, however, be increased much beyond the limit of 0.01-M. As an alternative, the ^{13}C isotope content in the solute can be increased over the natural abundance of 1.1%. In practice the gain in sensitivity by non-specific ^{13}C enrichment is limited by the appearance of ^{13}C—^{13}C spin–spin coupling patterns at higher levels of ^{13}C enrichment, and for many practical purposes also by the rather high costs of such experiments.

As an illustration, let us consider the C^α position of an amino acid, C^β—$C^\alpha(NH_3^\oplus)$—$COO^\ominus$. At a given ^{13}C enrichment of $A\%$, one has the probability A of finding ^{13}C in the position C^α, A^2 to have ^{13}C simultaneously in two neighboring positions, and A^3 to have ^{13}C simultaneously in all the three positions $COO^\ominus$, C^α, and C^β. At the natural ^{13}C abundance of 1.1%, the proton decoupled signal of C^α consists therefore of a singlet with 97.8% of the total intensity, of two doublets with 1.1% intensity arising from spin–spin coupling with $COO^\ominus$, and C^β, respectively, and a resonance of intensity

0.01%, which consists in general of four lines of equal intensity due to simultaneous spin–spin coupling with the ^{13}C nuclei in the positions $COO^{\ominus}$ and C^{β}. With increasing ^{13}C content, the peak intensities of the singlet, the doublet, and the four line resonance vary as $A(1-A)^2$, $\frac{1}{2}A^2(1-A)$, and $\frac{1}{4}A^3$, respectively. These functions are plotted in Fig. V-2.

Fig. V-2 shows that very high levels of ^{13}C enrichment will add to the complexity of the ^{13}C-NMR spectra[9]. When working with biopolymers, one will therefore in most cases be interested in the range of ^{13}C enrichment where the singlet resonance is strong both on the absolute scale, and relative to the corresponding multiplet lines. In practice the ^{13}C enrichment A will be chosen sufficiently small for the intensity of the four-line resonances (Fig. V-2, III) to be negligible. In this region one then selects A so that the difference between the intensities of the singlet resonances (Fig. V-2, I), and the doublet resonances (Fig. V-2, II) is maximum. The plot of $[A(1-A)^2 - \frac{1}{2}A^2(1-A)]$ vs. A (Fig. V-3) shows that this criterium is best met by ^{13}C enrichment of 0–30%. Combining the information of Fig. V-3 with economical considerations, one will therefore in non-specific isotope enrichment experiments aim for a ^{13}C content of 10–20%.

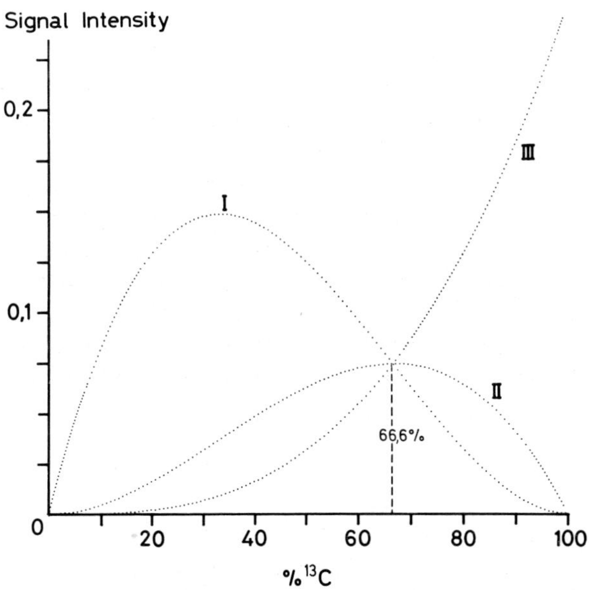

Fig. V-2. Plot vs. the ^{13}C isotope enrichment A of the functions $A(1-A)^2$ (I), $\frac{1}{2}A^2(1-A)$ (II), and $\frac{1}{4}A^3$ (III), which represent the intensities of the individual lines in the singlet, the doublet, and the four-line resonance of a C^{α} carbon atom, respectively (see text).

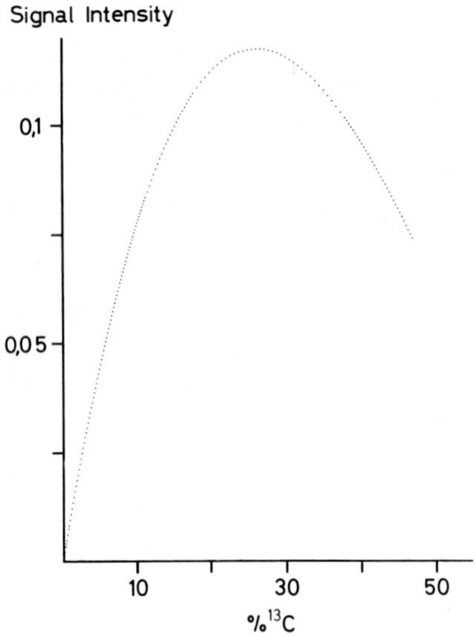

Fig. V-3. Plot vs. the ^{13}C isotope enrichment A of the function $[A(1-A)^2 - \frac{1}{2}A^2(1-A)]$, representing the intensity which the singlet resonance of a C^{α} carbon atom has above the level of the corresponding multiplet resonances (see text).

Reference Compounds

Throughout this book, ^{13}C chemical shifts are given with respect to the standard TMS, where positive shifts are to lower field. Literature values which had originally been related to some other standard, were converted to the TMS scale using one of the compounds in Table V-3 as a secondary standard. Similar procedures are now generally adopted for the presentation of ^{13}C chemical shift data[1, 2, 10]. Secondary standards are also used when TMS can for practical reasons not be employed as an internal reference, e.g. in aqueous media. Secondary standards in the literature include a variety of compounds in addition to the solvents in Table V-3, e.g. the C^{α} resonance of glycine[4].

Fourier Transform ^{13}C-NMR

Fourier transform techniques are at present the only relevant method for biological applications of ^{13}C-NMR. The commonly used spectrometers operate at field strengths H_0 of 14.1, 21.1, or 23.5 KG, which corresponds to ^{13}C resonance frequencies of 15.1, 22.6, and 25.1 MHz. Fourier transform ^{13}C-NMR spectrometers with superconducting solenoids producing fields H_0

of up to 85 KG are commercially available, but very few applications of high field ^{13}C-NMR have as yet been described. The standard equipment of the spectrometers allows for continuous or gated double resonance irradiation of a second nucleus, generally ^{1}H, during FT observation of ^{13}C, and field-frequency lock with a third nucleus, usually D. Partially relaxed ^{13}C-NMR spectra for studies of the T_1 relaxation times can readily be recorded, e.g. by a $180° - \tau - 90°$ pulse sequence, as illustrated in Fig. V-4.

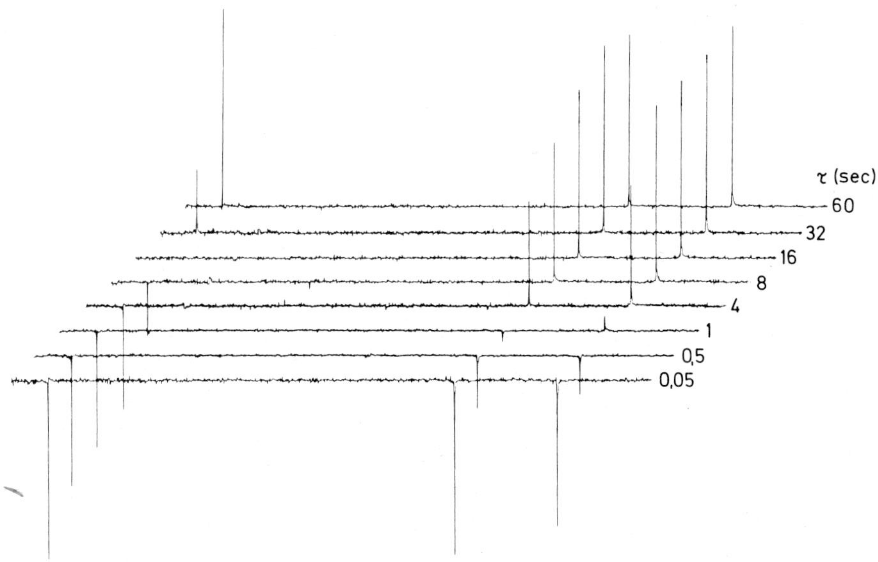

Fig. V-4. Partially relaxed ^{13}C-NMR spectra at 25.16 MHz of a ca 2-M solution of L-alanine in D_2O, pD = 10.5, obtained by the $180° - \tau - 90°$ pulse sequence and Fourier transformation. It is seen that T_1 increases from the methyl carbon ($T_1 = 1.9$ sec) to the C^{α} carbon ($T_1 = 3.5$ sec) to the carbonyl carbon ($T_1 \approx 28$ sec) (see Appendix V).

Heteronuclear ^{1}H—^{13}C Double Resonance Experiments

In Section II.4 we discussed the application of homonuclear double resonance experiments for purposes of resonance identification in ^{1}H NMR spectra. Similarly, specific double resonance irradiation of individual proton resonances is used to identify the resonance lines of the carbon atoms which are coupled to these protons (Fig. V-1C). However, an even more important type of double resonance experiment is the recording of proton noise-decoupled ^{13}C-NMR spectra, $^{13}C\{^{1}H\}$[11]. By noise modulation of the double irradiation radio frequency field, all the ^{1}H—^{13}C spin–spin couplings in a spectrum can simultaneously be decoupled, so that the ^{13}C-NMR of CH_3,

CH_2, CH, and quaternary carbons all appear as single lines (Fig. V-1A). The significance of the proton noise decoupling technique lies in the improved signal-to-noise ratio for the singlet ^{13}C resonances as compared to the corresponding multiplets in the undecoupled spectra (Fig. V-1D), and in the simplification of the spectra. These two features are both of particular importance for biological studies, where one has to work with low molar sample concentrations, and the spectra contain in general a large number of resonance lines. Most biological work so far has for practical reasons been restricted to studies of noise-decoupled spectra. It is quite obvious that much structural information contained in the ^{13}C—^{1}H spin–spin coupling constants is lost in these experiments.

The Nuclear Overhauser Enhancement

In addition to the improved signal-to-noise ratio which arises as a consequence of the addition of the intensities of the individual multiplet components into a single line, the so-called "nuclear Overhauser enhancement" (NOE) in proton noise-decoupled ^{13}C-NMR spectra yields a gain in signal intensity of up to threefold. The NOE results from a distortion of the Boltzmann populations of the spin levels (Equation I-1) by the double resonance irradiation of the protons. Its physical basis and practical applications have been treated extensively in a recent monograph by Noggle and Schirmer[12]. Suffice it here to mention that in the spin systems considered in the present context, the NOE is intimately linked to spin-lattice relaxation by ^{1}H—^{13}C dipole–dipole coupling. If T_1 of ^{13}C is caused entirely by dipolar coupling, and all the other relaxation mechanisms in Equation IV-24 are negligible, the NOE in the extreme motional narrowing situation is

$$\mathrm{NOE} = \frac{I}{I_0} = 1 + \frac{\gamma_H}{2\gamma_C} \qquad \text{(V-4)}$$

where I and I_0 are the ^{13}C-NMR signal intensities with and without double resonance irradiation of the protons. For the $^{13}C\{^{1}H\}$ experiment, Equation (V-4) shows that with a fully operative NOE, each carbon resonance will have a peak area 2.98 times that in the absence of ^{1}H irradiation. The NOE is decreased if sizeable contributions to the spin relaxation come from mechanisms other than dipole–dipole coupling, or if the extreme narrowing condition is not satisfied (see Section V.5). The NOE is clearly manifested in Fig. V-1, where the spectrum D represents the sensitivity in the absence of NOE, whereas the latter is operative in spectrum B, which was recorded with off-resonance proton irradiation.

In general, the contributions from the different relaxation mechanism to the overall relaxation rate, and the effective rotational correlation times are different for the individual carbon positions in a molecular structure and, as a consequence, the extent of the NOE may also vary for the different spins. Even if much care is taken to prevent saturation of individual resonance lines (Section IV.3), the relative integrated signal intensities in $^{13}C\{^{1}H\}$ spectra are thus often not proportional to the number of nuclei which they represent, so that noise-decoupled ^{13}C spectra cannot be used as a reliable quantitative measure of carbon atom concentrations.

Techniques for Resonance Identification

In practice, a combination of different observations will contribute to the identification of the resonances of the individual carbon positions. These may include comparison of the chemical shifts and T_1's with those of known molecular structures (Table V-1), selective irradiation of individual ones of the ^{1}H-NMR at low double resonance radio frequency power (Fig. V-1C), specific isotope substitutions, and investigations of the multiplet structures arising from ^{13}C—^{1}H spin–spin coupling. From the latter the carbon atoms with none, one, two, or three hydrogen atoms attached can readily be distinguished. For lack of sensitivity, the undecoupled spectra, which could also provide information on the number of protons attached to the neighboring carbon atoms (Fig. V-1D), cannot usually be obtained in biopolymers. Instead one uses a technique known as single frequency off-resonance decoupling, which allows recovery of some ^{13}C—^{1}H coupling information while retaining the NOE, and to some extent also the resolution of the noise-decoupled spectra. In an off-resonance decoupling experiment, a radio frequency field is applied at a high power level in a position several hundred Hz away from the protons to be irradiated. Under these conditions, the multiplet patterns arising from the one-bond spin–spin couplings are maintained with residual couplings J_R, i.e. the separations of the multiplet lines are smaller than those expected from $^{1}J_{^{13}CH}$, whereas the long range couplings are completely eliminated. As Fig. V-1B shows, C = 0, CH, and CH_3 carbon atoms (and in different examples of course also CH_2) are readily identified from the residual multiplet patterns in an off-resonance decoupling experiment.

The residual coupling J_R varies linearly with the separation of the double resonance proton irradiation frequency from the resonance frequency of the protons attached to the carbon atom under consideration. On this basis, different carbon atoms giving rise to the same multiplet pattern can then be distinguished from the residual couplings J_R. A graphical method has been developed to aid in the detailed interpretation of complex off-resonance decoupled spectra[13].

Alternative approaches using a pulse-modulated proton irradiation field in the place of continuous irradiation, enable one to obtain the complete fine structure of the ^{13}C lines (Fig. V-1D) without loosing the NOE[14]. On the basis of sensitivity criteria, and from considerations of the spectral resolution, it would appear that for the foreseeable future these methods should become more interesting for studies of small molecules than for investigations of biopolymers.

3. CARBON-13 NMR OF THE COMMON AMINO ACIDS

This section presents the ^{13}C-NMR parameters of the common amino acids. In the context of the main theme of this book, these data are primarily of interest as a basis for ^{13}C-NMR studies of peptides and proteins. As with the corresponding ^{1}H-NMR data, one is therefore really more directly interested in the spectra of amino acid residues in random coil linear oligopeptides than in those of the free amino acids.

Free Amino Acids in Aqueous Solution

^{13}C-NMR studies of amino acids have been reported by several different groups of authors[4, 9, 15–21]. Most of these investigations used compounds with natural abundance of ^{13}C, but some experiments were also done with ^{13}C enrichments of ca 15% (Refs. 4, 15), and others with 85% ^{13}C.[9] The ^{13}C chemical shift data for neutral aqueous solutions of the common amino acids are given in Table V-4.

Horsley et al.[4] showed that the amino acid ^{13}C chemical shifts can quite convincingly be predicted on the basis of the empirical additivity rules for substituent effects (Equation V-1)[3]. The chemical shift parameters for the different substituents in amino acids are given in Table V-5. In addition to these parameters, "refinement corrections"[3] must be applied in branched structures both at the position of the branching, and in α-position to it. In amino acids this concerns primary carbon atoms bound to tertiary carbons, which are denoted "1°(3°)", secondary carbons bound to tertiary positions 2°(3°), and correspondingly 3°(2°), and 3°(3°). Horsley et al.[4] showed that all the bonds with atoms other than hydrogen must be counted to define the character of the carbon atom, and that quite accurate predictions of the chemical shifts result for the amino acid side chains when the refinement corrections for hydrocarbons[3], which are also given in Table V-5, are used. For C^{α} carbons, an additional correction of -6.0 ppm has to be applied to obtain a satisfactory fit with the experiments. To illustrate these procedures, let us consider the situation in L-serine, $^{\oplus}H_3N$—$C^{\alpha}H(C^{\beta}H_2OH)$—$COO^{\ominus}$.

TABLE V-4

^{13}C Chemical shifts in ppm from TMS of the common amino acids in neutral D_2O*.

Amino acid	C=O	αC	βC	γC	δC	εC	ζC
Gly[a]	173.0	42.0					
Ala[a]	176.3	51.1	16.8				
Val[a]	174.8	61.1	29.7	18.6			
				17.4			
Ile[a]	174.7	60.4	36.6	25.2	11.8		
				15.4			
Leu[a]	176.1	54.2	40.5	24.9	22.7		
					21.6		
Ser[a]	172.6	56.9	60.8				
Thr[a]	173.5	61.0	66.6	20.0			
Pro[a]	174.1	61.1	29.2	23.9	46.0		
Asp[a]	175.0	52.7	37.1	178.3			
Glu[a]	175.1	55.2	27.6	34.0	181.8		
Lys[b]	174.9	54.8	30.2	21.9	26.7	39.5	
Arg[a]	174.7	54.6	28.0	24.4	41.0		157.0
Asn[c]	174.2	52.6	36.5	174.5			
Gln[c]	179.5	55.2	28.3	33.0	175.5		
Met[a]	174.8	54.7	30.5	29.6		14.7	
Cys[c, d]	174.2	57.3	26.0				
					Ring carbon atoms		
Trp[e]	175.2	56.1	27.4		C2 }	{ 126.0	
					C5 }	{ 122.9	
					C3	108.4	
					C4 }	{ 120.3	
					C6 }	{ 119.3	
					C7	112.8	
					C8	137.3	
					C9	127.5	
Phe[b]	174.5	56.8	37.0		C1	136.2	
					C2,6	130.3	
					C3,5	130.3	
					C4	128.6	
Tyr[a]	174.5	56.8	37.0		C1	(128.0)	
					C2,6	130.0	
					C3,5	117.0	
					C4	156.0	
His[a]	174.4	55.3	28.5		C2	136.7	
					C4	117.7	
					C5	131.0[f]	

*(a) From Ref. 4, using a chemical shift of 42.0 ppm for αC of Gly to convert to the TMS scale. (b) From Ref. 18, (c) From Ref. 17. (d) Cys was studied in the form H-L-Cys-$OCH_3 \cdot HCl$. (e) From Ref. 19, measured at pD 4.0. (f) From Ref. 21.

TABLE V-5

Chemical shift parameters A_j^X for the substituent groups in amino acids*.

Substituent group	Chemical shift parameters (ppm) A_α^X	A_β^X	A_γ^X	A_δ^X
$—CH_3$	9.1	9.4	−2.5	0.3
—OH	48.5	10.0	−5.7	0.5
$—NH_2$	29.0	11.4	−4.6	0.6
$—NH_3^{\oplus}$	26.0	7.5	−4.6	
—COOH	21.5	2	−2.5	0.6
$—COO^{\ominus}$	24.5	3.5	−2.5	

Refinement corrections for branched hydrocarbons:[3]

1° (3°)	−1.2 ppm
2° (3°)	−2.5 ppm
3° (2°)	−3.7 ppm
3° (3°)	−9.5 ppm
C^α correction shift:	−6.0 ppm

*A_j^X indicates the change in the chemical shift which arises when a proton in position j relative to the carbon atom i is replaced by the substituent X. From Ref. 4.

From Equation V-1, one has that

$$\delta^\beta = B + A_\alpha^{OH} + A_\alpha^{CH_3} + A_\beta^{NH_3^+} + A_\beta^{COO^-} + 2°(3°)$$
$$= -2.6 + 48.5 + 9.1 + 7.5 + 3.5 - 2.5$$
$$= 63.5 \text{ ppm}$$

$$\delta^\alpha = B + A_\alpha^{NH_3^+} + A_\alpha^{COO^-} + A_\alpha^{CH_3} + A_\beta^{OH} + 3°(2°) - 6.0 \text{ ppm}$$
$$= -2.6 + 26.0 + 24.5 + 9.1 + 10.0 - 3.7 - 6.0$$
$$= 57.3 \text{ ppm}$$

These values correspond indeed to a quite accurate prediction of the chemical shifts in Table V-4. From these empirical rules one would also expect that sequence effects on the chemical shifts in random coil peptide chains are quite small[4], a prediction which was confirmed by recent systematic studies of model peptides[18, 22–25].

The ^{13}C chemical shifts of amino acids show a marked pH dependence[15]. Successive deprotonation of the carboxylic acid and the amino group cause sizeable downfield shifts of the C=O, C^α and C^β NMR (Table V-6)[9, 26]. The chemical shifts of the species in acid or alkaline solution (Equation II-1) can

TABLE V-6

Chemical shifts arising from the deprotonation reactions II-1 in amino acids with non-ionizeable aliphatic side chains.

	$^{\oplus}H_3N{-}CH(R){-}COOH \rightarrow {}^{\oplus}H_3N{-}CH(R){-}COO^{\ominus}$	$^{\oplus}H_3N{-}CH(R){-}COO^{\ominus} \rightarrow H_2N{-}CH(R){-}COO^{\ominus}$
C=O	3.0 ± 0.5 ppm	9.0 ± 1.0 ppm
αC	2.0 ± 0.5 ppm	1.0 ± 0.5 ppm
βC	1.0 ± 0.5 ppm	4.0 ± 1.0 ppm
γC	< 0.7 ppm	< 0.1 ppm
δC	< 0.4 ppm	< 0.4 ppm

be obtained approximately by combining the data of Tables V-4 and V-6. At intermediate pH values, there is rapid exchange between the protonated and the deprotonated amino acids, so that the observed spectrum corresponds to the weighted average of those of the different species. A typical pH titration profile is shown in Fig. V-5. Similar titration shifts to those arising

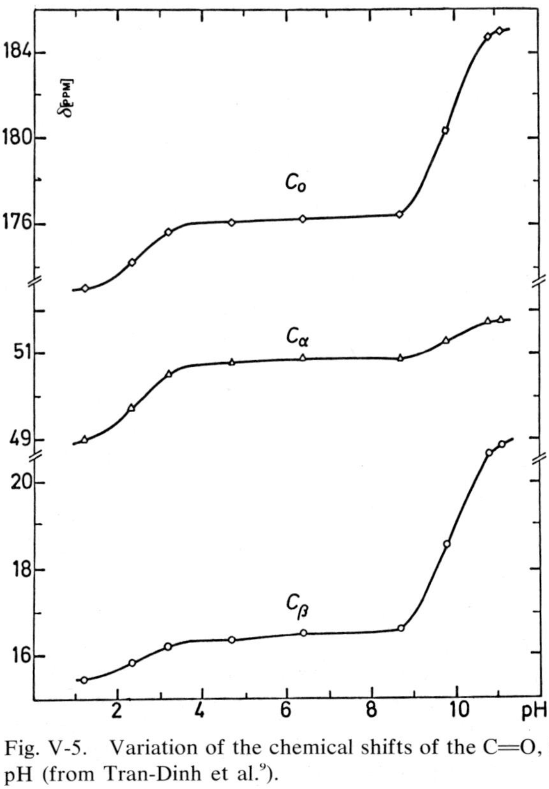

Fig. V-5. Variation of the chemical shifts of the C=O, C^{α} and C^{β} carbons of L-alanine with pH (from Tran-Dinh et al.[9]).

from the ionizeable backbone groups are found for side chain carboxylic acid and amino groups. The resulting titration curves have been analyzed in detail for several amino acids[26]. For histidine, the chemical shifts upon protonation of the imidazole ring are −2.4 ppm for C2, −4.8 ppm for C4, and 0.7 ppm for the C5 position[21].

Fig. V-6 shows the $^{13}C\{^1H\}$ spectrum of L-alanine enriched to 85% in ^{13}C.[9] As Fig. V-2 predicts, the multiplet resonances, and in particular the four-line resonance of C^{α}, are the dominant spectral features at this level of ^{13}C enrichment. The ^{13}C—^{13}C spin–spin coupling constants of L-alanine and L-leucine are given in Table V-7. The data are quite representative for amino acids with aliphatic side chains[9,27]. It is seen that only $^1J_{C^oC^{\alpha}}$ is markedly affected by pH variation. Because the other coupling constants are essentially constant throughout, they are most probably also valid for the corresponding amino acid residues in polypeptide chains.

T_1 measurements of free amino acids were reported by Pearson et al.[28]. Representative values for glycine in D_2O solution are $T_1 = 2.9$ sec for C^{α} and $T_1 = 84 \pm 5$ sec for the carboxyl carbon atom. While T_1 for C^{α} agrees rather well with the values for the protonated carbons in L-alanine, the relaxation time for the carboxyl carbon is approx. three times longer than that observed in Fig. V-4. This discrepancy arises most probably from the presence in the

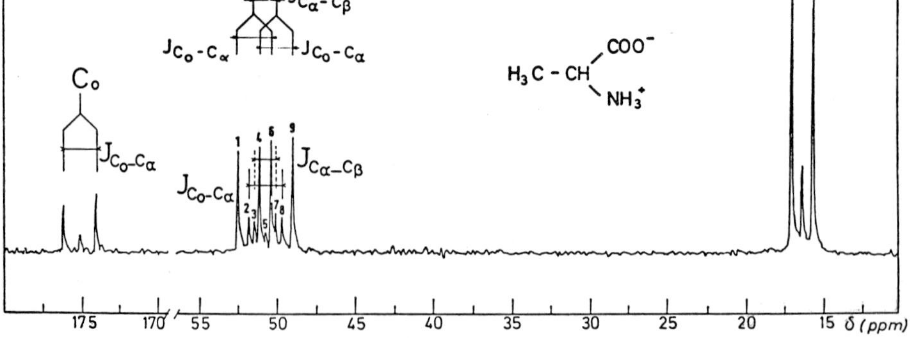

Fig. V-6. Fourier transform $^{13}C\{^1H\}$ NMR spectrum at 25.16 MHz of a 0.15-M solution in D_2O of L-alanine enriched to 85% in ^{13}C, pD = 7.6. The multiplet patterns resulting from ^{13}C—^{13}C spin–spin coupling are also indicated (from Tran-Dinh et al.[9]).

TABLE V-7

^{13}C—^{13}C spin–spin coupling constants (in Hz ± 0.5 Hz) obtained from amino acids enriched to 85% in ^{13}C (Ref. 9).

Amino acid	Carbon positions	pH 1.2	pH 6.4	pH 11.2
Alanine	$^1J_{C^oC^\alpha}$	59.2	54.1	52.7
	$^1J_{C^\alpha C^\beta}$	34.1	34.9	35.2
Leucine	$^1J_{C^oC^\alpha}$	59.5	53.6	52.6
	$^1J_{C^\alpha C^\beta}$	32.9	34.3	34.6
	$^1J_{C^\beta C^\gamma}$	31.7	31.3	31.7
	$^1J_{C^\gamma C^{\delta 1}}$	31.2	31.4	31.3
	$^1J_{C^\gamma C^{\delta 2}}$	35.3	34.9	34.7

sample used for Fig. V-4 of paramagnetic impurities, which can make a sizeable contribution to the carboxyl carbon relaxation[28].

Amino Acid Residues in Aqueous Solution

The ^{13}C chemical shifts for the common amino acid residues in neutral aqueous solution are given in Table V-8. These data were obtained from

TABLE V-8

^{13}C Chemical shifts of the common amino acid residues in neutral D_2O*.

Amino acid residue	C=O	αC	βC	γC	δC	εC	ζC
Gly[a]	172.7	43.5					
Ala[b]	175.8	50.8	17.7				
Val[a]	174.9	60.7	30.8	19.3			
				18.5			
Ile[a]	174.8	59.6	36.9	25.4	11.3		
				15.7			
Leu[a]	175.9	53.6	40.5	25.2	23.1		
					21.6		
Ser[b]	172.6	56.6	62.3				
Thr[b]	172.7	60.2	68.3	20.0			
Pro[b,c] trans	175.2	61.6	30.6	25.5	48.2		
Pro[b,c] cis		61.3	33.1	23.2	48.8		
Asp[b]	174.2	52.7	39.8	178.4			
Glu[b]	174.8	54.9	28.9	34.6	182.8		
Lys[b]	174.7	54.4	27.5	23.1	31.8	40.5	
Arg[a]	175.0	54.6	28.8	25.7	41.7		157.6
Asn[b]	173.1	51.5	37.7	175.6			
Gln[b]	174.0	54.1	28.1	32.2	179.0		
Met[a]	175.0	53.9	31.0	30.7		15.0	
Cys[d]	175.7	57.9	26.0				

TABLE V-8 (*contd.*)

Amino acid residue	C=O	αC	βC	γC	δC	εC	ζC
					Ring carbon atoms		
Trp[d]	176.7	56.7	27.4		C2 } C5 }	{126.0 {122.9	
					C3	108.4	
					C4 } C6 }	{120.3 {119.3	
					C7	112.8	
					C8	137.3	
					C9	127.5	
Phe[d]	176.0	57.4	37.0		C1	136.2	
					C2,6	130.3	
					C3,5	130.3	
					C4	128.6	
Tyr[d]	176.0	57.4	37.0		C1	(128.0)	
					C2,6	130.0	
					C3,5	117.0	
					C4	156.0	
His[b]	172.6	53.7	28.0		C2	135.2	
					C4	118.7	
					C5	130.3	

*The resonance positions are given with respect to TMS. (a) From Ref. 23–25. Measured in linear pentapeptides H-Gly-Gly-X-Gly-Gly-OH. To convert to the TMS scale, $\delta(CS_2) = 193.7$ ppm was used. (b) From Ref. 22. Measured in protected linear tetrapeptides TFA-Gly-Gly-X-L-Ala-OCH_3. To convert to the TMS scale, δ (internal dioxane) = 67.8 ppm was used. (c) In the prolyl peptide, trans- and cis-proline occurred with a relative abundance of approx. 4 : 1. (d) The chemical shifts of the aromatic ring carbon atoms are the same as in Table V-4, and those of C=O, C^α and C^β were adapted from Table V-4 with the peptide shifts for non-terminal amino acids in Table V-9.

studies of linear pentapeptides H-Gly-Gly-X-Gly-Gly-OH[23–25], and linear protected tetrapeptides TFA-Gly-Gly-X-L-Ala-OCH_3[22], where X stands for the different amino acid residues. As was to be expected from the additivity rules for substituent effects on the ^{13}C chemical shifts (Equation V-1 and Table V-5), the data in Table V-8 are only little different from those of the zwitterionic form of free amino acids (Table V-4).

Christl and Roberts[18] proposed a set of "peptide shift" parameters (Table V-9) to compute the ^{13}C chemical shifts in random coil peptide chains from the ^{13}C-NMR spectra of the constituent free amino acids. As a valuable complement of the data in Table V-8, these empirical rules can be quite helpful in the evaluation of the peptide chain termination effects on the ^{13}C-NMR spectra.

TABLE V-9

^{13}C Chemical shift changes (ppm) of amino acids on incorporation in peptides and titration shifts of peptides (from Christl and Roberts[18]).

Type	C=O	αC	βC	γC
Peptide shift of an N-terminal amino acid[(a)]	5.4 ± 1.0	1.3 ± 0.4 Gly 0.7 ± 0.2	− 0.2 ± 0.5	0.4 ± 0.5
Peptide shift of a C-terminal amino acid[(b)]	− 3.5 ± 1.0	− 1.0 ± 0.5 Gly − 2.0 ± 0.2	− 1.2 ± 0.6	− 0.4 ± 0.3
Peptide shift of a nonterminal amino acid[(c)]	1.5 ± 1.0	0.6 ± 0.4 Gly − 1.1 ± 0.1	0.0 ± 0.5	0.0 ± 0.1
Titration shift of N-terminal unit[(d)]	− 6.9 ± 1.0	− 1.0 ± 0.5 Gly − 2.6 ± 0.2	− 3.1 ± 1.0	− 0.6 ± 0.4
Titration shift of C-terminal unit[(e)]	3.0 ± 0.5	2.6 ± 0.5 Gly 2.2 ± 0.1	1.2 ± 0.2	0.3 ± 0.2

(a) Difference between the chemical shifts of a N-terminal amino acid in a zwitterionic peptide and the corresponding free, zwitterionic amino acid. (b) Difference between the chemical shifts of a C-terminal amino acid in a zwitterionic peptide and the corresponding free, zwitterionic amino acid. (c) Difference between the chemical shifts of a non-terminal amino acid in a zwitterionic peptide and the corresponding free amino acid. (d) Difference between the chemical shifts of a N-terminal amino acid in a peptide as anion and zwitterion. The values given correspond to a lower limit for the "titration shift of N-terminal units". (e) Difference between the chemical shifts of a C-terminal amino acid in a peptide as cation and zwitterion.

TABLE V-10

^{13}C Chemical shifts in ppm from internal TMS of the 20 common amino acid residues in d_6-DMSO solution*.

Amino acid residue	C=O	αC	βC	γC	δC	εC	ζC
Gly	168.7	42.1					
Ala	171.8	48.0	18.2				
Val	170.4	57.0	30.8	19.0 17.9			
Ile	170.6	56.2	37.1	24.2 15.1	11.0		
Leu	171.7	50.5	41.0	24.0	23.0 21.6		
Ser	169.7	54.8	61.6				
Thr	169.9	58.0	66.6	19.7			
Pro[(a)] trans		59.1	29.0	24.1	45.8		
Pro[(a)] cis		58.7	31.7	22.0	46.8		
Asp	170.3	49.2	36.3	171.5			
Glu	170.9	51.4	27.6	30.0	173.9		

TABLE V-10 (*contd.*)

Amino acid residue	C=O	αC	βC	γC	δC	εC	ζC
Lys[(b)]	171.2	51.8	26.6	22.1	31.5	38.6	
Arg(NO_2)[(c)]	171.0	51.7	29.5	24.7	40.3		159.3
Asn	170.9	49.3	37.1	171.2			
Gln	170.9	51.7	27.9	31.1	173.5		
Met	170.9	51.5	32.0	29.3		14.6	
| (Cys)$_2$[(d)]	169.7	51.6	40.3				
					Ring carbon atoms		
Trp	171.2	53.1	27.9		C2 } C5 }	{120.5 {123.5	
					C3	109.7	
					C4 } C6 }	{117.9 {118.2	
					C7	111.0	
					C8	135.9	
					C9	127.2	
Phe	170.9	53.6	37.7		C1	137.8	
					C2,6	127.9	
					C3,5	129.0	
					C4	126.1	
Tyr	171.0	53.9	36.9		C1	127.9	
					C2,6	130.1	
					C3,5	114.9	
					C4	155.5	
His[(b)]	170.0	51.8	27.6		C2	133.6	
					C4	116.8	
					C5	130.3	

*The data were obtained from studies of the protected tetrapeptides TFA-Gly-Gly-L-X-L-Ala-OCH_3. (a) in the prolyl peptide, trans- and cis-prolyl occurred with a relative abundance of ca. 2 : 1. (b) The peptides with Lys and His were used in the hydrochloride form. (c) Arg was studied in the form of the N^{ω} nitro derivative. (d) The cystine peptide was studied.

Amino Acid Residues in DMSO Solution

The ^{13}C chemical shifts for the 20 common amino acid residues in d_6-DMSO solution are reproduced in Table V-10. These data were obtained from investigations of the protected tetrapeptides TFA-Gly-Gly-X-L-Ala-OCH_3[22], where X stands for the different amino acids.

4. RANDOM COIL POLYPEPTIDE CHAINS

In much the same way as for ^{1}H-NMR studies (Chapter III), a detailed knowledge of the spectral properties of the "random coil" polypeptide chains is the basis for ^{13}C-NMR investigations of the molecular conformations in peptides and proteins. In this section, the relations between random coil ^{13}C-NMR spectrum, amino acid composition and sequence of polypeptide chains will be briefly surveyed. In this context, experiments for assigning individual resonance lines to particular amino acid residues in a peptide molecule are of much interest, and are therefore also discussed at this point.

^{13}C-NMR Spectra of Random Coil Polypeptide Chains

Analogous to the discussion of the situation encountered with ^{1}H-NMR (Chapter III), we compare the ^{13}C-NMR spectrum of the polypeptide hormone calcitonin M (Fig. III-3), which is known to occur predominatly in a non-globular extended solution conformation, with the computed hypothetical spectrum of the random coil form of this molecule. Taking into account that the resonance intensities in the experimental spectrum can deviate quite appreciably from those expected from the number of carbon atoms which they represent, one finds an almost complete coincidence between the observed and the computed spectrum in the regions from 0 to 40 ppm, and 100 to 160 ppm, which contain exclusively resonances of the amino acid side chains. From the data of Table V-10, most of the lines in these spectral regions can readily be assigned to the different types of amino acid side chains. On the other hand the resonances of the peptide backbone carbon atoms between 40 and 60 ppm, and 165 and 180 ppm are quite different in the spectra A and B of Fig. V-7. In particular, the resonances of the α-carbon atoms appear to be broader than what one would have expected from the spectra of the individual amino acid residues. This is quite typical for a number of different polypeptides in extended "random coil" forms[29]. As will be further discussed in Section V.5, it is quite likely that these are manifestations of local non-random conformational features of the peptide backbone, which reduce the segmental mobility.

The protected linear tetrapeptides TFA-Gly-Gly-X-L-Ala-OCH_3 were investigated in detail to obtain further insights into the ^{13}C-NMR of random coil peptide chains[22]. The random coil character of the peptide backbone in these molecules was evidenced in the ^{1}H and ^{19}F-NMR spectra[30]. The ^{13}C-NMR spectra for four tetrapeptides with different central amino acid residues X are shown in Fig. V-8. They are quite representative for flexible linear peptides. From the identification of the resonance lines of the

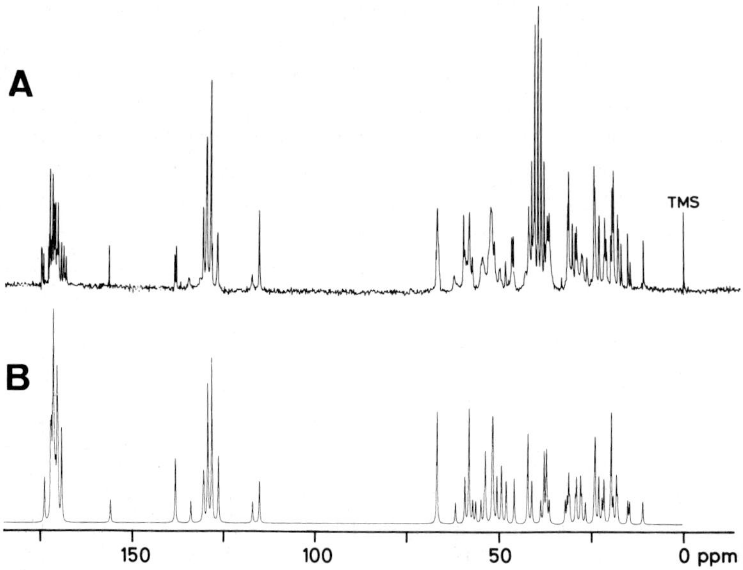

Fig. V-7. A. Fourier transform $^{13}C\{^1H\}$ NMR spectrum at 25.16 MHz of a 0.1 M solution in d_6-DMSO of the polypeptide hormone calcitonin M (Fig. III-3). The solvent resonance is centered at 39.6 ppm. B. Hypothetical spectrum for the random coil form of the polypeptide chain computed from the amino acid composition and the data of Table V-10.

individual amino acid residues, which will be described below, the influence of different amino acids X on the chemical shifts of the carbon resonances in the neighbouring residues Gly(2) and Ala(4) could be investigated. Overall, the amino acid sequence effects were found to be rather small, with the exception of those due to proline. Among the tetrapeptides with the other 19 common amino acids X, the relative chemical shifts observed in DMSO solution were 0.6 ppm for $C^{\alpha}(2)$, 0.4 ppm for C=O(2), 0.3 ppm for $C^{\alpha}(4)$, 0.2 ppm for $C^{\beta}(4)$, and 0.5 ppm for C=O(4). Similar sequence effects were found in experiments with different model peptides[18, 23–25].

Resonance Identification in Peptides

In Section V.2 we have seen that off-resonance proton decoupling experiments, combined with the application of the chemical shift parameters (Table V-5) and comparison with suitable organic molecules, can in most cases provide unambiguous resonance assignments in the individual amino acids. These methods will in general also suffice to differentiate between most of the resonances of the different types of amino acid residues in a polypeptide chain, in particular when the latter is in an extended random coil

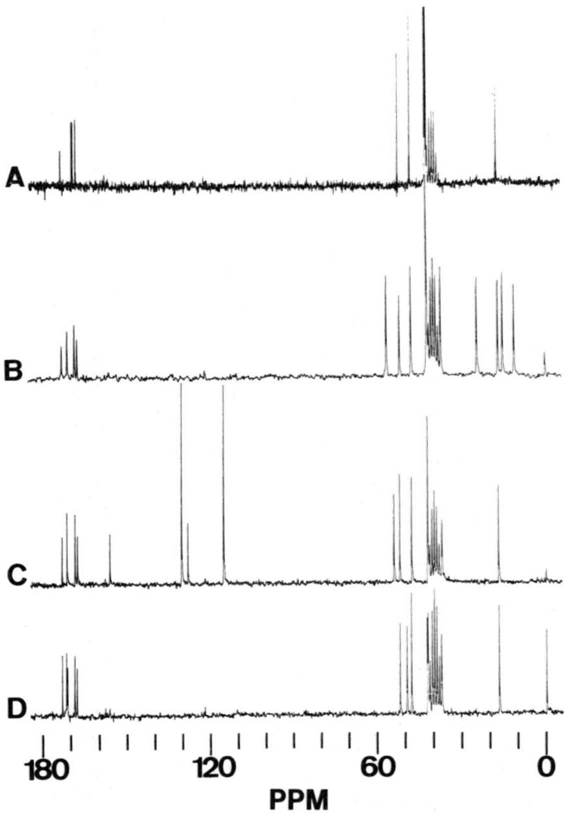

Fig. V-8. Proton noise-decoupled ^{13}C FT spectra at 25.16 MHz, of four peptides TFA-Gly-Gly-L-X-L-Ala-OCH_3. $T = 35°$, d_6-DMSO solvent resonance at 39.6 ppm: (A)X = Gly; (B)X = Ile; (C)X = Tyr; (D)X = Asn. In the spectra B, C, and D a trace of TMS had been added to the solutions (from Grathwohl and Wüthrich[22]).

form. In addition, however, much of the structural interpretation of the NMR spectra may depend on the differentiation between corresponding resonances of identical amino acid residues in different positions of the peptide chain. Several different approaches to the solution of this problem are briefly described in the following.

The introduction of ^{13}C enriched amino acid residues in certain positions of a peptide chain is in principle the most generally applicable procedure for the identification of individual carbon resonances. The isotope enrichment can either be in specific carbon atoms, e.g. the carbonyl carbon of an amino acid residue, or in the entire carbon skeleton of the amino acid. In the ^{13}C-NMR spectrum the resonances of the labelled carbon atoms can then readily be recognized from their increased intensities, and possibly also on the basis of the fine structure arising from $^{13}C—^{13}C$ spin–spin coupling (Fig. V-6). As the isotope enrichment over the natural abundance of ^{13}C can be up to 90-fold (Table I-1), there are no inherent difficulties from the spectroscopists point of view which would oppose the use of this technique even for high molecular weight proteins. On the other hand, the limited availability of suitably ^{13}C-labelled amino acids, the generally laborious chemical or biosynthetic procedures needed to introduce the labels into the desired positions of the peptide chains, and the rather high costs have so far limited the application of ^{13}C labelling to relatively few systems. Some examples will be described below.

The pH-dependence of the chemical shifts (Tables V-6 and V-9) can be used for the identification of the terminal amino acid residues in polypeptide chains[15,18]. For example in the above mentioned (Table V-8) model peptides H-Gly-Gly-X-Gly-Gly-OH, complete assignments of the resonances of the glycyl residues in positions 1, 2, 4, and 5 were derived from the pH-dependence in the region 1 to 10.[23–25]

In certain oligopeptides, the carbonyl carbon resonances could be assigned on the basis of selective $^{1}H—^{13}C$ double resonance experiments[31]. Since there are no protons bound to the carbonyl carbon atoms, the fine structure of the resonances in the absence of proton irradiation comes entirely from spin–spin coupling with the protons bound to the nearest and next nearest neighbor atoms, and the T_1-relaxation is mainly caused by dipole–dipole coupling with the protons bound to these nearby atoms. Selective proton double resonance irradiation can result in a collapse of the fine structure of individual carbonyl carbon resonances, and cause an increase of their intensities through nuclear Overhauser enhancement. The resonance assignments in the model peptides TFA-Gly-Gly-X-L-Ala-OCH_3 (Tables V-8 and V-10, Fig. V-8) provide an illustration. The spectra in Fig. V-9 show that the carbonyl carbon line of Gly(1) is at 167.3 ppm, because this resonance position is very little affected by the ionization state of the fragment TFA-Gly-Gly-OH. In Fig. V-10 the carbonyl carbon resonances of TFA-Gly-Gly-L-Val-L-Ala-OCH_3 recorded without proton irradiation are compared with those obtained with selective irradiation of the methyl protons of Ala(4). Note that for this experiment the amide protons had been replaced by deuterium. It is seen that only the line at 172.6 ppm is affected by the ^{1}H irradiation, and can therefore be assigned to Ala(4). In a corresponding experiment with TFA-Gly-Gly-L-Ala-L-Ala-OCH_3, the second line from the left was also affected by the irradiation of the methyl protons of alanine, and could thus be assigned to Ala(3), whence the line at 168.3 ppm had to come from Gly(2).

Further potentially useful methods for resonance assignments in pep-

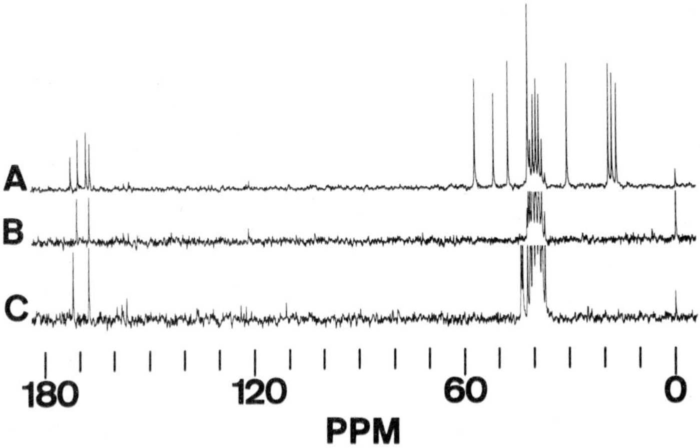

Fig. V-9. Proton noise-decoupled ^{13}C FT spectra at 25.16 MHz of three peptides in d_6-DMSO solution, $T = 35°$: (A) TFA-Gly-Gly-L-Val-L-Ala-OCH_3; (B) TFA-Gly-Gly-OH; (C) Same as B after addition of 1 equivalent of 4 M aqueous NaOD to the DMSO solution. A trace of TMS had been added to the solutions (from Grathwohl and Wüthrich[22]).

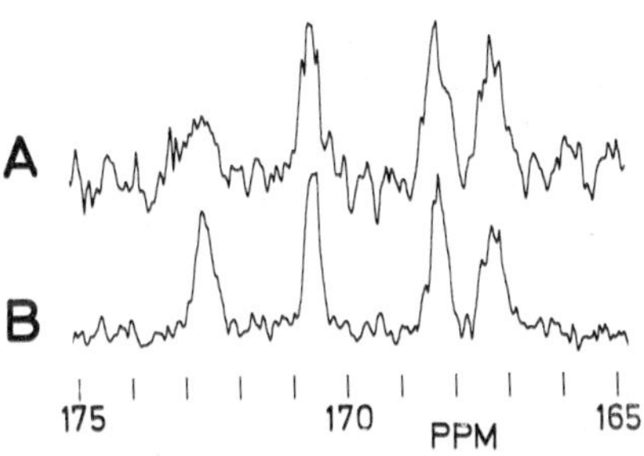

Fig. V-10. Carbonyl resonances in the ^{13}C spectrum of TFA-Gly-Gly-L-Val-L-Ala-OCH_3 (Fig. V-9) in a mixed solvent of d_6-DMSO and $D_2O = 9:1$, where the amide protons had been replaced by D: (A) Without proton irradiation; (B) Selective double resonance irradiation of the methyl protons of alanine (from Grathwohl and Wüthrich[22]).

tides include double resonance experiments in ^{15}N enriched compounds, and the use of paramagnetic shift or line broadening reagents, which will both be discussed in subsequent chapters.

Studies of the Amino Acid Composition and Amino Acid Sequence

The foregoing discussion of random coil polypeptide chains has shown that the amino acid composition is clearly evident in the ^{13}C-NMR spectra. Analogous to ^1H-NMR, ^{13}C-NMR can thus in principle be used for studies of the amino acid composition, in particular for the detection of certain specific amino acid residues in a peptide molecule (Tables V-8 and V-10). ^{13}C-NMR studies of the amino acid sequence in peptides so far relied on the pH-dependence of the resonances (Tables V-6 and V-9), and are therefore essentially restricted to the identification of the terminal and possibly the penultimate residues. Thus the complete sequence of linear di- and tripeptides, and in favorable cases even tetra- and pentapeptides can be determined by this non-destructive method[15,18,23]. The principal advantage of ^{13}C-NMR over ^{1}H-NMR for these experiments is in most cases the improved spectral resolution. Yet for routine experiments, ^{1}H-NMR will probably in general be more suitable, mainly because of the larger sensitivity.

5. CONFORMATIONAL STUDIES OF PEPTIDES AND PROTEINS

Over the past few years, an impressive collection of ^{13}C-NMR spectra has been accumulated which show that different molecular conformations of polypeptide chains can be manifested in readily observable differences of the ^{13}C-NMR parameters. Some of these conformation dependent spectral properties have already been related to particular features of the spatial molecular structures. The present section attempts to illustrate with selected examples the manifestations of particular peptide and protein conformations in the ^{13}C-NMR spectra, and to present a survey of the structural information which has so far been derived from ^{13}C-NMR data.

Polypeptide Backbone Conformation. Cis-, Trans- and Non-planar Peptide Groups

In Section III.2, we have seen how ^{1}H-NMR can be employed to investigate the backbone conformations in oligopeptides. These data can now be complemented with ^{13}C-NMR experiments. Thus, as was mentioned earlier, in ^{13}C enriched peptides the torsion angles ϕ_i (Figs. III-9 and III-10) can in principle independently be evaluated from measurements of the heteronuclear vicinal spin–spin coupling constants $^3J_{\mathrm{HN}^{13}\mathrm{C}'}$ and $^3J_{\mathrm{HN}^{13}\mathrm{C}^\beta}$ between the amide proton i and the carbon atoms C'_i or C^β_i, respectively, or $^3J_{^{13}\mathrm{C}'\mathrm{H}^\alpha}$ between C$'(i-1)$ and the C^α_i-proton (Fig. III-9).

^{13}C-NMR can also be used to delineate the internal hydrogen bonds in a peptide. These experiments are based on the observation that the peptide carbonyl carbon chemical shifts are quite markedly different in different solvents, in particular when one goes from a non-protic to a protic solvent[32]. For example the peptide carbonyl carbon resonances shift downfield by about 4 ppm on going from DMSO into 2,2,2-trifluoroethanol[32], 6 ppm on going from DMSO into TFA[33,34], and 3 ppm on going from DMSO into D_2O[32,33]. If the molecular conformation of a peptide chain involves peptide carbonyl groups which are shielded from contact with the solvent, e.g. by the formation of intramolecular hydrogen bonds, and this conformation is maintained in the different solvents, one might anticipate a less pronounced solvent dependence of the chemical shift for these carbonyl carbons than for the solvent accessible ones. This is illustrated in Fig. V-11 with the solvent titration dimethylsulfoxide-D_2O of the protected linear pentapeptide HCO-Val-Pro-Gly-Val-Gly-OCH_3. It is seen that on going from d_6-DMSO to D_2O, the carbonyl carbon line of Val(1) is shifted downfield to a lesser extent than those of the other amino acid residues. This was interpreted in terms of the formation of an intramolecular hydrogen bridge involving the N—H of Val(4), and the C=O of Val(1) (Fig. V-12). In a molecular structure with only one internal hydrogen bridge, the location of the latter can thus

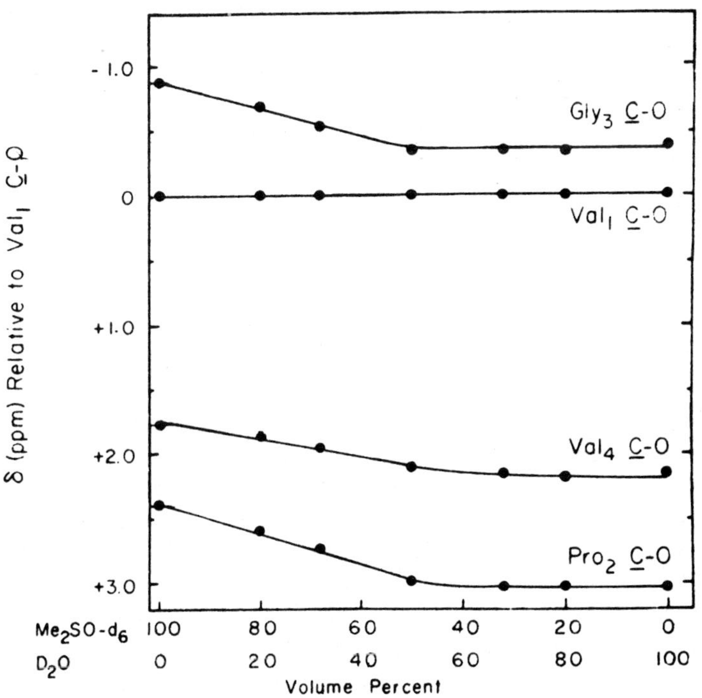

Fig. V-11. Solvent titration (d_6-DMSO-D_2O) of the carbonyl carbon resonances in the protected linear pentapeptide HCO-Val-Pro-Gly-Val-Gly-OCH_3. Val(1) is taken as the reference. The other peptide carbonyl carbon resonances are seen to shift downfield as the volume percent of D_2O is increased to about 50%. This result is interpreted to indicate that the carbonyl group of Val(1) is shielded from the solvent (from Urry et al.[32]).

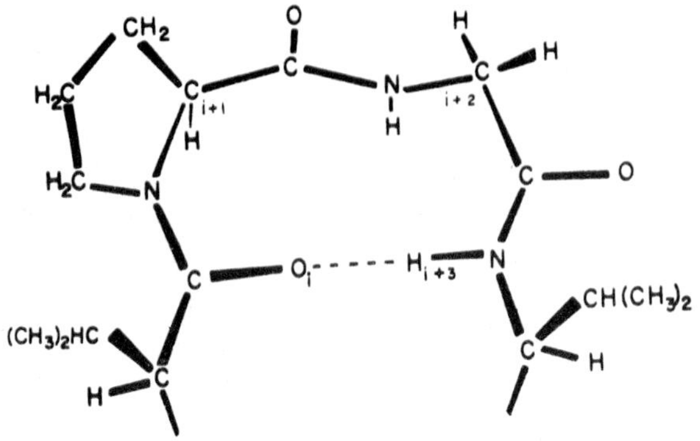

Fig. V-12. β-turn in the pentapeptide HCO-Val-Pro-Gly-Val-Gly-OCH$_3$ which would involve the formation of a hydrogen bond $C_i = O \cdots H\text{–}N_{(i+3)}$ (from Urry et al.[32]).

unambiguously be determined from the combination of the ^{1}H-NMR data on the amide protons and the ^{13}C-NMR data on the carbonyl carbons, provided the resonances of the individual amino acid residues have independently been identified[32]. It is quite apparent that even in more complex molecules with multiple internal hydrogen bonds, the combination of ^{1}H and ^{13}C-NMR can lead to a quite detailed description of the hydrogen bonding pattern, or at least enable one to restrict the number of possible types of conformations to relatively few which would be compatible with the NMR data.

As was already discussed in Chapter III, with the exception of the X-Pro peptide bonds, only trans peptide groups are usually encountered in polypeptide chains of the common amino acids (Fig. III-8). It is one of the major assets of ^{13}C-NMR for studies of peptide conformations, that the cis- and trans-forms of the X-Pro bonds can readily be distinguished on the basis of the chemical shifts of the proline ring carbon atoms[35–38]. In Fig. V-13 this is illustrated with three spectra of the linear dipeptide H-L-Ala-L-Pro-OH in basic, neutral, and acidic aqueous solution[33]. The resonances corresponding to the species containing respectively a cis or a trans Ala-Pro bond are identified by *c* and *t*. It is seen that the β- and γ-ring carbon resonances of proline are particularly sensitive to the cis-trans isomerization of the X-Pro peptide bond (see also Tables V-8 and V-10), whereas the relative chemical shifts of the other carbon resonances are rather small. A systematic study indicated that the chemical shift of the γ-carbon atom is probably the most reliable spectral evidence for the occurrence of the cis or trans form of the peptide bond involving the proline nitrogen[39]. The chemical shifts of both the β- and γ-ring carbons were found to be remarkably constant in different molecules. For example, in a series of model peptides including N-formylproline, N-acetylproline, glycylproline, and various chemically modified forms of these three compounds, the β-carbon resonance was found to be at 30.5 ± 0.6 ppm for the trans- and 32.2 ± 0.4 ppm for the cis-form, and the corresponding γ-carbon line at 25.1 ± 0.5 ppm and at 23.4 ± 0.3 ppm, respectively[39].

Fig. V-13 shows further that the cis-trans equilibrium of the peptide bond in H-Ala-Pro-OH is markedly dependent on pH. Because of the near identity of the covalent structure near the β- and γ-carbons in cis- and trans-proline, the equilibrium concentrations of the two forms can quite reliably be determined from the ^{13}C-NMR intensities, provided the segmental mobilities are not drastically different in the two species containing respectively cis- and trans-proline.

Additional ^{13}C-NMR features which could be useful for studies of the peptide bond conformations include the chemical shifts of the carbonyl carbon resonances. From comparison of corresponding amino acid residues in linear peptides and cyclodipeptides, it appears that the carbonyl carbon resonances of cis peptide groups are approximately 2 ppm to higher field than those of the corresponding trans peptide groups[33,34]. The chemical shifts

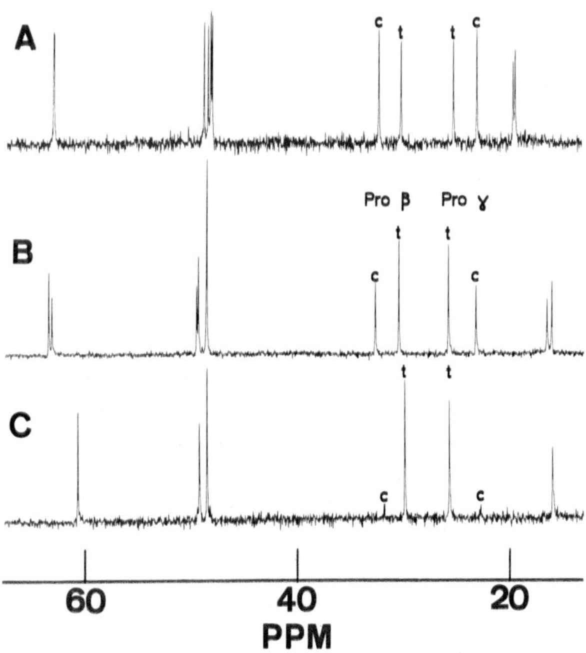

Fig. V-13. Spectral region from 0 to 60 ppm of the FT ^{13}C{^{1}H} NMR spectra at 25.16 MHz of an 0.1-M solution of H-L-Ala-L-Pro-OH in D_2O at A. pD = 12.2, B. pD = 5.4, C. pD = 1.4. The positions of the proline ring carbon atoms β and γ are indicated. *c* = cis-, *t* = trans-form of the peptide group (from Wüthrich et al.[33]).

of the carbonyl carbon resonances might also indicate the occurrence of "transoid" peptide groups, which would deviate from the planar form of the standard trans peptide group (see Section III.1). In one example, cyclo-tetraglycyl, a downfield shift of approximately 2 ppm relative to standard trans peptide groups was attributed to the occurrence of non-planar transoid peptide groups[33,34]. In ^{13}C enriched peptides, the vicinal spin–spin coupling constants $^3J_{^{13}C^{\alpha}NH}$ between $C^{\alpha}_{(i-1)}$ and the i-th amide proton, and $^3J_{^{13}C^{\alpha}\,^{13}C^{\alpha}}$[40] between the two α-carbon atoms bound to the peptide group could in principle also serve to investigate the dihedral angle ω about the peptide bond (Fig. III-9).

Spatial Arrangement of the Amino Acid Side Chains

As was discussed in Section III.1, the most direct means for determining the amino acid side chain conformations is to measure the torsion angles χ^j_i (Fig. III-10) by investigations of the vicinal spin–spin coupling constants. In addition to the proton–proton couplings (Equation III-8), one can for example employ the heteronuclear coupling constant $^3J_{^{13}C'C^{\beta}H}$ between the carbonyl carbon atom i and the protons attached to C^{β}_i. For three amino acids, $^3J_{^{13}C'C^{\beta}H}$ was measured at natural abundance of ^{13}C, and it was shown that with these complementary data a more definite description of the rotamer distribution about the C^{α}—C^{β} bond could be obtained than on the basis of the vicinal ^{1}H—^{1}H couplings $^3J_{HC^{\alpha}C^{\beta}H}$ alone[41].

Similar to the situation encountered with the ^{1}H-NMR, different polypeptide chain conformations can be manifested in quite sizeable chemical shifts of individual ^{13}C-NMR lines[19], which are an additional source of information on the environment of the amino acid side chains in interior parts of the molecules. Some illustrative examples will be described below in the section on globular proteins. The structural interpretation of these chemical shifts will in most cases follow similar lines as in ^{1}H-NMR. For example, the influence of the nearest neighbors on the pK-values of the ionizeable groups can be investigated in much the same way as with ^{1}H-NMR. Ring current effects upon the ^{13}C chemical shifts[42,43] would be expected to be less pronounced than for protons, mainly because of the different sterical situations with respect to near approach to neighboring aromatic rings (Fig. II-2). It has also been suggested that local electric field effects might cause conformation dependent ^{13}C-NMR chemical shifts[44].

Segmental Motions in Polypeptide Chains

Depending on the conformation adopted by a polypeptide chain in solution, its overall rotational motions may be characterized by sizeably different correlation times τ_R. While the rotations of a nearly spherical globular protein can to a good approximation be described by a single correlation time (Equation IV-31), the rotational motions of an extended polypeptide chain will in general be quite anisotropic, and at least two correlation times will be required to describe the motions about the different non-equivalent axes. In addition, the local mobility of individual segments of the polypeptide chain, and of specific amino acid side chains can be appreciably different depending on the molecular conformation. For example, one might expect that compared to the random coil form of linear peptides, the formation of local conformations of the type shown in Fig. V-12 would somewhat restrict the mobility of the backbone segment involved. Or, whereas the amino acid side chains are probably rather mobile in extended forms of a polypeptide chain, their mobility might be appreciably reduced in the interior of a globular protein. As we have seen in Sections IV.4 and V.1 (Equations IV-29 to IV-32, and V-3), the effective correlation times for intramolecular dipole–dipole relaxation are a combination of τ_R (Equation IV-32) and the correlation times for intramolecular rotational motions, and hence the relaxation times of protonated carbon atoms are suitable parameters for studies of the molecular dynamics.

The information content of the ^{13}C relaxation studies depends in each individual example on the spectral resolution, and the extent of the resonance identifications in the spectrum. Quite generally it will always be possible in studies of peptides and proteins, to differentiate between the relaxation times of the aliphatic amino acid side chains, the C^{α} carbon atoms, the aromatic side chains, and the carbonyl carbon atoms (Figs. V-1, V-7 and V-14, Table V-8), and hence to derive an overall picture of the dynamics of the molecule in different conformations. Once individual resolved resonances have been assigned to specific amino acid residues, more detailed information on specified regions of the polypeptide chains may be obtained. In this context, the introduction of ^{13}C-labelled marker groups into the macromolecules is of particular interest. Attractive applications may for example include the introduction of ^{13}C labels into the active sites of an enzyme, and to investigate the influence of enzyme interactions with substrates, inhibitors, or other effector molecules on its relaxation times.

As was pointed out in Sections IV.3 and V.1 (Fig. V-4), longitudinal relaxation times T_1 of ^{13}C can readily be measured by partially relaxed Fourier transform spectroscopy. An illustration of the application of this technique to globular proteins is given with the spectra of ribonuclease A in Fig. V-14[45]. However, when working with macromolecules, the extreme motional narrowing approximation (Equation IV-32) cannot a priori be applied, and for isotropic rotational motions the relaxation times arising

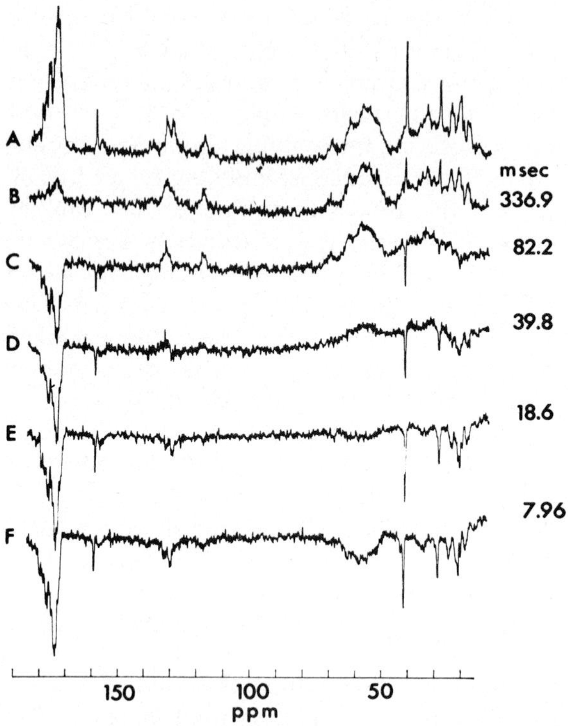

Fig. V-14. Proton noise-decoupled ^{13}C FT NMR spectra at 15.08 MHz of an 0.019-M aqueous solution of ribonuclease A at 45°, pH = 6.53. 16'384 scans had been accumulated, with a recycle time of 1.36 sec. A. Normal FT spectrum. B to F. Partially relaxed FT spectra with the τ-intervals (Equation IV-17) indicated in the figure (from Allerhand et al.[45]).

from dipolar coupling of two different spins H and C would be given by[46]

$$\frac{1}{T_1^C} = \frac{\hbar^2 \gamma_C^2 \gamma_H^2}{10 r^6} \left\{ \frac{\tau_c}{1+(\omega_H - \omega_C)^2 \tau_c^2} + \frac{3\tau_c}{1+\omega_C^2 \tau_c^2} + \frac{6\tau_c}{1+(\omega_H + \omega_C)^2 \tau_c^2} \right\} \quad \text{(V-5)}$$

$$\frac{1}{T_2^C} = \frac{\hbar^2 \gamma_C^2 \gamma_H^2}{20 r^6} \left\{ 4\tau_c + \frac{\tau_c}{1+(\omega_H - \omega_C)^2 \tau_c^2} + \frac{3\tau_c}{1+\omega_C^2 \tau_c^2} + \frac{6\tau_c}{1+\omega_H^2 \tau_c^2} + \frac{6\tau_c}{1+(\omega_H + \omega_C)^2 \tau_c^2} \right\} \quad \text{(V-6)}$$

The symbols had been explained when Equation IV-30 was introduced. In analogy to Equation IV-39, the treatment can be extended to carbon atoms carrying N protons ($N = 1, 2, 3$) in that the right hand sides of Equations V-5 and V-6 are multiplied by N. It is readily apparent, and will be confirmed by the discussion in the Appendix V (Figs. V-27 and V-28), that the behavior of the functions V-5 and V-6 is very similar to that for T_1 and T_2 in Fig. IV-12. Thus, two different values of the effective rotational correlation time τ_c correspond to any value of T_1 obtained in a macromolecule. Several additional experiments can be employed to choose the correct solution. First, the frequency dependence of T_1 would provide an unambiguous effective correlation time (Fig. V-27). Second, the combination of T_1 and T_2 would lead to the correct solution (Fig. V-28). Because of the experimental difficulties opposing accurate measurements of T_2 for ^{13}C, considerations of the line widths (Fig. I-2, Equation I-9) were in practice employed in this context[45]. Third, the nuclear Overhauser enhancement depends monotonously on τ_c[47] (Equation V-7 and Fig. V-15).

$$\mathrm{NOE}^{\mathrm{C}} = 1 + \frac{\gamma_{\mathrm{H}}}{\gamma_{\mathrm{C}}} \; \frac{\dfrac{6\tau_c}{1+(\omega_{\mathrm{H}}+\omega_{\mathrm{C}})^2\tau_c^2} - \dfrac{\tau_c}{1+(\omega_{\mathrm{H}}-\omega_{\mathrm{C}})^2\tau_c^2}}{\dfrac{\tau_c}{1+(\omega_{\mathrm{H}}-\omega_{\mathrm{C}})^2\tau_c^2} + \dfrac{3\tau_c}{1+\omega_{\mathrm{C}}^2\tau_c^2} + \dfrac{6\tau_c}{1+(\omega_{\mathrm{H}}+\omega_{\mathrm{C}})^2\tau_c^2}} \qquad \text{(V-7)}$$

It is seen that under conditions where the ^{13}C relaxation is entirely by ^{13}C—^{1}H dipole–dipole coupling, the NOE is 2.988 in the extreme motional narrowing limit given by

$$(\omega_{\mathrm{C}} + \omega_{\mathrm{H}})\tau_c \ll 1 \qquad \text{(V-8)}$$

where Equation V-7 adopts the form V-4. On the other hand, the NOE decreases to 1.153 when the rotational motions are slow enough so that $\omega_{\mathrm{C}}\tau_c \gg 1$. Hence the combination of T_1 and the NOE, which can both be obtained from ^{13}C{^{1}H}FT experiments, leads to the correct solution for τ_c. A detailed analysis of the frequency dependence of T_1^{C}, T_2^{C}, and NOE^{C} for systems with and without intramolecular reorientations was presented by Doddrell et al.[47]

Flexible Linear Peptides

With the definition of a "random coil polypeptide chain" adopted in Chapter III (footnote p. 66), any extended flexible form of a linear peptide will in general qualify for this term, and the ^{1}H and ^{13}C-NMR of the amino acid side chains will correspond quite closely to the spectrum expected for the random coil form (Figs. III-2 and V-7). This does not preclude the occurrence of local non-random, energetically favored backbone conformations, e.g. by formation of ring structures through hydrogen bonds (Fig. V-12). This view is also supported by the observation that denatured forms of proteins with intact disulfide bonds, e.g. lysozyme, ribonuclease A, and the basic pancreatic trypsin inhibitor (Figs. III-20 to III-22) give rise to typical random coil spectra of the amino acid side chain resonances. The above mentioned conformation dependent ^{13}C-NMR features of polypeptide chains enable one to obtain more direct information on specific peptide backbone conformations than could be obtained by ^{1}H-NMR alone, or by ^{1}H and ^{13}C-NMR observation of the amino acid side chains, and the analysis of the data can be extended to a more subtle distinction between random coil and non-random spatial arrangements of the polypeptide chains.

The experiments to be discussed here should be of particular interest for investigations of medium size linear polypeptides, where the folding of the molecule is relatively little restricted by the covalent structure. On the basis of general experience, a model assuming that the peptide groups are in the standard planar forms and the occurrence of cis peptide groups is limited to N-substituted amino acids, is adequate for this class of compounds. For a given sequence of the common amino acid residues, the sterically allowed backbone conformations can then be generated by variations of the torsion angles ϕ_i and ψ_i about the single bonds linking the peptide groups, ω about X-Pro, and χ_i^j about the single bonds of the amino acid side chains (Fig. III-10). With the exception of the interconversion between the cis- and trans-forms of the X-Pro bonds, for which the free energy of activation $\Delta G^{\ddagger}$ is of the order 20 Kcal · mole^{-1}, the low energy barriers for rotational motions about the individual single bonds will allow for rapid interconversion among the different rotational states. Therefore, the NMR spectra of random coil polypeptide chains correspond to the average over all the rotational states about the individual single bonds populated according to the Boltzmann law, where the ensemble of conformations to be included would to a good approximation be given by Ramachandran ϕ-ψ plots for the isolated dipeptide segments (Ref. 5 of Chapter III), and by certain equilibrium ratios of the cis- and trans- forms of the X-Pro peptide bonds. ^{13}C-NMR manifestations of the formation of energetically favored conformations in which the equilibrium populations of the rotational states deviate from the random coil behavior, include variations of the relaxation times caused by reduced segmental mobility of the backbone, spectral features typical for the formation of intramolecular hydrogen bonds, and variations of the cis:trans ratio of the X-Pro bonds.

Studies of the segmental motions have for example been reported for aqueous solutions of the proline peptides poly(L-prolyl), poly(hydroxyl-L-prolyl), poly(L-prolyl-glycyl), poly(hydroxyl-L-prolyl-glycyl), and poly-(glycyl-glycyl-L-prolyl-glycyl), with molecular weights in the range 7000 to 13000[48]. The spin-lattice relaxation times T_1 measured in these molecules are given in Table V-11, and the corresponding effective correlation times for rotational motions, τ_c, are presented in Table V-12. The authors[48] concluded that the five polypeptides studied are not highly ordered in aqueous solution. The effective correlation times of 0.1 to 0.9 nanoseconds

TABLE V-11

Spin-lattice relaxation times T_1 (sec) of the carbon atoms in polypeptide chains of L-proline, hydroxyl-L-proline (Hyp), and glycine in aqueous solution at 32°. $\bar{M}_w$ = weight average molecular weight (from Torchia and Lyerla[48]).

		Poly(Pro)	Poly(Hyp)	Poly-(Pro-Gly)	Poly-(Hyp-Gly)	Poly(Gly(1)-Gly(2)-Pro(3)-Gly(4))
$\bar{M}_w$		11500	9000	13200	10500	6700
Pro or Hyp	C^α	0.077	0.069	0.129	0.129	0.188
	C^β	0.079	0.058	0.111	0.092	0.156
	C^γ	0.109	0.113	0.118	0.177	0.188
	C^δ	0.044	0.041	0.082	0.076	0.115
	C'	1.6	1.5	1.5	1.5	2.7
Gly	C^α			0.067	0.069	0.105
	C'			1.5	1.6	3.4
Gly(1)C^α						0.105
Gly(4)C'						2.2

TABLE V-12

Effective correlation times τ_c (nanoseconds) for rotational motions of the carbon atoms in polypeptide chains of L-proline, hydroxyl-L-proline (Hyp), and glycine in aqueous solution at 32°.
$\bar{M}_w$ = weight average molecular weight (from Torchia and Lyerla[48]).

		Poly(Pro)	Poly(Hyp)	Poly-(Pro-Gly)	Poly-(Hyp-Gly)	Poly(Gly(1)-Gly(2)-Pro(3)-Gly(4))
$\bar{M}_w$		11500	9000	13200	10500	6700
Pro or Hyp	C^α	0.61	0.68	0.36	0.36	0.26
	C^β	0.30	0.41	0.21	0.26	0.15
	C^γ	0.22	0.42	0.20	0.26	0.13
	C^δ	0.53	0.58	0.29	0.31	0.21
	C'	0.8	0.9	0.60	0.61	0.33
Gly	C^α			0.35	0.34	0.22
	C'			0.61	0.57	0.26
Gly(1)C^α						0.22
Gly(4)C'						0.31

appear to be determined essentially entirely by segmental motions, and are little affected by the molecular weight differences in the range 7000 to 13000. It is clear from Table V-12 that the polypeptide C^α correlation times decrease as the fraction of pyrrolidine residues in the polymers decreases.

This is a consequence of the reduced steric interactions when Pro or Hyp are replaced by Gly. The numbers given in the Tables V-11 and V-12 for the three copolypeptides of Gly and Pro, or Hyp, respectively, are quite representative for flexible extended polypeptide chains. It is further worth noting that the approximately equal C^α correlation times observed for Pro, Hyp, and Gly indicate that the segmental mobility of the backbone is highly cooperative, since quite different rotational potential functions are obtained for isolated Gly and Pro residues. For the ring carbon atoms of Pro and Hyp, the values for τ_c seem to indicate that there are rapid ring motions involving primarily C^β and C^γ, with the Pro ring being more mobile than the Hyp ring. In a different study, the rotational mobilities of backbone and side chain carbons of poly(γ-benzyl-L-glutamate) in the helical and random coil states were investigated[49]. For the α-carbons, the T_1 relaxation times were found to be nearly equal in the two conformation states, e.g. at 15.18 MHz, T_1's of 0.039 sec and 0.070 sec were observed for the helical and the random coil forms, respectively, of a sample with molecular weight 17000. The NOE was 1.2 for the helix, and 2.6 for the coil, which showed that the effective correlation times for rotational reorientation of the C^α's decrease by a factor of about 30 when going from the helix to the random coil form, i.e. from ca. 25 nsec to 0.8 nsec. (Fig. V-15). The near equality of the T_1's in the two conformations arises because of their particular positions on different sides of

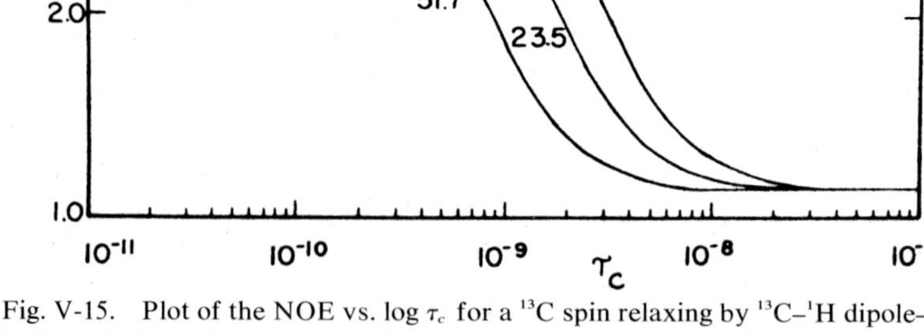

Fig. V-15. Plot of the NOE vs. log τ_c for a ^{13}C spin relaxing by ^{13}C–^1H dipole–dipole coupling. In the calculation it was assumed that the rotational tumbling of the spin system was restricted to isotropic rotations with the velocity indicated by τ_c, and that there was complete proton decoupling. Results are shown for $H_0 = 14.1$ kG ($\omega_C = 9.48 \times 10^7$ rad sec^{-1}, $\omega_H = 3.77 \times 10^8$ rad sec^{-1}), $H_0 = 23.5$ kG ($\omega_C = 1.58 \times 10^8$ rad sec^{-1}, $\omega_H = 6.29 \times 10^8$ rad sec^{-1}), and $H_0 = 51.7$ kG ($\omega_C = 3.48 \times 10^8$ rad sec^{-1}, $\omega_H = 1.38 \times 10^9$ rad sec^{-1}) (from Doddrell et al.[47]).

the minimum in the functional relation (Equation V-5) between T_1 and τ_c (Fig. V-27). In both helical and random coil poly (γ-benzyl-L-glutamate), there is a progressive increase of T_1 of the protonated carbons of the side chains when going away from the polymer backbone. This is a consequence of the segmental mobility of the side chains, which is quite typical for non-globular conformations of polypeptide chains.

A ^{13}C-NMR investigation of hydrogen bond formation in linear peptides was reported for the repeat peptides of elastin, e.g. H-Val-Pro-Gly-Gly-OH, H-Val-Pro-Gly-Val-Gly-OH, and H-Ala-Pro-Gly-Val-Gly-Val-OH[50]. Complete resonance assignments were derived from a suitable set of ^{13}C-labelled analogue peptides, and the position of the single hydrogen bridge in these peptides (Fig. V-12) determined by solvent titrations, as outlined in Fig. V-11.

On the following basis, studies of the cis:trans ratio of the X-Pro peptide bonds provide a means to demonstrate the occurrence of energetically favored peptide conformations[51]. $\Delta G°$ for the equilibrium between the cis- and trans- forms of X-Pro is of the order -2.0 to $+2.0$ Kcal · mole^{-1} in a variety of linear oligopeptides, so that the two forms are simultaneously present at ambient temperature. Investigations of a series of model peptides H-(Ala)$_m$-Pro-(Ala)$_n$-OH, with $m = 1–6$, $n = 0–2$, and analogues of these peptides with different amino acids preceding Pro[33, 51, 52], indicated that $\Delta G°$ for the cis-trans equilibrium of a single prolyl residue in a polypeptide chain can be decomposed into two terms ΔG°_{XP}, and $\Delta G^{\circ}_{conf.}$. The structural term ΔG°_{XP} is typical for a sequence -CO-X-Pro-NH-, where X is one of the common amino acids. Deviations from ΔG°_{XP} of the proline cis:trans ratio arise when the populations of the rotational states described by the torsion angles ϕ_i, ψ_i, and χ_i^j are different in the molecular species which contain cis- or trans- proline, respectively. The occurrence of energetically non-equivalent manifolds of flexible polypeptide conformations can thus be manifested in the cis-trans equilibrium of X-Pro, even when the free energy stabilizing the preferred species, $\Delta G^{\circ}_{conf.}$, is as small as several tenth of one Kcal · mole^{-1}. Applications of these principles to a series of oligopeptides H-(Ala)$_n$-X-Pro-OH (e.g. Fig. V-13) revealed the occurrence of energetically favored conformations including an eight-membered ring structure closed by a hydrogen bond between the terminal carboxylic acid group, and the amide proton of the amino acid preceding proline[52]. Studies of biological peptides include for example investigations of calcitonin M and partial sequences of this polypeptide hormone (Fig. V-16), the thyrotropin-releasing factor, angiotensin, and the melanocyte-stimulating hormone release-inhibiting factor[53]. With the introduction of ^{13}C enriched Pro into the polypeptide chains, this technique should be applicable to quite large molecules.

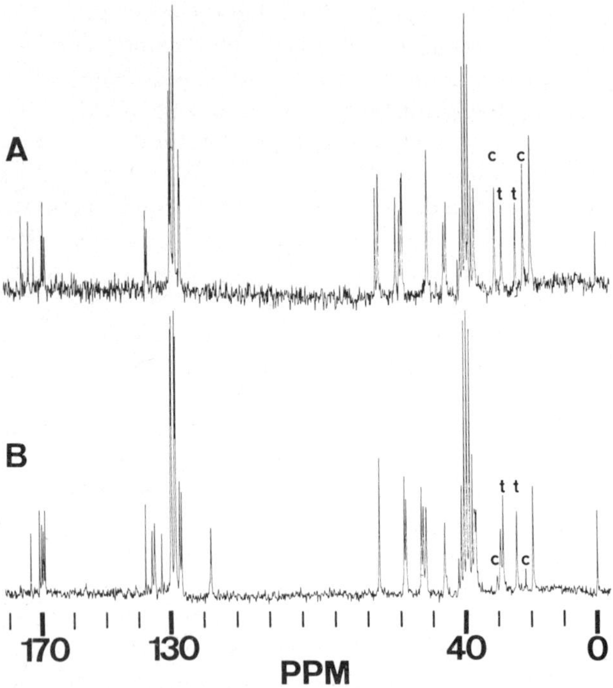

Fig. V-16. Proton noise-decoupled ^{13}C-NMR spectra of solutions in d_6-DMSO of two linear proline-containing peptides, which correspond to the segments 21-23 and 19-23 of calcitonin M (Fig. III-3). A. H-L-Thr-L-Phe-L-Pro-OH. B. H-L-Phe-L-His-L-Thr-L-Phe-L-Pro-OH. The letters *c* and *t* identify the resonances of the β- and γ-ring carbons of cis and trans proline, respectively. It is seen that the elongation of the peptide chain at the N-terminus sizeably affects the cis:trans ratio of the Phe-Pro peptide bond, even though it is separated from the latter by two amino acid residues. No evidence of cis-proline could be observed for the entire calcitonin M sequence (from Wüthrich and Grathwohl[51]).

Cyclopeptides

In this section the ^{13}C-NMR spectra of cyclo-glycyl-L-prolyl-glycyl-glycyl-L-prolyl-glycyl and cyclo-glycyl-L-alanyl-glycyl-glycyl-L-prolyl will be presented, which had served in Section III.3 to illustrate the use of ^1H-NMR for conformational studies of peptides. This will be followed by a survey of selected investigations of cyclo-peptide conformations.

c-(-Gly-L-Pro-Gly-)$_2$ and the selectively deuterated analogue *c*-(Gly-L-Pro-Gly(d_2)-) were studied in d_6-DMSO solution, and in a mixed solvent of d_6-DMSO and D_2O. The ^{13}C-NMR spectra needed to ascertain the resonance assignments given in Table V-13 are presented in Figs. V-17 and V-18. The

TABLE V-13

^{13}C Chemical shifts of *c*-(-Gly-L-Pro-Gly-)$_2$ in d_6-DMSO solution at 35° (from Grathwohl et al.[31]).

Carbon assignment			δ (ppm)
CH_2	Pro	C^γ	24.7
		C^β	28.8
	Gly (1,4) Gly (3,6)	C^α	42.5
	Pro	C^δ	45.8
CH	Pro	C^α	60.9
C=O	Gly (1,4)		166.7
	Gly (3,6)		169.2
	Pro		172.4

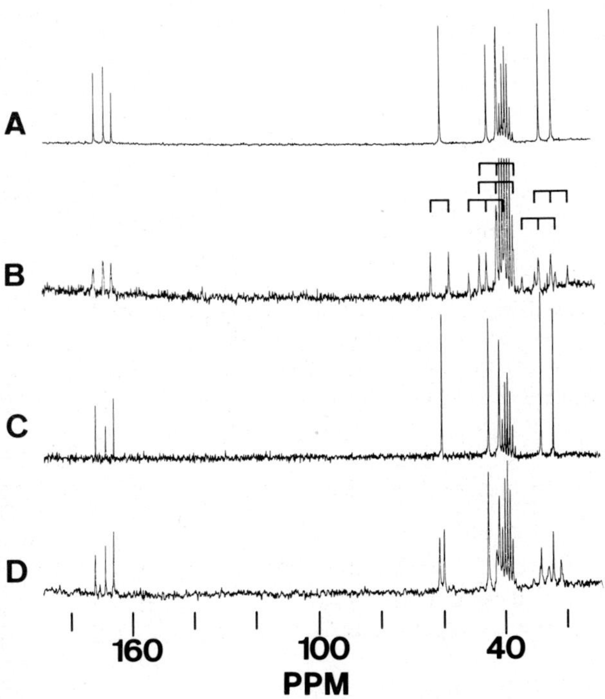

Fig. V-17. ^{13}C FT NMR spectra at 25.16 MHz, $T = 35°$, d_6-DMSO solvent resonance at 39.6 ppm. A and B. *c*-(-Gly-L-Pro-Gly-)$_2$; A. proton noise-decoupled; B. No proton irradiation. C and D. *c*-(-Gly-L-Pro-Gly(d_2)-)$_2$; C. Proton noise-decoupled; D. Double resonance irradiation at a position approx. 20 Hz upfield from the δ-methylene proton resonances of proline (b in Fig. III-13) (from Grathwohl et al.[31]).

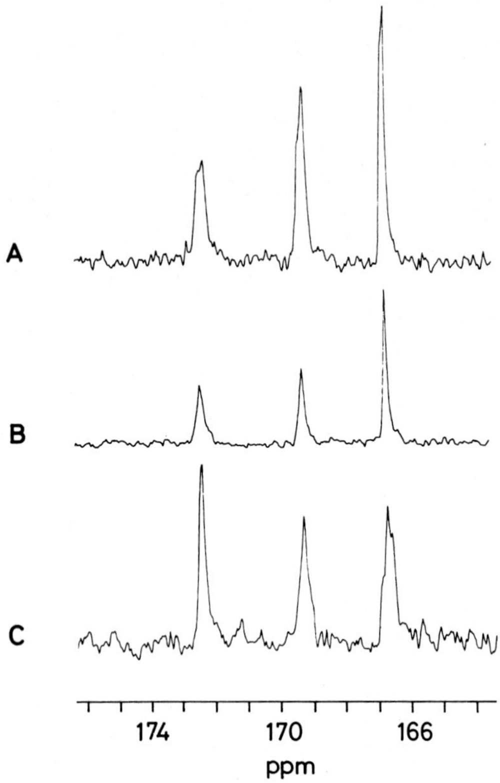

Fig: V-18. Carbonyl carbon ^{13}C-NMR lines of A. *c*-(-Gly-L-Pro-Gly-)$_2$ in d_6-DMSO, double resonance irradiation of the C^{α}-proton resonances of Gly(1) and (4) (resonance d in Fig. III-13). B and C. *c*-(-Gly-L-Pro-Gly(d_2)-)$_2$ in d_6-DMSO + 4% D_2O; B. Same conditions as A; C. Double resonance irradiation of the β- and γ-methylene protons of proline (resonance a in Fig. III-13) (From Grathwohl et al.[31]).

methylene, methine, and carbonyl carbon resonances expected from the covalent structure

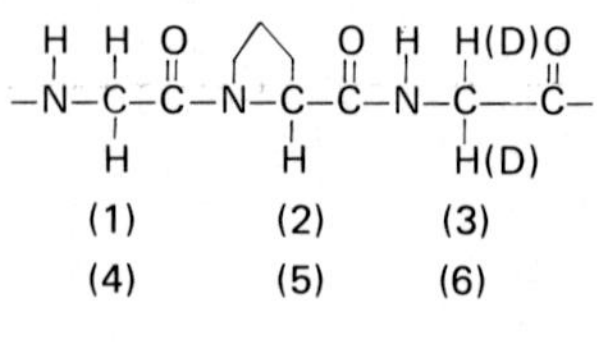

could be distinguished from the fine structure of the resonances in spectrum B of Fig. V-17, and with Table V-10 the lines at 24.7, 28.8, and 60.9 ppm could be assigned to trans-proline. The assignment of the resonances at 42.5

and 45.8 ppm to C^{α} of Gly and C^{δ} of Pro, respectively could be confirmed by the observation in spectrum C of Fig. V-17, that deuteration of Gly(3) affected only the intensity of the C^{α} resonance of Gly, and by the specific ^{1}H—^{13}C double resonance experiment in spectrum D of Fig. V-17. The individual carbonyl carbon resonances were assigned from observations made after selectively removing individual long range C′-proton spin–spin interactions through the introduction of Gly(d_2) in positions 3 and 6, replacement of the amide protons by D from a trace of D_2O added to the d_6-DMSO solutions, and specific double resonance irradiation of individual ^{1}H-NMR lines (Fig. III-13). Compared to the spectrum of the fully protonated peptide obtained without ^{1}H irradiation (Fig. V-17, B), decoupling of the C^{α}-protons of Gly(1,4) affects mainly the C═O line at highest field (Fig. V-18A), and decoupling of the proline C^{β} and C^{γ} methylene protons causes an increase of the intensity of the lowest field line (Fig. V-18C). Deuteration of C^{α} of Gly(3,6) causes a reduced intensity of the line at 169.2 ppm in the decoupled spectra (Figs. V-17C and V-18B). These observations led to the resonance identifications of Table V-13, which have also been confirmed by studies of ^{13}C-enriched peptide analogues[54]. The ^{13}C-NMR parameters (Table V-13) support the molecular conformation derived from the ^{1}H-NMR data (Tables III-3 and III-4) of *c*-(-Gly-L-Pro-Gly-)$_2$ (Fig. III-15). In particular, the C_2-symmetry of the molecule is readily apparent in the ^{13}C-NMR, and the trans conformation of the Gly-Pro peptide bond could unambiguously be derived from the C^{β} and C^{γ} carbon chemical shifts of Pro. An additional interesting observation emanating from these experiments concerns the influence of intramolecular hydrogen bond formation on the carbonyl carbon chemical shift. In *c*-(-Gly-L-Pro-Gly-)$_2$, the hydrogen bonded C═O(1,4) (Fig. III-15) is observed at higher field than the other carbonyl lines[31,54]. This contrasts with observations in different molecules, where hydrogen bonding of C═O with N—H groups was found to cause sizeable downfield shifts of the ^{13}C resonances[55], e.g. also in enterobactin, where the selective decoupling techniques illustrated in Figs. V-10 and V-18 had also been employed to assign the C═O resonance lines[56].

The ^{13}C-NMR spectrum of *c*-(-Gly-L-Ala-Gly-Gly-L-Pro-) is shown in Fig. V-19, and the ^{13}C chemical shifts are given in Table V-14[35]. The resonance assignments in Table V-14 were derived from off-resonance ^{1}H—^{13}C double resonance experiments, and comparison with model peptides (Table V-10). The two molecular conformations *M* and *m* observed in the ^{1}H-NMR (Figs. III-16 to III-18, Table III-5) are readily recognized also in the ^{13}C-NMR spectrum. It is seen that the conformational differences are manifested in chemical shift differences of all the carbon atoms. On the basis of the C^{β} and C^{γ} resonances of proline, the cis- and trans- forms of the Gly-Pro peptide bond could be assigned to the species *m* and *M*, respectively. The

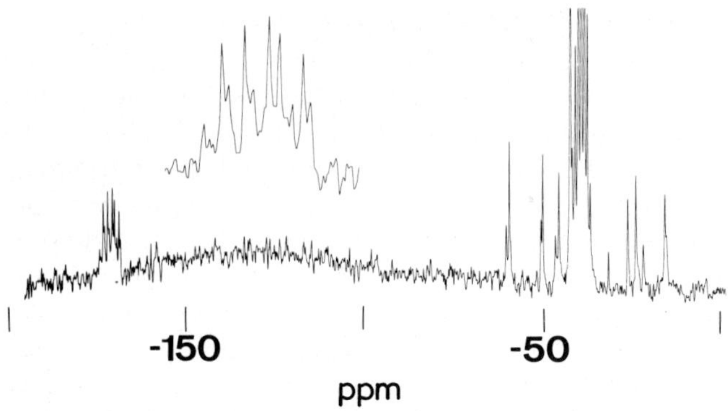

Fig. V-19. $^{13}C\{^1H\}$ FT NMR spectrum at 25.16 MHz of *c*-(-Gly-L-Ala-Gly-Gly-L-Pro-) in d_6-DMSO at 28°. The carbonyl carbon region from 165 to 175 ppm is also shown on an expanded scale (from Wüthrich et al.[35]).

TABLE V-14

^{13}C Chemical shifts (ppm) of the two molecular conformations *M* (65%) and *m* (35%) of *c*-(-Gly-L-Ala-Gly-Gly-L-Pro-) which occur simultaneously in a solution in DMSO at 28° (from Wüthrich et al.[35]).

	Carbon assignment		Conformation *M*	Conformation *m*
CH_3	Ala		16.3	15.8
CH_2	Pro	C^γ	24.3	22.2
		C^β	26.7	32.1
	Gly(1), Gly(3), Gly(4)	C^α	42.8	(40.6)
	Pro	C^δ	46.0	46.8
CH	Ala	C^α	50.4	51.0
	Pro	C^α	59.6	60.5
C=O	Gly(1), Ala, Gly(3), Gly(4), Pro		168.6, 169.9, 170.5, 171.9, 173.2	168.2, 169.2, 171.4, 172.8, 174.2

conformation dependent chemical shift differences of the carbonyl carbons, which have not been individually assigned in this molecule, appear to be quite sizeable, whereas rather small effects of less than 1 ppm are observed for the other carbon positions.

^{13}C-NMR was used for studies of a variety of synthetic cyclic model peptides. In *c*-(-L-Pro-Gly-)$_3$, the ^{13}C spectra revealed multiple solvent dependent equilibria between symmetric and asymmetric conformations with different combinations of trans- and cis- Gly-Pro peptide bonds[57]. In *c*-(L-Pro-Gly-)$_2$, the ^{13}C-NMR showed that the molecular conformation includes two cis Gly-Pro peptide bonds[58]. Furthermore, evidence was found for an equilibrium between two molecular species containing respectively cis′ and trans′ proline, where the rotation about the C^{α}—C′ single bond of proline was sufficiently slow, so that at low temperatures separate resonances could be observed for the two species.

^{13}C-NMR has also been applied for conformational studies of natural cyclopeptides. The investigations of the cyclodecapeptide antibiotic gramicidin S-A, cyclo-(-D-phenylalanyl-L-prolyl-L-valyl-L-ornithyl-L-leucyl-)$_2$, are among the outstanding examples. The resonances in the ^{13}C-NMR spectrum of gramicidin S-A[59] were identified through studies of peptide analogues in which individual amino acid residues were enriched to 10% in ^{13}C.[60] The selective incorporation of ^{13}C enriched amino acids was obtained biosynthetically by growing *Bacillus brevis* from a spore suspension on a minimal medium supplemented with equal amounts of each of the amino acids found in gramicidin S. In each experiment, one of the amino acids was added in the ^{13}C enriched form. Besides the ^{13}C-NMR work, the increased intensity of the ^{13}C satellites of the enriched amino acids was also used to confirm the earlier ^{1}H-NMR assignments of gramicidin S-A. Overall, the results of these experiments nicely confirmed the molecular conformations of gramicidin S-A proposed from earlier X-ray and ^{1}H-NMR studies (Ref. 33 and 83–86 of Chapter III). Gramicidin S-A was also used to test the delineation of intramolecular hydrogen-bonded carbonyl groups by solvent titration (Fig. V-11)[32], and the segmental mobility of this peptide was investigated by T_1-measurements of ^{13}C (Ref. 61). Typical T_1 values at 15.18 MHz in a methanol solution of gramicidin S-A are 1.6 sec for the quaternary carbons, 0.14 sec for the α-carbons, and 0.1 to 0.6 sec for the protonated carbon atoms of the amino acid side chains. Other biological cyclopeptides investigated with ^{13}C-NMR include the neurohypophyseal hormones oxytocin and lysine-vasopressin, for which chemical shifts and longitudinal relaxation times of ^{13}C have been investigated in detail[62–66], and various metal ion specific compounds[56], some of which will be further discussed below.

Alkali Ion Specific Cyclic Peptides and Depsipeptides

The ability of certain cyclic peptides and depsipeptides to form stable alkali ion complexes, e.g. the decapeptide antamanide and the depsipeptide antibiotics valinomycin and the enniatins, has aroused much interest in recent years, and these compounds are being widely used as tools for investigations of processes associated with ion transport through membranes. It was previously established by various techniques, including ^{1}H-NMR (Ref. 89–93 of Chapter III), that in these compounds a number of oxygen atoms of the backbone are so arranged in space that they can replace the hydration shell of a cation. ^{13}C-NMR has recently been advanced as a complementary technique for structural studies in these systems[67–70].

As compared to the ^{13}C-NMR properties of peptides discussed up to this point, a new feature in these systems is that ion-dipole interactions cause downfield shifts of the order 4 to 6 ppm in carbonyl groups bound to univalent diamagnetic metal ions. Thus the metal binding sites in the peptides can quite readily be delineated from the chemical shift differences between the free ligand and the metal complex. This is illustrated in Fig. V-20 by the comparison of the ^{13}C-NMR spectra of valinomycin* and its

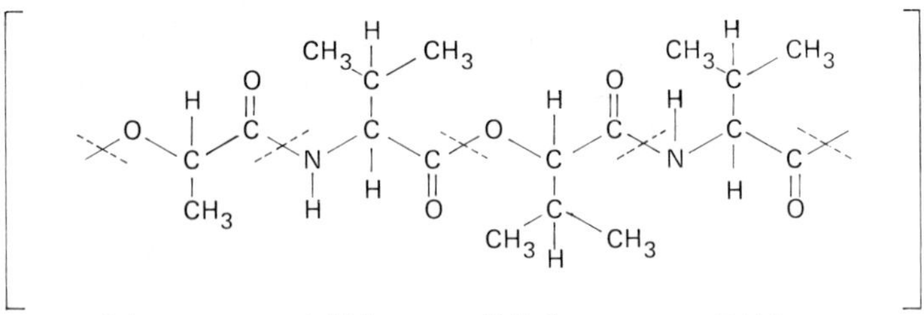

potassium complex. Each of the four carbonyl carbon lines of valinomycin corresponds to three identical residues. Upon complexation with K^+, two of the carbonyl lines, which had tentatively been assigned to L-Val and D-Val[67, 68], are shifted downfield by 4.6 ppm, and 3.9 ppm, respectively, and were therefore assigned to the six coordination sites of $K^{\oplus}$. The much smaller shifts of the other two carbonyl resonances were explained by the combined influence of long range ion-dipole interactions with the bound K^+ ion, and conformational rearrangement of the depsipeptide moiety in the K^+ complex.

Globular Proteins

The globular arrangement of polypeptide chains is quite readily apparent in the overall features of the ^{13}C-NMR spectra. As a consequence of the different microenvironments for individual amino acid residues, and the restricted segmental mobility in interior parts of globular proteins, the chemical shifts and the longitudinal relaxation times of ^{13}C can differ

*L-Lac = L-lactic acid, D-HyIv = D-α-hydroxy-isovaleric acid.

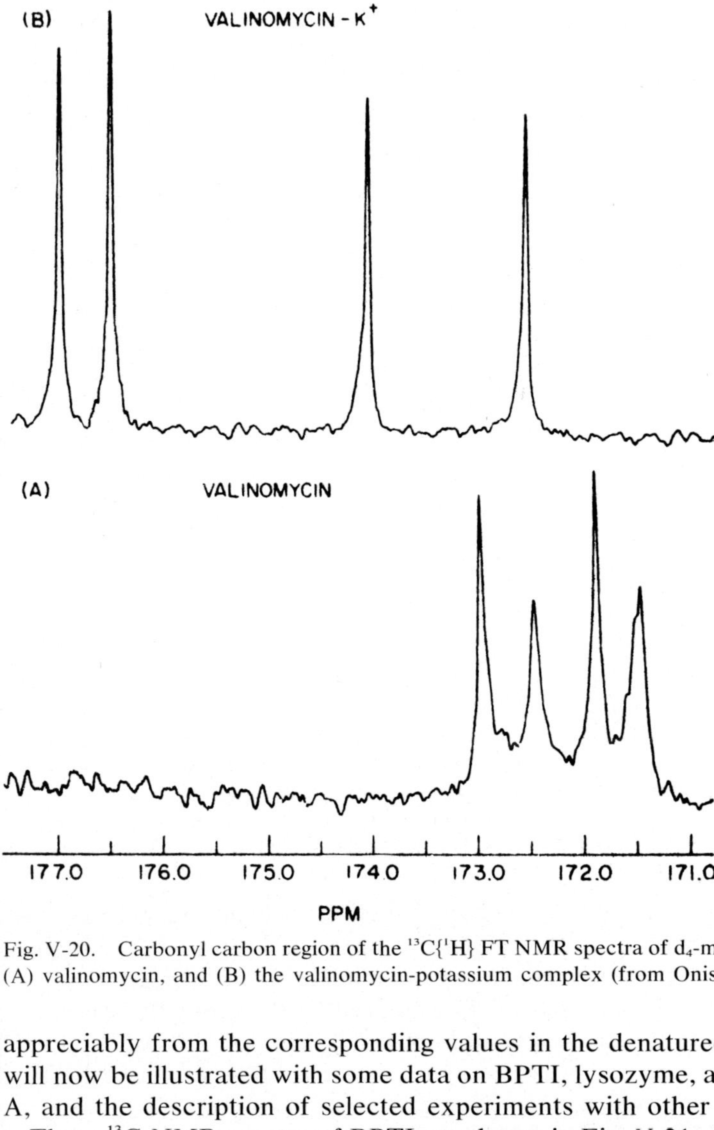

Fig. V-20. Carbonyl carbon region of the $^{13}C\{^1H\}$ FT NMR spectra of d_4-methanol solutions of (A) valinomycin, and (B) the valinomycin-potassium complex (from Onishi et al.[68]).

appreciably from the corresponding values in the denatured proteins. This will now be illustrated with some data on BPTI, lysozyme, and ribonuclease A, and the description of selected experiments with other proteins.

Three ^{13}C-NMR spectra of BPTI are shown in Fig. V-21. As we have seen in Chapter III, a quite typical random coil ^{1}H-NMR spectrum is obtained when this protein is dissolved in DMSO (Fig. III-22). Correspondingly, the ^{13}C-NMR lines of the amino acid side chains of BPTI in d_6-DMSO, which are mostly in the range from 0 to 40 ppm, and 100 to 165 ppm (Fig. V-21C), correspond quite closely to the spectrum expected for the random coil polypeptide chain (Fig. V-21B). The spectrum of the native globular

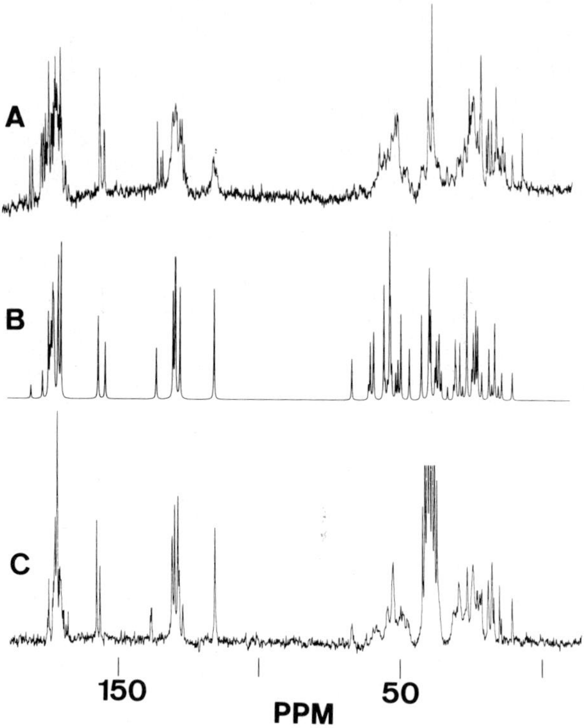

Fig. V-21. Proton noise-decoupled FT ^{13}C-NMR spectra at 25.16 MHz of the basic pancreatic trypsin inhibitor: A. Native BPTI in D_2O solution; B. Hypothetical spectrum for a DMSO solution of the random coil form of the polypeptide chain of BPTI computed from the data in Table V-10; C. Denatured BPTI in d_6-DMSO solution (from Wüthrich[71]).

molecule (Fig. V-21A) is on the other hand quite markedly different. There is a dispersion of the chemical shifts of identical amino acid residues in different positions of the polypeptide chain, and also an increase of the line width of certain resonances. This is most readily apparent in the aromatic region of the spectrum, where e.g. the resonances of the 3,5 ring carbons of the four tyrosines are spread over a range of approximately 4 ppm in native BPTI, whereas in the denatured protein they give rise to a single sharp line at 115 ppm. ^{13}C-NMR was employed to determine pK-values for the individual tyrosyl residues in native BPTI[72]. In these studies the pH titration behavior of the C_4 ring carbon resonances was followed in the spectral region between 150 and 160 ppm (Fig. V-21), which contains otherwise only the C^{ζ} resonances of the 6 arginyl residues (Table V-8).

^{13}C-NMR spectra of lysozyme are shown in Fig. V-22. They were recorded at 15.18 MHz using sample tubes with 20 mm diameter[19], which allows the observation of single carbon resonances in the spectra of proteins. As in

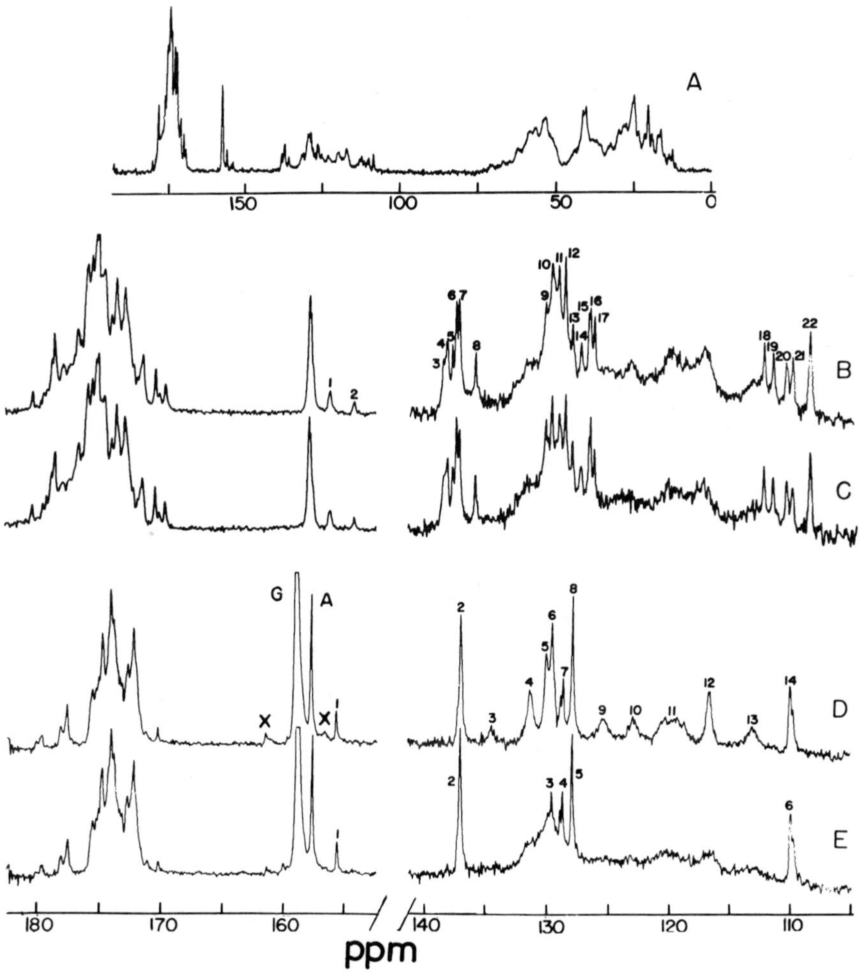

Fig. V-22. ^{13}C FT NMR spectra of hen egg-white lysozyme at 15.18 MHz in a 20 mm sample tube. A. Proton noise-decoupled spectrum of native lysozyme about 20% w/v in 0.1 M NaCl, pH 4.1, 45°. B. Region from 110 to 180 ppm of spectrum A on an expanded scale. Peaks 1-22 arise from the 28 non-protonated aromatic carbons. C. Same as B, but with noise-modulated off-resonance proton decoupling. The decoupling frequency was centered at about −6 ppm (Fig. III-12A). The random-noise modulation bandwidth was about 300 Hz. D. Region from 110 to 180 ppm of the proton noise-decoupled ^{13}C-NMR spectrum of guanidine-denatured lysozyme, about 25% w/v in 0.1 M NaCl and 6.2 M guanidinium chloride, pH 4.3, 45°. The resonance of the 6.2 M guanidinium ion (denoted by G) is the only signal that is strong enough to yield detectable spinning side bands (denoted by X). Peak A is the resonance of the ζ carbons of the 11 arginine residues. Peaks 1-14 arise from the 87 aromatic carbons. E. Same as D, but with noise-modulated off-resonance proton decoupling. Decoupling conditions were the same as for native lysozyme in C. Peaks 1–6 arise from the 28 nonprotonated aromatic carbons. The authors point out that with their home built apparatus, the spectra A–E can be recorded in 4, 13, 13, 7, and 7 hours, respectively (from Allerhand et al.[19]).

BPTI, the conformational differences between native and denatured lysozyme are clearly manifested in the ^{13}C-NMR (compare the spectra B and D of Fig. V-22). The ^{13}C chemical shifts of the non-protonated aromatic carbons in native and denatured lysozyme were identified on the basis of the smaller line widths as compared to those of the protonated aromatic carbons. In the latter, ^{13}C—^{1}H dipole–dipole coupling is an effective line broadening mechanism (Equation V-6), which leads to line widths of the order 36 Hz in native lysozyme. Because of the r^{-6}-dependence of dipolar interactions, the corresponding line widths of the quaternary carbons are only a few Hz.[19] The assignment of the non-protonated carbon lines was confirmed by means of noise-modulated off-resonance proton decoupling, which selectively broadens protonated carbons (spectra C and E of Fig. V-22)[73]. It can be seen that the 28 non-protonated aromatic carbon atoms of the 6 Trp, 3 Phe, 3 Tyr, and 1 His residues give rise to 22 sharp lines in the spectrum of native lysozyme. In the guanidinium chloride denatured protein, rapid segmental motions cause some of the protonated carbon resonances to sharpen appreciably as compared to the native molecule. Noise-modulated off-resonance decoupling broadens these lines again, and shows that the 28 quaternary carbons give rise to 6 lines in denatured lysozyme (spectra D and E of Fig. V-22). Chemical shifts and assignments for these resonances are given in Table V-15. The table shows that the conformation dependent chemical shifts of these resonances are as large as 2.5 ppm. Fig. V-22 shows further that the carbonyl carbon resonances in native lysozyme are spread out over an appreciably larger spectral region than in the denatured protein. The strong line at 157.9 ppm corresponds to C^{ζ} of the 11 arginyl residues in lysozyme.

In ribonuclease A, tentative assignments of individual lines in the ^{13}C-NMR spectrum (Fig. V-14) have been presented, and the longitudinal relaxation times were measured in the native and the denatured oxidized protein[45,74]. Some typical values for T_1, and the corresponding effective correlation times for rotational reorientation are: α-carbons, $T_1 = 0.04$ sec and $\tau_c = 30$ nanoseconds in native ribonuclease, $T_1 = 0.12$ sec and $\tau_c = 0.4$ nanoseconds in the denatured protein. β-carbons of threonine, $T_1 = 0.04$ sec, $\tau_c = 30$ nanoseconds, and $T_1 = 0.10$ sec, $\tau_c = 0.5$ nanoseconds, respectively. ϵ-carbons of lysine, $T_1 = 0.33$ sec and $\tau_c = 0.07$ nanoseconds, and $T_1 = 0.31$ sec and $\tau_c = 0.08$ nanoseconds, respectively. The introduction of ^{13}C enriched His and Gly as NMR probes into semisynthetic ribonuclease S′ has also been described[75,76].

Applications of ^{13}C-NMR to other proteins include investigations of various lipoproteins[77–79]. ^{13}C-NMR appears to be a particularly suitable non-destructive method for investigations of the lipid : protein ratios in these molecules, and for studies of the segmental mobility of the lipid chains in the

TABLE V-15

^{13}C Chemical shifts of non-protonated aromatic carbons in lysozyme (from Allerhand et al.[19])*.

Native lysozyme Line Nr.[(a)]	δ (ppm)	Denatured lysozyme Line Nr.[(a)]	δ (ppm)	Resonance assignment[(b)]	
1	156.2	1	155.6	Tyr	C4
2	154.1				
3	138.7	2	137.2	Phe	C1
4	138.5			Trp	C8
5	138.0				
6	137.7				
7	137.4				
8	136.1				
9	130.3	3	129.7	His	C5
10	129.8	4	128.7	Tyr	C1
11	129.3	5	128.0	Trp	C9
12	128.7				
13	128.2				
14	127.5				
15	126.9				
16	126.7				
17	126.4				
18	112.3	6	109.9	Trp	C3
19	111.6				
20	110.5				
21	110.0				
22	108.5				

*(a)See Fig. V-22; (b) see Tables V-8 and II-6.

lipoproteins. The covalent attachment of ^{13}C enriched NMR labels to specific amino acid side chains was for example described for myoglobins, which were carboxymethylated with [2-^{13}C] bromoacetate[80].

6. BIBLIOGRAPHY

1 Stothers J. B., Carbon-13 NMR Spectroscopy, Academic Press, New York, 1972

2 Levy G. C. and Nelson G. L., Carbon-13 Nuclear Magnetic Resonance for Organic Chemists, Wiley-Interscience, New York, 1972

3 Grant D. M. and Paul E. G., J. Amer. Chem. Soc. *86*, 2984, (1964)

4 Horsley W., Sternlicht H. and Cohen J. S., J. Amer. Chem. Soc. *92*, 680 (1970)

5 Maciel G. E., in Nuclear Magnetic Resonance Spectroscopy of Nuclei other than Protons, T. Axenrod and G. A. Webb (Eds.), Wiley-Interscience, New York, 1974, p. 187

6 Jaeckle H., Haeberlen U., and Schweitzer D., J. Magn. Res. *4*, 198 (1971)

7 Doddrell D. and Allerhand A., J. Amer. Chem. Soc. *93*, 1558 (1971)
8 Allerhand A., Childers R. F., Goodman R. A., Oldfield E. and Ysern X., Amer. Lab. *4*, 19 (1972)
9 Tran-Dinh S., Fermandjian S., Sala E., Mermet-Bouvier R., Cohen M., and Fromageot P., J. Amer. Chem. Soc. *96*, 1484 (1974)
10 Levy G. C. and Cargioli J. D., J. Magn. Res. *6*, 143 (1972)
11 Ernst R. R., J. Chem. Phys. *45*, 3845 (1966)
12 Noggle J. H. and Schirmer R. E., The Nuclear Overhauser Effect, Academic Press, New York, 1971
13 Birdsall B., Birdsall N. J. M. and Feeney J., Chem. Comm. *1972*, 316
14 Freeman R. and Hill H. D. W., J. Magn. Res. *5*, 278 (1971)
15 Horsley W. J. and Sternlicht H., J. Amer. Chem. Soc. *90*, 3738 (1968)
16 Horsley W., Sternlicht H. and Cohen J. S., Biochem. Biophys. Res. Comm. *37*, 47 (1969)
17 Voelter W., Jung G., Breitmaier E. and Bayer E., Z. Naturforsch. *26b*, 213 (1971)
18 Christl M. and Roberts J. D., J. Amer. Chem. Soc. *94*, 4565 (1972)
19 Allerhand A., Childers R. F. and Oldfield E., Biochemistry *12*, 1335 (1973)
20 Bradbury J. H. and Norton R. S., Biochim. Biophys. Acta *328*, 10 (1973)
21 Reynolds W. F., Peat I. R., Freedman M. H. and Lyerla J. R., J. Amer. Chem. Soc. *95*, 328 (1973)
22 Grathwohl C. and Wüthrich K., J. Magn. Res. *13*, 217 (1974)
23 Keim P., Vigna R. A., Marshall R. C., and Gurd F. R. N., J. Biol. Chem. *248*, 6104 (1973)
24 Keim P., Vigna R. A., Morrow J. S., Marshall R. C., and Gurd F. R. N., J. Biol. Chem. *248*, 7811 (1973)
25 Keim P., Vigna R. A., Nigen A. M., Morrow J. S. and Gurd F. R. N., J. Biol. Chem. *249*, 4149 (1974)
26 Quirt A. R., Lyerla J. R., Peat I. R., Cohen J. S., Reynolds W. F. and Freedman M. H., J. Amer. Chem. Soc. *96*, 570 (1974)
27 Sogn J. A., Craig L. C. and Gibbons W. A., J. Amer. Chem. Soc. *96*, 4694 (1974)
28 Pearson H., Gust D., Armitage I. M., Huber H., Roberts J. D., Stark R. E., Vold R. R. and Vold R. L., Proc. Nat. Acad. Sci. *72*, 1599 (1975)
29 Wüthrich K. and Baumann R., unpublished data
30 Bundi A., Grathwohl C., Hochmann J., Keller R. M., Wagner G. and Wüthrich K., J. Magn. Res. *18*, 191 (1975)
31 Grathwohl C., Schwyzer R., Tun-Kyi A. and Wüthrich K., FEBS Letters *29*, 271 (1973)
32 Urry D. W., Mitchell L. W. and Ohnishi T., Proc. Nat. Acad. Sci. *71*, 3265 (1974)
33 Wüthrich K., Grathwohl C. and Schwyzer R., in Peptides, Polypeptides and Proteins, E. R. Blout, F. A. Bovey, M. Goodman and N. Lotan (Eds.), J. Wiley, New York, 1974, p. 300
34 Grathwohl C., Tun-Kyi A., Bundi A., Schwyzer R. and Wüthrich K., Helv. Chim. Acta *58*, 415 (1975)
35 Wüthrich K., Tun-Kyi A. and Schwyzer R., FEBS Letters *25*, 104 (1972)
36 Thomas W. A. and Williams M. K., Chem. Comm. *1972*, 994
37 Bovey F. A., in Chemistry and Biology of Peptides, ed. J. Meienhofer, Ann Arbor Sci. Publ., Ann Arbor 1972, p. 3
38 Smith I. C. P., Deslauriers R. and Walter R., in Chemistry and Biology of Peptides, ed. J. Meienhofer, Ann Arbor Sci. Publ., Ann Arbor 1972, p. 29
39 Dorman D. E. and Bovey F. A., J. Org. Chem. *38*, 2379 (1973)
40 Doddrell D., Burfitt I., Grutzner J. B. and Barfield M., J. Amer. Chem. Soc. *96*, 1241 (1974)
41 Feeney J., Hansen P. E. and Roberts G. C. K., Chem. Comm. *1974*, 465
42 Levin R. H. and Roberts J. D., Tetrahedron Lett. *1973*, 135
43 Du Vernet R. and Boekelheide V., Proc. Nat. Acad. Sci. *71*, 2961 (1974)

44 Batchelor J. G., Prestegard J. H., Cushley R. J. and Lipsky S. R., J. Amer. Chem. Soc. *95*, 6358 (1973)
45 Allerhand A., Doddrell D., Glushko V., Cochran D. W., Wenkert E., Lawson P. J. and Gurd F. R. N., J. Amer. Chem. Soc. *93*, 544 (1971)
46 Solomon I., Phys. Rev. *99*, 559 (1955)
47 Doddrell D., Glushko V. and Allerhand A., J. Chem. Phys. *56*, 3683 (1972)
48 Torchia D. A. and Lyerla J. R., Biopolymers *13*, 97 (1974)
49 Allerhand A., and Oldfield E., Biochemistry *12*, 3428 (1973)
50 Urry D. W., Mitchell L. W. and Onishi T., Biochemistry *13*, 4083 (1974)
51 Wüthrich K. and Grathwohl Ch., FEBS Lett. *43*, 337 (1974)
52 Grathwohl Ch. and Wüthrich K., to be published
53 Deslauriers R., Walter R., and Smith I. C. P., Biochem. Biophys. Res. Comm. *53*, 244 (1973)
54 Pease L. G., Deber Ch. M., and Blout E. R., J. Amer. Chem. Soc. *95*, 258 (1973)
55 Bovey F. A., Nuclear Magnetic Resonance Spectroscopy, Academic Press, New York, 1969, p. 233
56 Llinás M., Wilson D. M. and Neilands J. B., Biochemistry *12*, 3836 (1973)
57 Madison V., Atreyi M., Deber C. M., and Blout E. R., J. Amer. Chem. Soc. *96*, 6725 (1974)
58 Deber C. M., Fossel E. T. and Blout E. R., J. Amer. Chem. Soc. *96*, 4015 (1974)
59 Gibbons W. A., Sogn J. A., Stern A., Craig L. C. and Johnson L. F., Nature *227*, 840 (1970)
60 Sogn J. A., Craig L. C. and Gibbons W. A., J. Amer. Chem. Soc. *96*, 3306 (1974)
61 Allerhand A., and Komoroski R. A., J. Amer. Chem. Soc. *95*, 8228 (1973)
62 Lyerla, J. R., and Freedman M. H., J. Biol. Chem. *247*, 8183 (1972)
63 Deslauriers R., Walter R. and Smith I. C. P., Biochem. Biophys. Res. Comm. *48*, 854 (1972)
64 Walter R., Prasad K. U. M., Deslauriers R., and Smith I. C. P., Proc. Nat. Acad. Sci. US *70*, 2086 (1973)
65 Deslauriers R., Walter R. and Smith I. C. P., Proc. Nat. Acad. Sci. USA *71*, 265 (1974)
66 Deslauriers R., Smith I. C. P. and Walter R., J. Amer. Chem. Soc. *96*, 2289 (1974)
67 Bystrov V. F., Ivanov V. T., Koz'min S. A., Mikhaleva I. I., Khalilulina K. and Ovchinnikov Y. A., FEBS Letters *21*, 34 (1972)
68 Ohnishi M., Fedarko M. C., Baldeschwieler J. D. and Johnson L. F., Biochem. Biophys. Res. Comm. *46*, 312 (1972)
69 Grell E., Funck T. and Sauter H., Eur. J. Biochem. *34*, 415 (1973)
70 Patel D. J., Biochemistry *12*, 496 (1973)
71 Wüthrich K., Pure Appl. Chem. *37*, 235 (1974)
72 Maurer W., Haar W. and Rüterjans H., Z. Phys. Chemie *93*, 119 (1974)
73 Wenkert E., Clouse A. O., Cochran D. W. and Doddrell D., J. Amer. Chem. Soc. *91*, 6879 (1969)
74 Glushko V., Lawson P. J. and Gurd F. R. N., J. Biol. Chem. *247*, 3176 (1972)
75 Chaiken I. M., J. Biol. Chem. *249*, 1247 (1974)
76 Chaiken I. M., Cohen J. S. and Sokoloski E. A., J. Amer. Chem. Soc. *96*, 4703 (1974)
77 Hamilton J. A., Talkowski C., Williams E., Avila E. M., Allerhand A., Cordes E. H. and Camejo G., Science *180*, 193 (1973)
78 Hamilton J. A., Talkowski C., Childers R. F., Williams E., Allerhand A. and Cordes E. H., J. Biol. Chem. *249*, 4872 (1974)
79 Assmann G., Highet R. J., Sokoloski E. A. and Brewer H. B., Proc. Nat. Acad. Sci. *71*, 3701 (1974)
80 Nigen A. M., Keim P., Marshall R. C., Morrow J. S., Vigna R. A., and Gurd F. R. N., J. Biol. Chem. *248*, 3724 (1973)
81 Deslauriers R., Walter R. and Smith I. C. P., FEBS Lett. *37*, 27 (1973)
82 Wüthrich K. and Wagner G., FEBS Lett. *50*, 265 (1975)

APPENDIX V

INTERPRETATION OF NMR RELAXATION STUDIES. NUMERICAL EXAMPLES

In the Chapters I, IV, and V, some of the fundamental aspects of nuclear spin relaxation have been introduced, and the principles underlying the experimental measurements of relaxation times were outlined. Furthermore, the impact of the spin relaxation times on certain technical aspects of NMR experiments with biopolymers, and the interpretation of the relaxation times in terms of the molecular dynamics have been illustrated with various studies of peptides and proteins. In the following Chapters VI and VIII, the treatment of spin relaxation phenomena will be extended to include paramagnetic species, and systems with chemical exchange of nuclear spins. It is the purpose of this appendix to complement these discussions with a description of numerical examples illustrating how the relaxation times are actually obtained from the NMR measurements.

First, the longitudinal relaxation times T_1 for ^1H and ^{13}C in L-alanine will be computed from the partially relaxed FT NMR spectra presented in Figs. IV-10 and V-4. This will be followed by an account of how the longitudinal relaxation times for ^{13}C nuclei in a protein may be obtained, using the basic pancreatic trypsin inhibitor (BPTI, Fig. V-21) as an illustration. Finally, the correlation times τ_c and τ_R for the rotational molecular motions will be evaluated from the combined experimental observations on T_1, T_2, and the nuclear Overhauser enhancement (NOE). In the description of these examples, the emphasis is on the qualitative aspects. Therefore, a mathematical error analysis is not included in the discussion.

^1H Relaxation Times T_1 in a D_2O Solution of L-Alanine

The intensities of the corresponding resonances in the individual partially relaxed FT NMR spectra in Fig. IV-10 are related by

$$M_z = M_0(1 - 2e^{-\tau/T_1}) \tag{V-9}$$

where τ is the delay time in the $180° - \tau - 90°$ pulse sequence, M_0 the thermal equilibrium magnetization, and M_z the z-component of the magnetization after the delay time τ (see Section IV.2). It is convenient to rewrite this relation in the form

$$\ln(M_0 - M_z) = -\frac{\tau}{T_1} + C \qquad \text{(V-10)}$$

where C is a constant. Equation V-10 comprises three significant experimental variables, M_0, M_z, and τ. The delay time τ may in all these experiments be predetermined as one of the instrument settings, whereas M_0 and M_z are evaluated from the spectra. With the modern spectrometers the scaling of successive spectra may be held constant, so that relative values of M_0 and M_z can readily be obtained. M_0 corresponds to the signal intensity in the spectrum obtained after a delay of $\tau \geqslant 5T_1$.

To evaluate the intensities of the individual lines, the ^{1}H-NMR spectra of L-alanine in Fig. IV-10 were integrated (Fig. V-23), and the integrals taken as relative values of M_z. Values for T_1 were then evaluated by a least squares fit of these numbers to Equation V-10. $T_1 = 2.0$ sec for the methyl protons, 5.9 sec for the C^α-proton, and ≈ 12 sec for the solvent HDO were thus obtained. The slopes in the plots of $\ln(M_0 - M_z)$ vs. τ in Fig. V-24 correspond approximately to the values for $-1/T_1$ obtained from the computer fit (Equation V-10). A check of these T_1-values is provided by the relation indicated in Fig. IV-5, that M_z vanishes after a waiting period $\tau = T_1 \ln 2$. From Fig. IV-10 one would thus estimate that $T_1 \approx 2$ sec for the methyl protons, $T_1 \approx 6$ sec for the C^α-proton, and $T_1 \approx 12$ sec for HDO, which coincides quite satisfactorily with the solutions represented in Fig. V-24.

^{13}C Relaxation Times T_1 in a D_2O Solution of L-Alanine

In principle the situation encountered in Fig. V-4 is very similar to that with the ^{1}H-NMR spectra. Reliable integrated intensities of the ^{13}C-NMR lines are, however, in practice not readily obtained. Because the width of the corresponding lines in the different partially relaxed spectra is essentially identical, one therefore often chooses to take the signal heights as a measure of the resonance intensities. Thus, the signal height of the resonances recorded after $\tau = 60$ sec was taken to represent M_0 (note that a delay time of approximately 150 sec is required to obtain a reliable value of M_0 for the carboxylic acid carbon atom). Values for M_z after different delay times τ were derived in an analogous way from the other spectra in Fig. V-4. From a least squares fit of the data to equation V-10, $T_1 = 1.9$ sec for C^β, 3.5 sec for C^α, and ≈ 28 sec for the carboxyl carbon was obtained.

From the τ values at which M_z vanishes (Fig. V-4), one estimates that $T_1 \approx 1$ sec, 2 sec, and 23 sec for C^β, C^α, and $COO^\ominus$, respectively. These

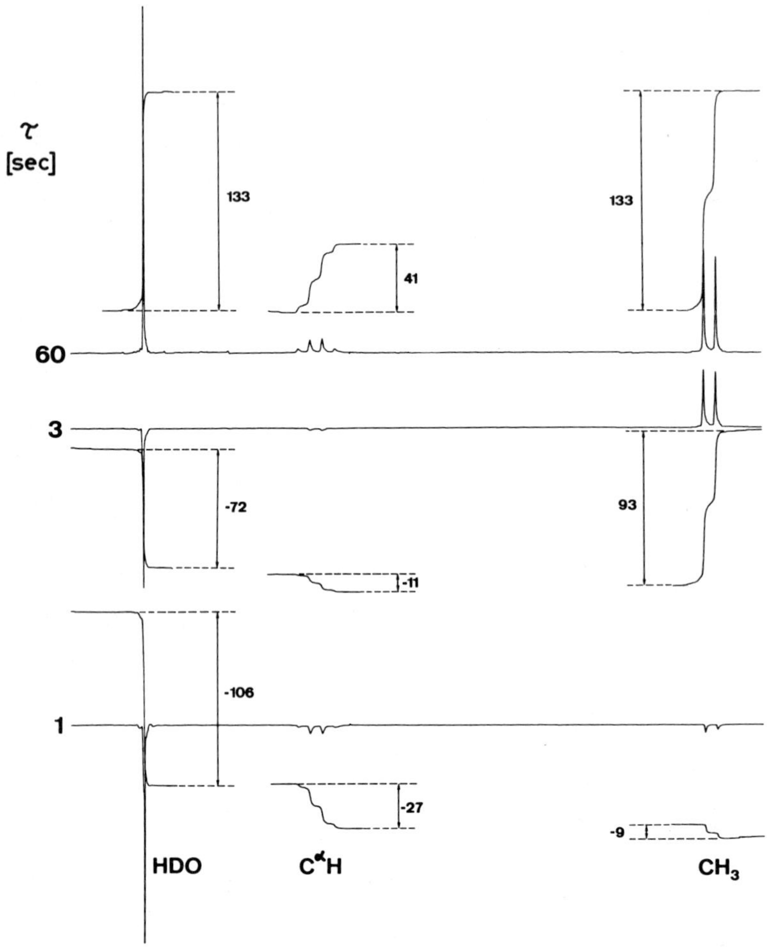

Fig. V-23. Partially relaxed FT ^{1}H-NMR spectra of a D_2O solution of L-alanine. The three traces with delay times in the $180° - \tau - 90°$ pulse sequence of $\tau = 1$, 3, and 60 sec have been selected from among those in Fig. IV-10, and completed with the integrated spectra. The relative signal intensities measured in mm on the original spectra are also indicated.

values are rather short, indicating that it is not advisable in studies of T_1, to rely solely on measurements of the null-point of M_z. Nevertheless, within the accuracy that is inherent in this approximate evaluation, it confirms the numbers obtained from the computer fit. The T_1's for C^β and C^α are in good agreement with the values quoted in the literature, i.e. 2.1 and 3.4 sec[81]. As was discussed on p. 174, the relaxation time for the carboxyl carbon is rather short compared e.g. with the value for glycine[28].

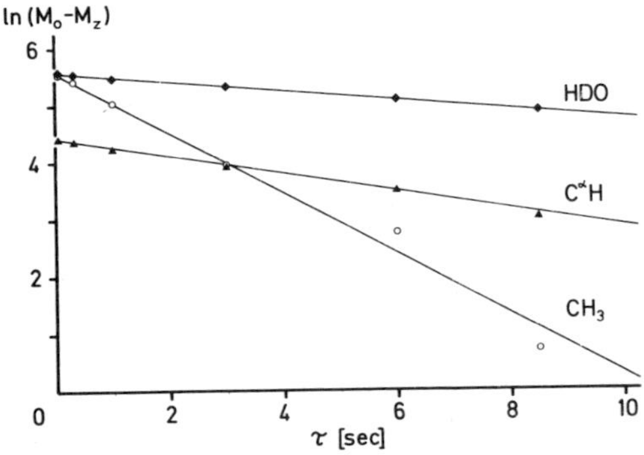

Fig. V-24. Representation of the experimental signal intensities in the partially relaxed FT ^{1}H-NMR spectra of a D_2O solution of L-alanine (Fig. IV-10) in the form of a plot of $\ln(M_0 - M_z)$ vs. τ. The integrated signal intensities of the spectrum obtained with $\tau = 60$ sec were used as M_0, and corresponding values for M_z after variable time intervals τ were obtained from the integrated intensities obtained in the other spectra (see Fig. V-23). From a least squares fit of these data to Equation V-10, $T_1 = 2.0$ sec for the methyl protons, $T_1 = 5.9$ sec for the C^α-proton, and $T_1 \approx 12$ sec for the solvent resonance was obtained. The straight lines in the figure represent Equation V-10, with the slopes $= -1/T_1$ corresponding approximately to the above values for T_1.

^{13}*C Relaxation Times* T_1 *in a* D_2O *Solution of BPTI*

Fig. V-25 shows a series of partially relaxed ^{13}C-NMR spectra of this protein. As indicated in the top trace of the figure, six resonances have been selected which will now be further discussed. Three of these are well resolved, prominent sharp lines which can from their chemical shifts be assigned to C^δ of Ile (◆), C^ϵ of lysine (▲), and C^ζ of Arg (○). In addition, three quite broad absorption bands were chosen to represent the α-carbon resonances (△), the 3,5 ring carbons of tyrosine (+), and the protonated ring carbons of phenylalanine (∗). The height of these signals in the top trace was taken to represent M_0, and correspondingly, values for M_z after different time intervals τ were obtained from the signal heights in the other spectra. These resonance intensities are plotted in Fig. V-26, and the relaxation times obtained from a least squares fit to Equation V-10 are given in the caption to Fig. V-25. Again, checks with the τ-values for which M_z vanishes, give a quite satisfactory agreement with the T_1's obtained from the computer fit.

Overall, considerably shorter relaxation times are observed in the protein than in the individual amino acids. Among the six selected resonances, the longest T_1 was found for the unprotonated carbon atom of the guanidinium group of arginine, i.e. $T_1 \approx 500$ msec. Inspection of Fig. V-25 shows that this T_1 is quite similar to that for the peptide carbonyl carbon resonances between 170 and 180 ppm. Relatively long relaxation times of 300 msec, and

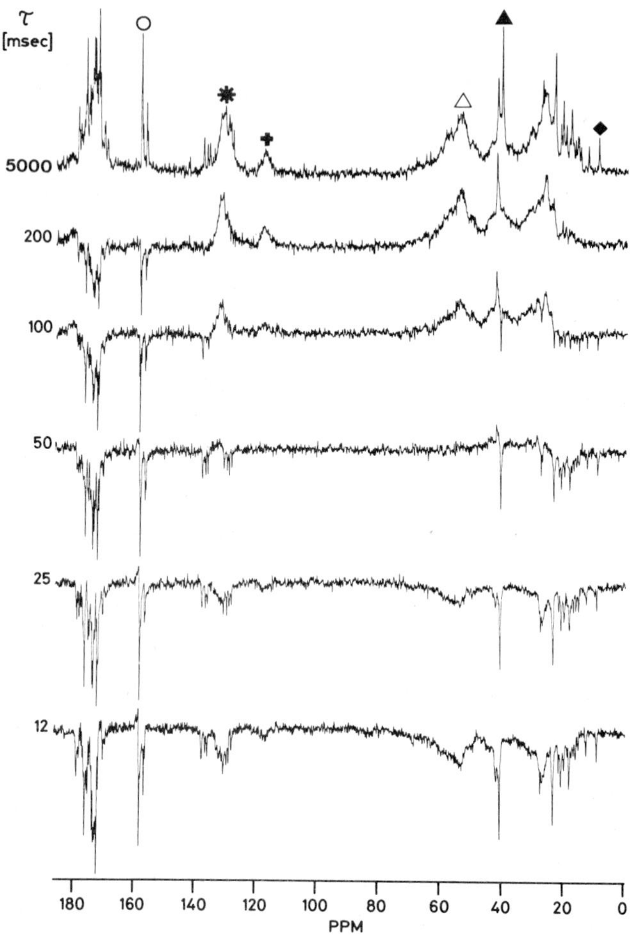

Fig. V-25. Partially relaxed proton noise-decoupled FT ^{13}C-NMR spectra at 25.16 MHZ of a 0.05-M solution of the basic pancreatic trypsin inhibitor (BPTI) in D_2O, pD = 4.7, $T = 35°$. The τ-intervals in the $180° - \tau - 90°$ pulse sequence are also indicated. The following resonances are identified in the top trace: ◆ C^{δ} of Ile at 9.2 ppm. $T_1 \approx 300$ msec; ▲ C^{ϵ} of lysine at 40.5 ppm. $T_1 \approx 230$ msec; △ Resonance position at 53.5 ppm selected to represent the α-carbon resonances. $T_1 \approx 75$ msec; + 3,5-ring carbon atoms of tyrosine at 117.0 ptm. $T_1 \approx 75$ msec; * Resonance position at 130.5 ppm selected to represent the protonated ring carbons of phenylalanine, and the 2,6-ring carbons of tyrosine. $T_1 \approx 85$ msec; ○ C^{ζ} of Arg at 158.2 ppm. $T_1 \approx 500$ msec.

230 msec, respectively, are observed for the protonated terminal carbon atoms of the side chains of isoleucine and lysine, whereas $T_1 \approx 75$ to 85 msec is typical for the bulk of the α-carbons, and the protonated aromatic ring carbon atoms of tyrosine and phenylalanine.

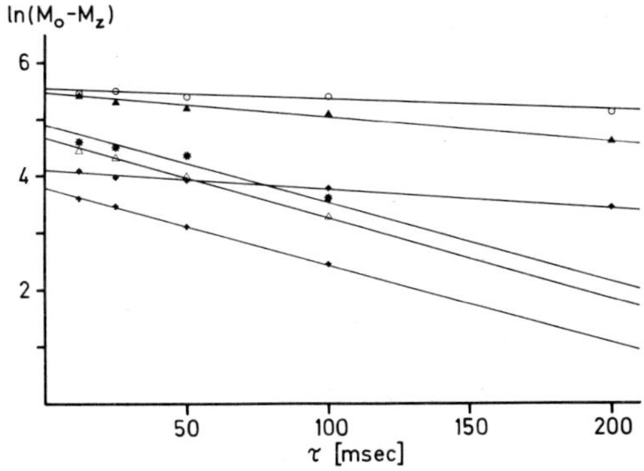

Fig. V-26. Experimental signal intensities in the partially relaxed FT ^{13}C-NMR spectra of BPTI (Fig. V-25). The values of T_1 obtained from a computer fit of these data to Equation V-10 are given in the figure caption V-25. The curves represent Equation V-10, with the slope = $-1/T_1$ corresponding approximately to the T_1's obtained from the least squares fit.

Correlation Times for the Rotational Motions of L*-Alanine*

The ^{13}C relaxation of the α-carbon atom in L-alanine comes essentially entirely from dipole–dipole coupling with the directly attached proton. Furthermore, it would appear from the molecular structure that the rotational motions of the relaxation vector $\boldsymbol{r}_{IS}$ linking the $^{13}C^{\alpha}$ atom with the C^{α}-proton are restricted to the overall rotations of the molecule. Hence in Equation V-5, we have for the α-carbon atom that $\tau_c \approx \tau_R$ (Equation IV-31). The reciprocal of the function V-5, representing the dependence of T_1 on τ_c, is plotted in Fig. V-27 for three different fields H_0. The two correlation times which correspond to the T_1 value observed for C^{α}, i.e. $T_1 = 3.5$ sec at 23.5 KG are $2 \cdot 10^{-11}$ sec, and $1 \cdot 10^{-6}$ sec. We know from experience that only the shorter one of the two solutions represents a reasonable description for alanine. Let us nevertheless check how this intuitive choice can be substantiated with the measurements of the spin–spin relaxation time T_2, and the NOE. The function V-6, representing the dependence on τ_c of the full line width at half height in Hz, $1/\pi T_2$ (Fig. I-2), is plotted in Fig. V-28. The line widths corresponding to $\tau_R = 2 \cdot 10^{-11}$ sec and $1 \cdot 10^{-7}$ sec are, respectively, ≈ 0.1 Hz and > 100 Hz. Inspection of Fig. V-4 then clearly settles the issue, since the line width of the C^{α} resonance is less than 1 Hz. Similarly, the NOE would be 1.15 for $\tau_R = 1 \cdot 10^{-7}$ sec, and 2.98 for $\tau_R = 2 \cdot 10^{-11}$ sec (Equation V-7 and Fig. V-15). Obviously, a sizeable NOE is observed in D_2O solutions of L-alanine (Fig. V-1), which gives yet another independent clue from which to decide that $\tau_R \approx 2 \cdot 10^{-11}$ sec.

Let us now examine the relaxation times T_1 of the nuclei in the methyl group on the basis of this rotational correlation time. From the geometry of

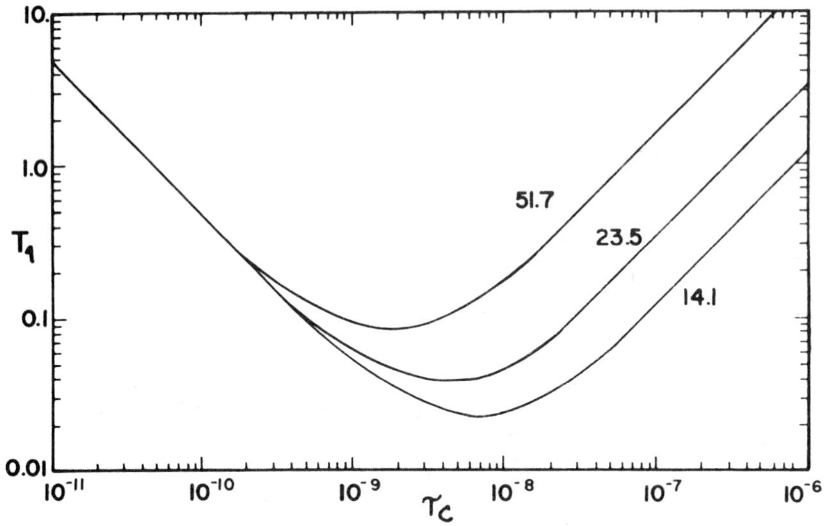

Fig. V-27. Log-log plot of T_1 vs. τ_c (both in seconds) for a ^{13}C spin relaxing by a dipolar interaction with a single covalently bonded proton, i.e. $|\boldsymbol{r}_{IS}| = 1.09$ Å (Equation V-5). In the computation of these curves, which are given for the three field strength $H_0 = 14.1$, 23.5, and 51.7 KG, it was assumed that the rotational reorientation of the relaxation vector $\boldsymbol{r}_{IS}$ (Fig. IV-11) is isotropic and can be described by one correlation time τ_c, and that one has complete proton decoupling (from Doddrell et al.[47]).

the methyl group, $r_i = 1.8$ Å for each pair of protons, and each individual proton is relaxed by the two neighboring protons. First, let us assume that the rotational motions of the methyl protons are restricted to the τ_R rotations of the entire molecule. The form of Equation IV-40 for this limiting case corresponds to the functions in Fig. IV-12, except that the T_1-values have to be divided by two, to account for the presence of two nearby protons. Thus, T_1 for the methyl protons of L-alanine is found to be 1.3 sec. Second, if the methyl group is assumed to spin rapidly about the C^α—C^β bond, with $\tau_{Me} \ll \tau_R$, one finds from Equation IV-40 that $T_1 \approx 5.2$ sec. Of these two models, the first one agrees quite closely with the experimental value $T_1 = 2.0$ sec, indicating that the intramolecular rotation of the methyl group about the C^α—C^β-bond in L-alanine is quite severely restricted, with $\tau_{Me} > 10^{-11}$ sec. This may independently be checked by an analogous consideration of the $^{13}C^\beta$-relaxation. With the effective correlation time $\tau_c = \tau_R$, and taking into account that $^{13}C^\beta$ is relaxed by 3 protons (Equation V-5), one finds that T_1 for $^{13}C^\beta$ should be $\frac{1}{3}T_1$ for $^{13}C^\alpha$, i.e. 1.2 sec. For rapid spinning of the methyl group, i.e. $\tau_{Me} \ll \tau_R$, the effective correlation time in Equation V-5 becomes $\approx \frac{1}{9}\tau_R$[12], and hence T_1 for $^{13}C^\beta$ would be ≈ 10 sec. Again, the experimentally observed value of $T_1 = 1.9$ sec implies that the intramolecular rotation of the methyl group is hindered by a sizeable energy barrier.

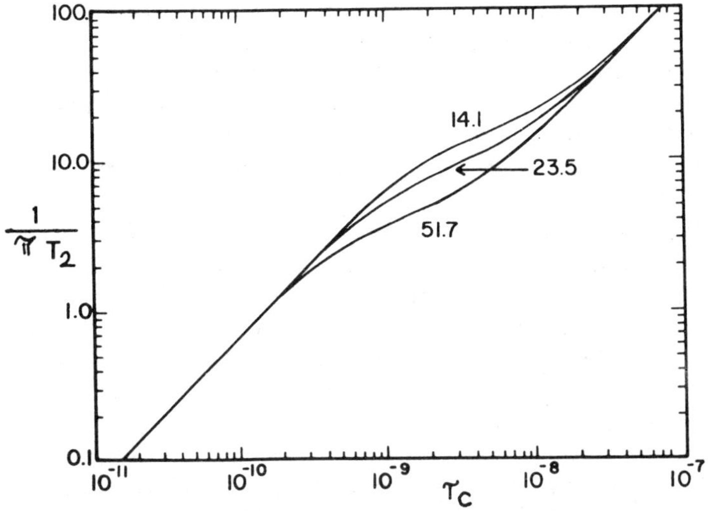

Fig. V-28. Log-log plot of $1/\pi T_2$, which is the full width in Hz at half height of the resonance line (Fig. I-2), vs. the correlation time τ_c (in sec) for isotropic rotational tumbling of the relaxation vector r_{IS}. The curves were computed with Equation V-6 for a identical system as that considered in Fig. V-27, and are given for $H_0 = 14.1$, 23.5, and 51.7 KG (from Doddrell et al.[47]).

Rotational Correlation Times for BPTI

The ^{1}H-NMR studies of BPTI had shown that for four aromatic rings of phenylalanine and tyrosine the correlation times for rotational motions about the C^β—C^γ bond are longer than 10^{-3} sec at 35° (Section III.4, Fig. III-35)[82]. This information combined with an inspection of the molecular model built from the X-ray data, indicates that for all the eight aromatic rings in BPTI, i.e. 4 Phe and 4 Tyr, the intramolecular rotations should be rather slow, with correlation times $\gg 10^{-8}$ sec. Therefore, a reliable value for τ_R of BPTI may be obtained from T_1 of the aromatic carbon atoms. The two correlation times corresponding to the measured value of $T_1 \approx 80$ msec are 7×10^{-10} sec, and 2×10^{-8} sec (Fig. V-27). The respective line widths are ≈ 5 Hz, and ≈ 30 Hz (Fig. V-28). Considering the quite dramatic line broadening observed when going from the denatured form to the globular form of BPTI (Fig. V-21), one obtains an indication that $\tau_R \approx 2 \times 10^{-8}$ sec gives a more realistic description of the BPTI solution. This is then confirmed by the experimental observation that the NOE for the aromatic carbons is of the order 1.1 (Fig. V-29). This is what one obtains from Fig. V-15 for a rotational correlation time of 2×10^{-8} sec, whereas at 23.5 KG $\tau_R = 7 \times 10^{-10}$ sec would correspond to a NOE of ≈ 2.7. Considering the size of the BPTI molecule ($M = 6000$), the correlation time τ_R is rather long, which is probably a consequence of the high protein concentration used for these experiments (Fig. V-25).

Of the other resonances of protonated carbons which were analyzed in the spectra of Fig. V-25, those of the C^α-carbons show a very similar behavior to that of the aromatic carbons. This indicates that for the bulk of the polypeptide backbone, the rotational motions are essentially restricted to the overall rotational tumbling of the protein molecule. On the other hand, appreciable freedom for intramolecular segmental motions is evidenced in the relaxation times of the terminal side chain carbon atoms of lysine and isoleucine. In the following, we assume that the rotational motions for these carbon atoms are isotropic, and can be described by a single effective correlation time τ_c. Furthermore, we take into account that these carbon atoms carry two and three hydrogen atoms, respectively. For C^ϵ of lysine, the correlation times corresponding to $NT_1 = 460$ msec (Fig. V-27) are 1×10^{-10} sec, and 2×10^{-7} sec. Since for $\tau_c = 2 \times 10^{-7}$ sec, the corresponding line width would be > 100 Hz (Fig. V-28), and the NOE = 1.1 (Fig. V-15), this correlation time can be excluded from further consideration (Figs. V-25 and

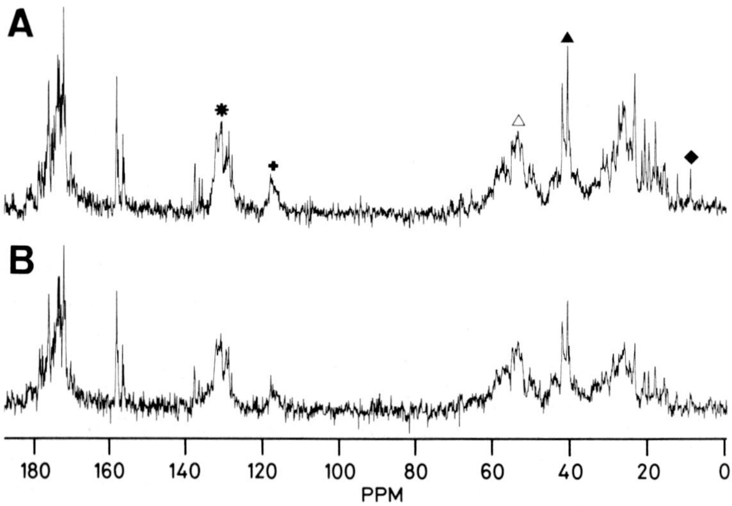

Fig. V-29. FT ^{13}C-NMR spectra at 25.16 MHz of a 0.03-M solution of BPTI in D_2O, pD = 4.7, $T = 35°$. A. Proton noise-decoupled spectrum. B. Spectrum recorded with gated proton double resonance irradiation, i.e. the H_2-field was turned on only during the time interval used for signal acquisition. In this experiment, one obtains full ^{1}H—^{13}C decoupling. The NOE, on the other hand, is eliminated, since it would only arise if the ^{1}H double resonance irradiation were maintained for a time period $\gtrsim T_1$ of ^{13}C.
The following are values for the NOE in spectrum A, which were estimated from comparing the spectra A and B. ◆ C^δ of Ile, NOE $\not\approx 2.3$; ▲ C^ϵ of Lys, NOE ≈ 1.6; △ C^α carbon atoms, NOE ≈ 1.2; + C3,5 of Tyr, NOE ≈ 1.1; $*$ C2-6 of Phe and C2,6 of Tyr, NOE ≈ 1.2.

V-29), and hence we have that $\tau_c \approx 1 \times 10^{-10}$ sec for C^{ϵ} of lysine. In an analogous way, $\tau_c \approx 6 \times 10^{-11}$ sec is found for C^{δ} of isoleucine. Overall, a comparison with the data on other polypeptides and proteins presented in Section V.5 shows that the situation encountered with BPTI is quite representative for globular proteins.

CHAPTER VI

PARAMAGNETIC CENTERS IN PROTEINS

So far we have focused our attention on diamagnetic amino acids, peptides and proteins, i.e. molecules which contain no unpaired electrons. While diamagnetic materials constitute the bulk of matter in biological systems, many biological processes involve also paramagnetic molecules. In paramagnetic metalloproteins the unpaired electron is usually located on or near the metal ions. There are also short lived organic free radicals which appear as intermediate reaction products, e.g. in photosynthesis, radiation damage, biological aging, and have also been postulated in the early stages of cancerous growth. Among the principal techniques for investigations of paramagnetic centers are electron paramagnetic resonance (EPR) spectroscopy, NMR, and measurements of the macroscopic paramagnetic susceptibility. Adhering to the main theme of this book, we shall in the following mainly concentrate on the applications of NMR experiments. Because of the dominant magnetic moment of the unpaired electron, which exceeds the proton magnetic moment about 650-fold, the electron-nucleus interactions cause extensive variations of the NMR chemical shifts and relaxation times in paramagnetic molecules, which opens new avenues for NMR investigations of the molecular structures. In addition to their use with the rather limited selection of natural paramagnetic species, these techniques have become applicable to a wide variety of diamagnetic biological systems, which are being labelled with artificial paramagnetic probes prior to the NMR measurements. Prominent among these technical aids are the nitroxide "spin labels", and transition metal or rare earth complexes as "NMR shift probes", and "NMR relaxation probes".

The experiments described in the present chapter include typical high resolution NMR studies of natural paramagnetic proteins, and the use of paramagnetic probes in high resolution NMR investigations of diamagnetic peptides and proteins. As in the foregoing Chapters II, III, and V, the data thus obtained will bear mainly on the molecular conformations of the molecules studied, and in addition also on the electronic structures of natural paramagnetic centers in metalloproteins. The latter aspect is treated in some detail in the Appendix VI. In Chapter IX, additional applications of paramagnetic NMR probes will be described which bear mainly on structural and kinetic features of intermolecular interactions in multicomponent biological systems, and include primarily investigations of intermolecular NMR relaxation enhancement effects (Section I.3).

The subject of the present chapter has been treated in numerous review articles and monographs. These include surveys of experimental and theoretical aspects of NMR studies in paramagnetic molecules in general[1–3], of the biological applications of nitroxide spin labels[4,5], and of NMR investigations of natural and synthetically labelled biopolymers[6–8]. A thorough discussion of the EPR spectra of the paramagnetic centers is beyond the scope of this book, but frequent reference to EPR data will be made throughout this chapter. For further reference on EPR spectroscopy, many excellent reviews and textbooks are available, e.g. Refs. 4, 5, 9–11.

1. ELECTRON PARAMAGNETISM

Electron Paramagnetic Resonance

Electrons have a spin $S = 1/2$ (Table VI-1), and hence for a system of one unpaired electron we encounter a situation which is qualitatively very similar to that of a nuclear spin $I = 1/2$ (Section I.2). In the absence of an

TABLE VI-1

Magnetic properties of the unpaired electron

Electron spin	$S = \frac{1}{2}$
Bohr magneton $\left(=\frac{eh}{4\pi mc}\right)$	$\beta = 0.92731 \times 10^{-20}$ erg · sec^{-1} · Gauss^{-1}
g-value of free electron	$g_e = 2.0023$
Gyromagnetic ratio of free electron	$\gamma_e = 1.76 \times 10^7$ rad · sec^{-1} · Gauss^{-1}
Electron–proton moment ratio	$\frac{\gamma_e}{\gamma_H} = 658.23$

external magnetic field, an equal number of electrons are in the $m_s = 1/2$ and $m_s = -1/2$ spin states. When the sample is exposed to a static external magnetic field H_0, the energetic degeneracy of the two spin states is lifted (Fig. VI-1), and their population is then given according to the Boltzmann distribution law by

$$\frac{n_\alpha}{n_\beta} = e^{-\Delta E/kT} \qquad \text{(VI-1)}$$

Compared to the nuclear spin systems, the energy difference ΔE at a given field H_0 is approximately 10^3 times larger for the electron spin, and correspondingly the sensitivity for observation of electron paramagnetic resonance (EPR) signals is higher than for NMR. The resonance condition is

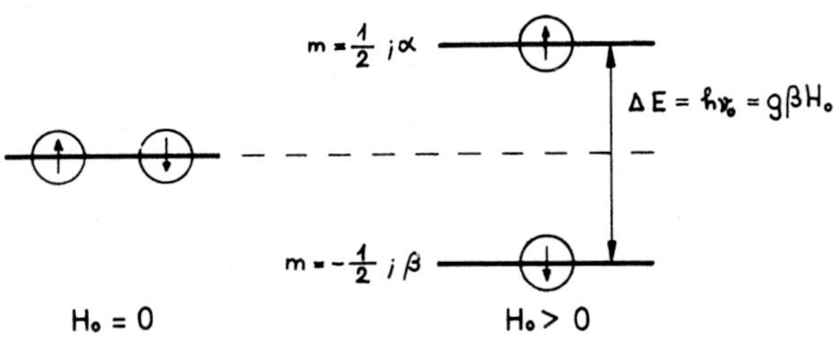

Fig. VI-1. Magnetic energy levels for an electron spin $S = 1/2$ with and without application of an external magnetic field.

usually written as

$$h\nu = g\beta H_0 \tag{VI-2}$$

where the resonance frequency ν is in the microwave region, β is the Bohr magneton, and g is the "g-factor".* $g = h\nu/\beta H_0$ is a characteristic quantity for a given molecule, and depends on the molecular structure. In an analogous way as the chemical shifts in NMR spectroscopy, the g-factors are used to indicate the positions of the individual signals in the EPR spectra. If the unpaired electron is located in a molecule which also contains nuclear spins, electron–nuclear spin–spin coupling can produce a so-called "hyperfine structure" of the EPR signal, which is also an important source of structural information. Through spin–spin coupling with a nuclear spin I, an EPR signal is typically split into $(2I+1)$ lines, in analogy to the interactions causing the multiplet patterns in NMR (Section II.2). This is illustrated in Fig. VI-2d with the EPR spectrum of di-t-butyl nitroxide.

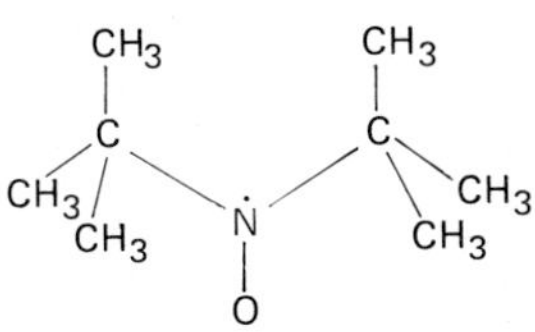

(VI-3)

In this molecule, which is the parent compound for a wide variety of nitroxide spin labels[4,5], the unpaired electron is predominantly located in the N-O group. Therefore the EPR signal shows a three-line hyperfine structure which arises from the interactions with the ^{14}N ($I = 1$) nucleus. The isotropic EPR spectrum, which is typical for solutions of low molecular weight nitroxide spin labels, can be described by the spin Hamiltonian

$$\mathcal{H}_s = g\beta S_z H_0 + ha\boldsymbol{S}\cdot\boldsymbol{I} \tag{VI-4}$$

---

*The resonance condition for a nuclear spin can be written correspondingly as $h\nu = g_N\beta_N H_0$, where β_N is the nuclear magneton, and g_N the g-factor for the nucleus.

Here, g is the isotropic electronic g-factor, β the Bohr magneton, S_z the component of the electronic spin operator along the external magnetic field $\boldsymbol{H}_0$, a the isotropic hyperfine coupling constant, and $\boldsymbol{S}$ and $\boldsymbol{I}$ are the electron and nuclear spin operator, respectively. The EPR parameters for di-t-butyl nitroxide are given in Table VI-2. It is seen that the g-factor is only slightly different from the free electron value (Table VI-1). This is quite typical for organic free radicals, whereas largely different g-factors can be observed for paramagnetic metal ions[9–11].

TABLE VI-2

EPR parameters for di-t-butyl nitroxide[5]

Single crystal spectrum	$g_x = 2.0089$
	$g_y = 2.0061$
	$g_z = 2.0027$
	$\|A_x\| \approx 14\text{–}17$ MHz
	$\|A_y\| \approx 14\text{–}17$ MHz
	$\|A_z\| = 87$ MHz
Solution spectrum	$g = 2.0060$
	$\|a\| = 41$ MHz

Anisotropic Effects in the EPR Spectra

It is a general feature of the EPR parameters that they are anisotropic in the sense that they vary with the orientation of the paramagnetic molecule in the static magnetic field H_0. This is illustrated in Fig. VI-2 with the single crystal spectra of di-t-butyl nitroxide. In the single crystal, the radicals are held in a fixed orientation with respect to the external magnetic field as described in the figure caption VI-2, and it is readily apparent that both the center of the spectrum, and the extent of the hyperfine splitting depend on the orientation of the crystal. The anisotropic EPR spectrum can be described by the spin Hamiltonian

$$\mathcal{H}_s = \beta \boldsymbol{S} \cdot \tilde{g} \cdot \boldsymbol{H} + h\boldsymbol{S} \cdot \tilde{A} \cdot \boldsymbol{I} \qquad \text{(VI-5)}$$

where $\tilde{g}$ is the g-tensor, and $\tilde{A}$ the hyperfine tensor that includes both the anisotropic dipolar interaction and the isotropic Fermi contact interaction between the electronic and nuclear spins. The principal tensor elements for di-t-butyl nitroxide are presented in Table VI-2. More pronounced anisotropic effects are observed for paramagnetic metal ions. As we will see throughout this chapter, the anisotropy of the EPR parameters can greatly affect the NMR spectra of paramagnetic molecules. Fig. VI-2 and Table VI-2 show further that the parameters of the solution spectrum of di-t-butyl nitroxide lie between the extreme values for the single crystal. Because of

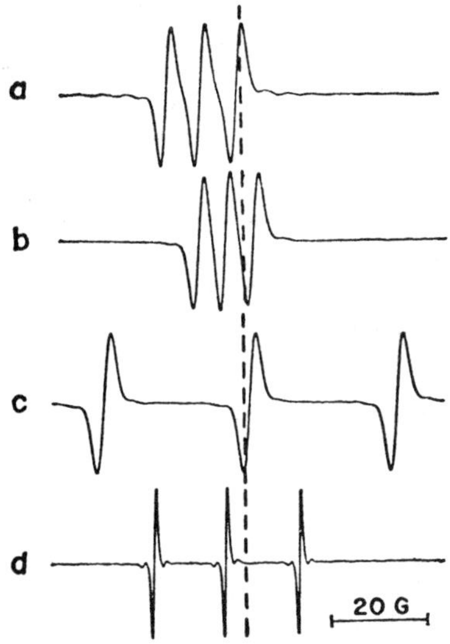

Fig. VI-2. X-band first-derivative EPR spectra of di-t-butyl nitroxide illustrating the anisotropic effects in EPR. *a*, *b*, and *c* are the single crystal spectra obtained from a tetramethyl-1,3-cyclo-butanedione host crystal doped with small amounts of the nitroxide radical. The host crystal orients the nitroxide radicals so that their respective molecular axes (the *x* axis is parallel to the N—O bond, the *z* axis parallel to the non-bonding nitrogen *p*-orbital, and the *y* axis perpendicular to the *xz* plane) are aligned in certain directions relative to the crystal axis. The crystal has been rotated to align the external magnetic field H_0 with the *x*, *y*, and *z* axes of the oriented radicals producing the spectra *a*, *b*, and *c*, respectively. The principal values of the tensors $\tilde{g}$ and $\tilde{A}$ are given in Table VI-2. Note that the hyperfine interactions of 87 MHz, and 14–17 MHz yield multiplet line separations in a field swept spectrum of 32 *G*, and 6 ± 1.5 *G*, respectively. Spectrum *d* was obtained from a 10^{-5}-M solution of di-t-butyl nitroxide in di-t-butyl ketone. The vertical dotted line represents the reference position $g = 2.0036$ (from Griffith and Waggoner[5]).

the rapid isotropic rotational tumbling in the solution, the parameters of the solution spectrum correspond simply to the average of the single crystal data, with $g = 1/3\ (g_x + g_y + g_z)$, and $a = 1/3\ (A_x + A_y + A_z)$.

Electron Spin Relaxation

During an EPR observation, the thermal populations of the electronic spin states (Fig. VI-1) are perturbed. In analogy to the formalism introduced for NMR (Sections I.2 and IV.1), the relaxation processes which reestablish the equilibrium situation may be described by the longitudinal or spin-lattice electronic relaxation time T_{1e}, and the transverse or spin–spin relaxation time T_{2e}. The latter is related to the line width of the EPR signal in the same way as indicated in Fig. I-2 for the nuclear spin systems.

On the basis of the electron spin relaxation times, one can readily appraise whether the EPR signal of a given radical will be observable in solution, and what the principal impact of the unpaired electron on the NMR of the molecule will be. For systems with $T_{1e} \approx T_{2e} \lesssim 10^{-10}$ sec, the EPR lines will be so broad that the EPR signal cannot readily be detected, while on the other hand the NMR line widths in these species will be only little affected by the unpaired electron. For species with electron relaxation times $> 10^{-10}$ sec, the EPR spectrum in solution is usually not difficult to observe, while under these conditions the NMR lines are drastically broadened by the interactions with the electron spin (see Section VI.3). Representative values for T_{1e} in solutions of different paramagnetic species are given in Table VI-3.

TABLE VI-3

Electronic spin–lattice relaxation times T_{1e}*.

	T_{1e} (sec)
Organic free radicals (including nitroxide spin labels)	$> 1 \times 10^{-8}$
Mn^{2+}	$\approx 3 \times 10^{-9}$
Cu^{2+}	$\approx 2 \times 10^{-8}$
VO^{2+}	$\lesssim 1 \times 10^{-8}$
Ni^{2+}	$\lesssim 1 \times 10^{-11}$
Co^{2+}	$< 5 \times 10^{-12}$
Fe^{2+}	$< 1 \times 10^{-11}$
Fe^{3+} ($S = 5/2$)	$\lesssim 1 \times 10^{-10}$
Fe^{3+} ($S = 1/2$)	$\approx 2 \times 10^{-12}$
Gd^{3+}	$\approx 1 \times 10^{-10}$
Other lanthanide(III) ions	$\lesssim 1 \times 10^{-12}$

*Most of the values given in this table have been derived from NMR observations in aqueous solutions of the metal ions[12], those for Fe^{3+} from studies of hemoproteins[6]. While this table gives an indication of the T_{1e}'s to be expected under the conditions of a typical NMR experiment, T_{1e} for a given paramagnetic center can vary appreciably with the molecular environment, the magnetic field strength H_0, and the temperature.

Following the above qualitative criteria, the EPR signals in dilute solutions of organic free radicals, Mn^{2+}, Cu^{2+}, VO^{2+}, high spin Fe^{3+}, and Gd^{3+} will in general not be difficult to detect, and at the same time these paramagnetic centers are typical "NMR relaxation probes" (Sections VI.3 and VI.5). On the other hand, in solutions of Ni^{2+}, Co^{2+}, Fe^{2+}, low spin Fe^{3+}, and lanthanide

(III) ions other than Gd^{3+}, the EPR lines are very broad, and since these paramagnetic species have little impact on the NMR line widths, they represent typical "NMR shift reagents" (Sections VI.2, VI.4, and VI.5).

In view of the interpretation of the NMR spectra of paramagnetic molecules, it is important to note that even for systems with long electronic relaxation times, only the isotropic EPR parameters can be obtained from solution studies. Information on the anisotropic EPR features have to be derived from EPR experiments in single crystals or in frozen solution,* i.e. under conditions which are obviously different from those of the NMR experiments to be considered in this context. The electronic relaxation times in solids at low temperatures are in most cases much longer than those in solutions at ambient temperature, so that the EPR parameters can be determined for most of the species in Table VI-3.

2. NMR HYPERFINE SHIFTS IN PARAMAGNETIC MOLECULES

The electron–nuclear interactions in paramagnetic molecules can be manifested in extensive variations of the NMR chemical shifts. This is illustrated in Fig. VI-3, where the ^1H-NMR spectrum of the paramagnetic low spin Fe^{3+} complex with porphin is compared with that of the corresponding diamagnetic Zn^{2+} complex. In both spectra the protons in positions 1 to 8 (a), and the four mesoprotons (b) are equivalent, as one might have anticipated from the D_{4h} symmetry of the molecular geometry (Fig. VI-3). In Fe^{3+} (Porphin) $(CN)_2$, the resonance b is ca. 9 ppm, and the resonance a ca. 24 ppm to higher field than in Zn^{2+} (Porphin), and the resonance a is also markedly outside the spectral region for diamagnetic compounds (Table II-1). The chemical shifts in the paramagnetic molecule can be expressed as the sum of two terms

$$\delta_{obs} = \delta_d + \delta_{hf} \qquad \text{(VI-6)}$$

where δ_d is the chemical shift in the absence of the unpaired electron, and δ_{hf} the hyperfine shift arising from electron–nucleus interactions. If we assume in the example of Fig. VI-3 that the chemical shifts of Zn^{2+} (Porphin) are representative for a hypothetical diamagnetic Fe^{3+}-porphin complex, δ_{hf} can thus be obtained through Equation VI-6. We note further that the NMR line

*It might be added at this point that in analogy to EPR, anisotropic NMR parameters may be obtained from studies of solids. The shielding constant σ in Equation I-4 would then have to be replaced by a shielding tensor, and the occurrence of anisotropic spin–spin couplings is also typical for NMR spectra of solids.

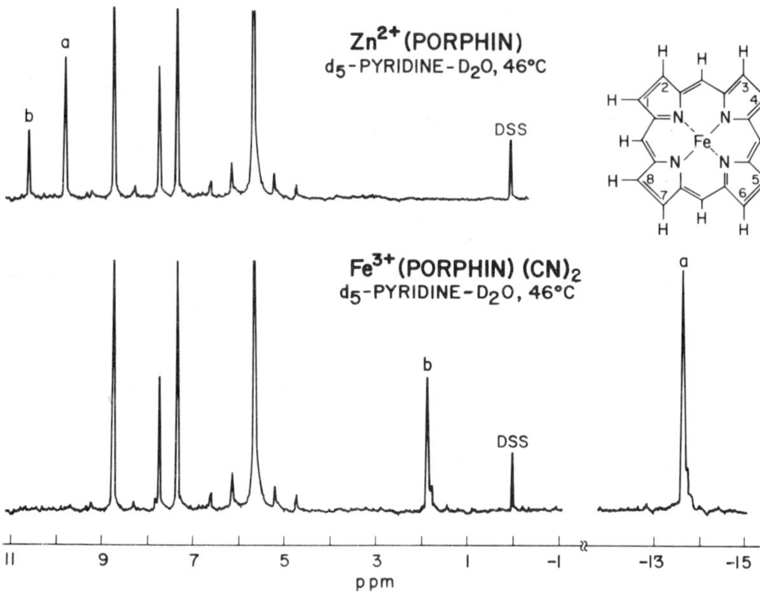

Fig. VI-3. ^{1}H-NMR spectra at 220 MHz of the paramagnetic cyanoferriporphin complex and the diamagnetic Zn^{2+}-porphin complex. The resonances *a* and *b* were identified from their relative intensities. DSS is the resonance line of the internal reference. The lines between 4.5 and 9.5 ppm correspond to the solvent resonances and their spinning side bands (from Wüthrich[6]).

widths are not markedly affected by the unpaired electron in Fe^{3+} (Porphin) $(CN)_2$ (see Section VI.3).

The term "hyperfine shift" implies that δ_{hf} and the hyperfine multiplet patterns in the EPR spectra (Fig. VI-2) have a common origin. The different manifestations of the electron–nucleus spin–spin interactions in the EPR and NMR spectra are a consequence of the dynamics of the spin systems. On the one hand, the nuclear spin relaxation is slow on the EPR time scale. Hence the electronic spins see a static and almost uniform distribution of the nuclear spins among the $(2I+1)$ spin states, and as a result the EPR signal is split into a field independent hyperfine multiplet with $(2I+1)$ lines. On the other hand, the electron spin relaxation is rapid on the NMR time scale. Therefore the doublet structure of the NMR lines which one would expect to arise from the interaction with the electron spin $S = 1/2$, collapses to an average resonance line (see Fig. II-10, where the exchange rate ν_e would be determined by T_{1e}). The hyperfine shift in ppm of the average resonance is

$$\delta_{hf} = \frac{a_N(n_\alpha - n_\beta)}{2(n_\alpha + n_\beta)\nu_0} \qquad \text{(VI-7)}$$

where n_α and n_β are the populations of the electronic spin states (Equation VI-1), a_N is the extent of doublet splitting of the NMR line by the electron in Hz, and ν_0 the NMR frequency. Equation VI-7 shows that δ_{hf} is proportional to the population difference between the electronic spin states. From Equation VI-1 and Fig. VI-1 one then has in the high temperature approximation[10], which applies to all the experiments discussed in this book, that δ_{hf} is proportional to both the external magnetic field H_0, and the reciprocal of the absolute temperature $1/T$.

Porphin (Fig. VI-3) is the parent compound of the natural heme groups, which form an integral part of the active sites in the hemoproteins[6]. In the following the NMR spectra of natural hemes will be used to further illustrate the hyperfine shift phenomenon. We shall then see that two types of electron–nucleus interactions can contribute to δ_{hf}, i.e. dipole–dipole coupling which produces the so-called "pseudocontact shifts" δ_{pc}[13], and Fermi contact coupling which produces the "contact shifts" δ_c[14].

^{1}H and ^{13}C-NMR Hyperfine Shifts in Natural Hemes

Protoheme IX, the complex formed between an iron ion and protoporphyrin IX, is the prosthetic group in hemoglobin, myoglobin, cytochromes of class b, and other hemoproteins. The structure of this complex is shown in Fig. VI-4. The iron ion is bound to the four nitrogen atoms of the planar conjugated porphyrin ring system. Three different types of side chains are attached to the ring carbon atoms 1–8, the so-called "β-carbon atoms": the ring methyls in positions 1, 3, 5, and 8, the vinyl groups 2 and 4, and the propionic acid side chains 6 and 7. All the four meso carbon atoms b have a hydrogen atom attached. In the following we are concerned with low spin ferriprotoheme IX (Fe^{3+}, $S = 1/2$) obtained when two cyanide ions are coordinated to the axial positions of the heme iron. This compound will be referred to as Fe^{3+}(PROTO)$(CN)_2$. In addition, low spin ferrideuteroheme IX, Fe^{3+}(DEUT)$(CN)_2$, which differs from protoheme IX in that the 2,4-vinyl groups are replaced by hydrogen atoms (Fig. VI-5), and the corresponding diamagnetic Zn^{2+} complexes, Zn^{2+}(PROTO) and Zn^{2+}(DEUT), will also be included in the discussion.

In the ^{1}H-NMR spectrum of Fe^{3+}(PROTO)$(CN)_2$ (Fig. VI-4), it is readily apparent that the resonances a, f, and g are outside the usual spectral region for diamagnetic molecules (Table II-1). Compared to the diamagnetic Zn^{2+}(PROTO) (Fig. VI-5), the ring methyl resonances a and the vinyl methine proton resonances f are shifted downfield by approximately 12 ppm, and 4 ppm, respectively, while the meso-protons b and the vinyl methylene protons g experience an upfield shift by the unpaired electron of approximately 10 ppm. In the spectra of Zn^{2+}(DEUT) and Fe^{3+}(DEUT)$(CN)_2$ (Figs.

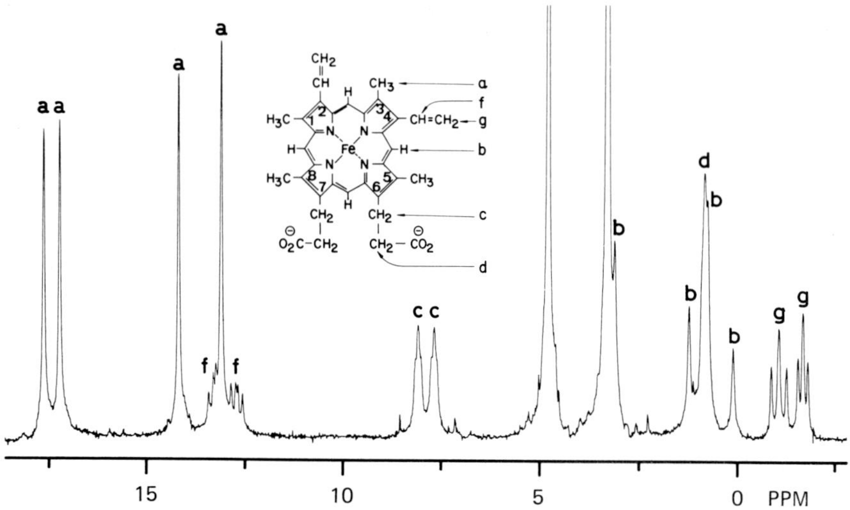

Fig. VI-4. Fourier transform ^{1}H-NMR spectrum at 100 MHz of Fe(III)(PROTO)$(CN)_2$ in CD_3OD. $T = 29°$. The structure of the complex, where the axial cyanide ligands have been omitted, is also indicated. The resonance assignments indicated by the letters *a* to *g* were derived from the resonance intensities, the fine structure arising from proton–proton spin–spin coupling, and in the case of the resonances *c* and *d* also from ^{1}H—^{13}C decoupling experiments (from Wüthrich and Baumann[15]).

VI-5 and VI-6), the chemical shifts of the resonances a, b, c, and d (and, where applicable, of the methylester lines e) are very similar to those in the corresponding protoporphyrin IX compounds. In the place of the resonances f and g of protoporphyrin IX, one has in deuteroporphyrin IX the two lines f^D of the 2,4-protons. For the ensuing discussions it will be of considerable interest that the resonances f^D are shifted to an extreme high field position by the unpaired electron.

In the ^{13}C-NMR spectrum of Zn^{2+}(PROTO) (Fig. VI-7), all the resonances fall within the usual spectral region for diamagnetic molecules (Table V-1). In the three ^{13}C-NMR spectra of the corresponding paramagnetic complex Fe^{3+}(PROTO)$(CN)_2$ (Figs. VI-8 and VI-9), all the 34 resonances expected from the molecular structure (Fig. VI-4) are observed. Those resonance lines which are hidden by the solvent in Fig. VI-8 can readily be detected in Fig. IV-9, and v.v. Comparing the spectra in Figs. VI-7 to VI-9, we note that in Fe^{3+}(PROTO)$(CN)_2$ there are high field hyperfine shifts of approximately 50 ppm for the methyl carbons a and the methylene carbons c, 65 ppm for the vinyl carbons f, 20 ppm for the β-carbons 1–8, and 110 ppm for the quaternary carbons q, whereas the resonances d and g are shifted downfield by 50 ppm, and 40 ppm, respectively, by the unpaired electron.

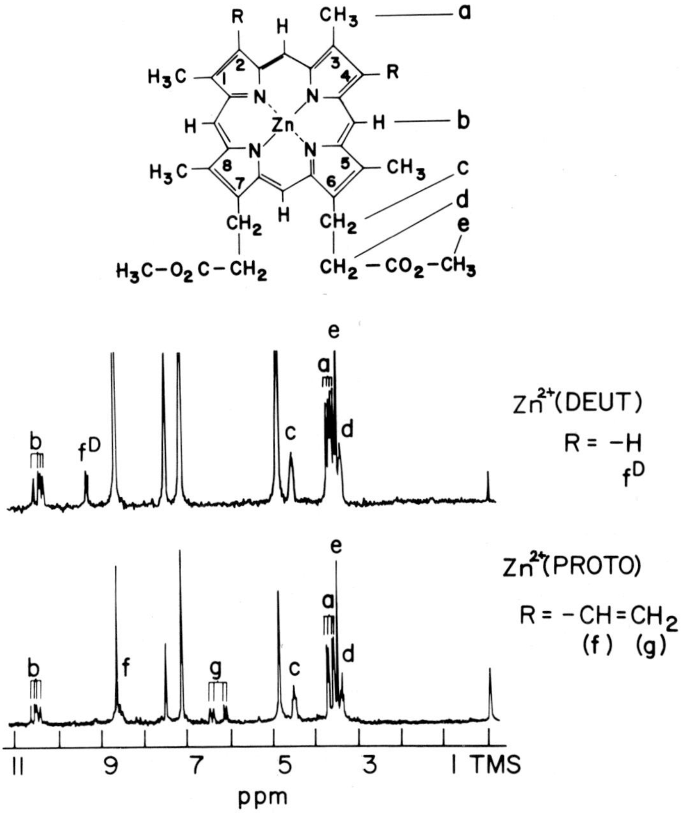

Fig. VI-5. ¹H-NMR spectra at 220 MHz of Zn^{2+}(PROTO) and Zn^{2+}(DEUT) in d_5-pyridine solution, $T = 25°$. The resonance assignments indicated by the letters a to g are based on the relative intensities and the fine structure multiplets from proton–proton spin–spin coupling[16]. The strong resonances between 7 and 9 ppm come from d_5-pyridine, and the line at 4.9 ppm from a trace of water (from Wüthrich[6]).

Analogous to the situation encountered in the ¹H-NMR spectra, the ¹³C-NMR of Zn^{2+}(DEUT) and Fe^{3+}(DEUT)$(CN)_2$ are overall very similar to those of the corresponding protoporphyrin IX complexes. Again, the 2,4-carbons are of particular interest. The hyperfine shifts for these two lines were found to be -32.8 ppm and -22.8 ppm, respectively[15], which is of the same order as the hyperfine shifts of the carbon atoms 1–8 in Fe^{3+}(PROTO)$(CN)_2$.

The ¹H and ¹³C hyperfine shifts for the symmetrical compound Fe^{3+}(Porphin)$(CN)_2$[20] (Fig. VI-1) are compiled in Table VI-4, those for Fe^{3+}(PROTO)$(CN)_2$ and the 2,4-positions of Fe^{3+}(DEUT)$(CN)_2$[15] in Table VI-5. The following are two general features in these tables:

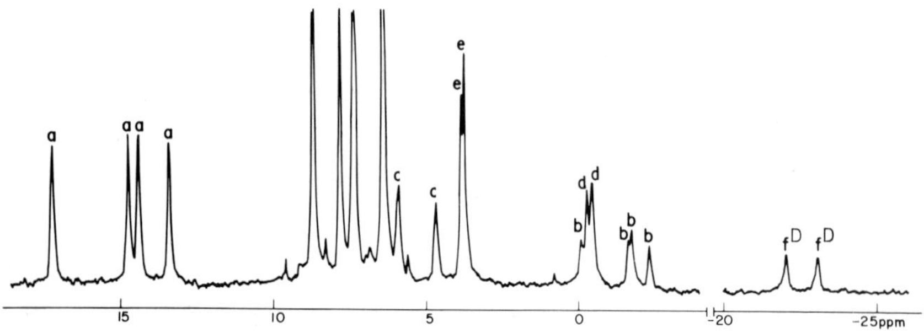

Fig. VI-6. ^{1}H-NMR spectrum at 220 MHz of the dimethylester of Fe^{3+}(DEUT)$(CN)_2$ in a mixed solvent of d_5-pyridine and D_2O 4:1, $T = -9°$. The resonance assignments by the letters *a* to *f* refer to the structure of the corresponding Zn^{2+} compound in Fig. VI-5. The unmarked lines between 5 and 10 ppm correspond to the resonances of d_5-pyridine and D_2O, and their spinning side bands (from Wüthrich, Shulman, Wyluda and Caughey[17]).

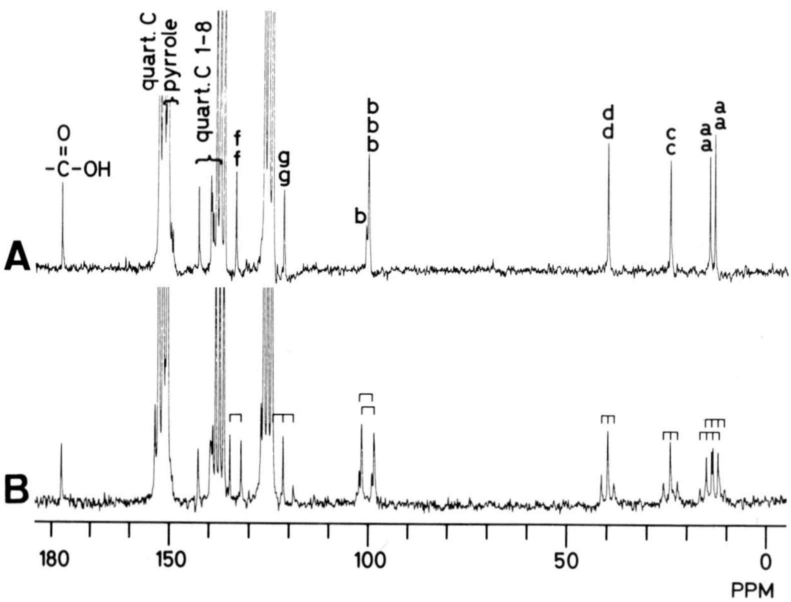

Fig. VI-7. FT ^{13}C-NMR spectra at 25.16 MHz of a ca. 0.01-M solution of Zn^{2+}(PROTO) in d_5-pyridine, $T = 29°$. A. proton noise decoupled, B. proton off-resonance irradiation at -5 ppm (see Fig. VI-5). The resonance assignments refer to the molecular structure in Fig. VI-5, with the only difference that the free acid rather than the dimethylester had been used here. From the off-resonance proton double resonance irradiation experiment B, and with the empirical additivity rules V-1, the resonance assignments were quite straightforward. They had also been ascertained by comparison with Zn^{2+}(DEUT)[15], and with the metal free porphyrins[18]. The solvent resonances are at 124, 137, and 150 ppm (from Wüthrich and Baumann[15]).

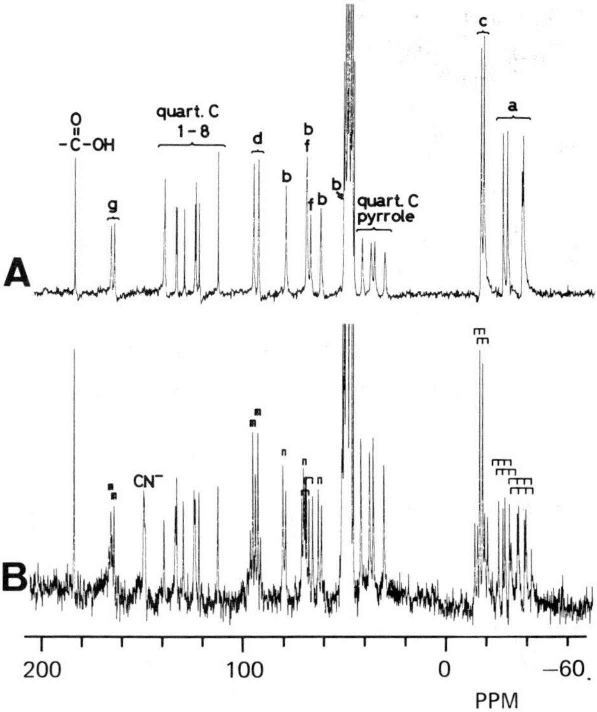

Fig. VI-8. FT ^{13}C-NMR spectra at 25.16 MHz of Fe(III)(PROTO)$(CN)_2$ in CD_3OD, $T = 29°$. The strong resonance at 49 ppm comes from the solvent. A. proton noise decoupled. B. off-resonance proton irradiation at −5 ppm. Because the ^{1}H-NMR lines are nicely separated (Fig. VI-4), all the resonance identifications indicated in the figure could be derived from the residual ^{13}C—^{1}H spin–spin couplings in spectrum B. In selective ^{1}H—^{13}C double resonance experiments, the corresponding ring methyl ^{1}H and ^{13}C lines were also identified[19] (from Wüthrich and Baumann[15]).

(i) There is no simple relation between the distance from the heme iron to the nuclei, and the sign and extent of the hyperfine shifts.

(ii) The sign of the hyperfine shifts can be different for nuclei in nearly identical locations. For example, the resonances of the protons attached to the β-carbons -2 and -4 in $Fe^{3+}(DEUT)(CN)_2$, and all the ring methyl ^{13}C resonances are shifted upfield, whereas the ring methyl ^{1}H-NMR lines are shifted downfield by the unpaired electron.

As will be discussed in the following, these qualitative features of the hyperfine shifts are indicative of a situation where the observed hyperfine shifts contain sizeable contributions from Fermi contact coupling.

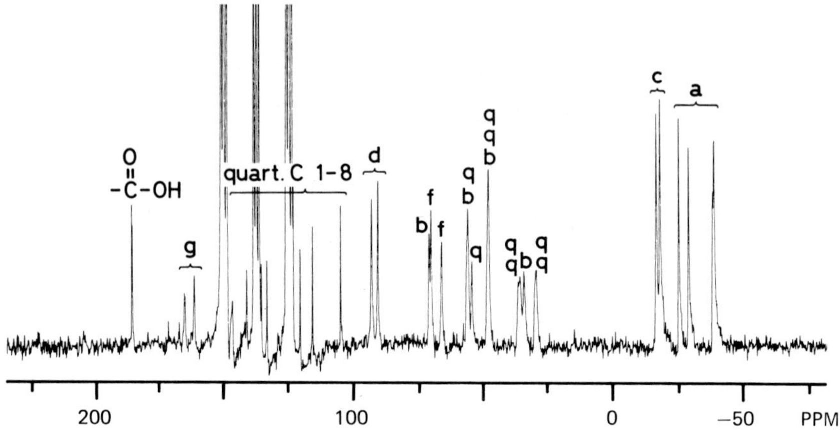

Fig. VI-9. ^{1}H noise decoupled FT ^{13}C-NMR spectrum of Fe(III)(PROTO)$(CN)_2$ in a mixed solvent of d_5-pyridine and D_2O 4:1, $T = 29°$. The solvent resonances are at 124, 137, and 150 ppm. The resonance assignments are in the nomenclature of Fig. VI-4, and *q* stands for the quaternary carbons next to the pyrrole nitrogen atoms (from Wüthrich and Baumann[15]).

TABLE VI-4

^{1}H and ^{13}C-NMR hyperfine shifts in Fe^{3+}(Porphin)$(CN)_2$ in a mixed solvent of d_5-pyridine and D_2O 4:1, $T = 34°$*.

Atom position (Fig. VI-3)	δ^H_{hf}(ppm)	δ^C_{hf}(ppm)
1–8 (a)	−24.6	−40.6
Meso- (b)	−9.3	−49.7
Quaternary carbons		−101.4

*Positive numbers indicate shifts to lower field at fixed frequency (from Ref. 20).

Pseudocontact Shifts

Pseudocontact shifts result from the dipolar interactions between electron and nuclear spins. They can only arise in systems where the magnetic susceptibility of the paramagnetic center is anisotropic, and hence the electron–nucleus dipole–dipole coupling does not average to zero under the rotational molecular tumbling in solution. As a consequence of the small anisotropy of the *g*-factor (Table VI-2), pseudocontact shifts are as a rule negligibly small in organic free radicals[21]. On the other hand they can present a sizeable, or even dominant contribution to the hyperfine shifts in certain transition metal compounds and, as discussed in Section VI.5, also in rare earth compounds.

TABLE VI-5

^1H and ^{13}C-NMR hyperfine shifts in Fe^{3+}(PROTO)$(CN)_2$ in a mixed solvent d_5-pyridine—D_2O 4:1, $T = 29°$†.

Atom position (Fig. VI-4)		δ^H_{hf}	δ^C_{hf}
Ring methyls	(a)	13.3 } *	* { −52.7
		12.8 }	{ −52.3
		9.1 } *	* { −43.1
		7.4 }	{ −39.4
Propionates	(c)	2.6	−42.4
		2.3	−40.9
	(d)	−2.9	51.0
		−2.9	53.8
Vinyls	(f)	3.2	−66.8
		2.4	−62.7
	(g)	−8.5 }	{ 40.3
		−8.2 }	{ 43.9
		−8.1 }	
		−7.6 }	
β-Carbons	(1–8)		−35
			−25
			−20
			−7
			−4
			1
			6
			...
Quaternary carbons next to the pyrrole nitrogens (*q* in Fig. VI-8)	(q)		−120
			−120
			−114
			−113
			−102
			−102
			−96
			−94
Meso-	(b)	−11.2	−65.5
		−10.2	−51.9
		−9.7	−43.9
		−7.1	−28.7
Fe^{3+}(DEUT)$(CN)_2$ (Fig. VI-5)			
2,4β-positions (f^D)		−25.5	−32.8
		−25.3	−22.8

†Positive numbers indicate shifts to lower field at fixed frequency. The correspondence between ^1H and ^{13}C-NMR of the ring methyls indicated in the table (*) was established by selective ^1H—^{13}C spin decoupling[19]. In the bottom line, the data for the 2,4-ring positions of Fe^{3+}(DEUT)$(CN)_2$ are also given (from Refs. 6 and 15).

Quantitative expressions describing the sign and the magnitude of the pseudocontact shifts in transition metal compounds have been derived by different authors[22–24]. They have the general form

$$\delta_{pc} = \frac{\beta^2 S(S+1)}{3kT} F\left(\tilde{g}, \frac{\Omega, \theta}{r^3}\right) \qquad \text{(VI-8)}$$

β is the Bohr magneton, S the total electronic spin, k the Boltzmann constant, T the absolute temperature, and F is an algebraic function of the g-tensor anisotropy, and the geometrical factors Ω, θ and r. r is the distance from the electron spin to the nuclear spin, Ω the polar angle of r, and θ the angle between the projection of r in the xy plane and the y axis (Fig. VI-10). For the case of rhombic symmetry, Equation VI-8 takes the form[24]

$$\delta_{pc} = \frac{\beta^2 S(S+1)}{9kTr^3} \left\{\left[g_z^2 - \frac{1}{2}(g_x^2 + g_y^2)\right](3\cos^2\Omega - 1) + \frac{3}{2}(g_y^2 - g_x^2)\sin^2\Omega\cos 2\theta\right\} \qquad \text{(VI-9)}$$

which reduces to the form of Equation VI-10 for axial symmetry, i.e. $g_z \equiv g_{\parallel}$, $g_x = g_y \equiv g_{\perp}$

$$\delta_{pc} = \frac{\beta^2 S(S+1)}{9kTr^3}(g_{\parallel}^2 - g_{\perp}^2)(3\cos^2\Omega - 1) \qquad \text{(VI-10)}$$

In deriving Equations VI-9 and VI-10[22–24], it was assumed that the system contained only one thermally populated energy level in the absence of the magnetic field H_0, and that T_{1e} is very short, which is the most interesting situation for high resolution NMR studies of hyperfine shifts. Experience indicates that Equations VI-9 and VI-10 provide an adequate description for many transition metal compounds.

It is readily apparent from Equations VI-9 and VI-10 that the pseudocontact shifts for all the nuclei in a paramagnetic molecule can be computed if the molecular geometry, the principal g-values, and the orientation of the g-tensor with respect to the molecular structure are known. The potential of the pseudocontact shifts for investigations of the molecular conformations lies in their intimate relation with r, Ω, and θ. In principle, if the g-tensor in a paramagnetic molecule is known from EPR experiments, the molecular structure can in favorable cases be determined solely on the basis of the pseudocontact shifts.

Let us now briefly return to the low spin ferric porphyrin complexes, and in particular to the symmetrical porphin complex (Fig. VI-3). Low temperature EPR studies showed that if the z-axis is perpendicular to the porphyrin ring (Fig. VI-10), one has that $g_z > g_x$ or g_y. For solutions of Fe^{3+}(Porphin)$(CN)_2$ at ambient temperature, the NMR spectra show further that the g-tensor has axial symmetry. Application of Equation VI-10 with

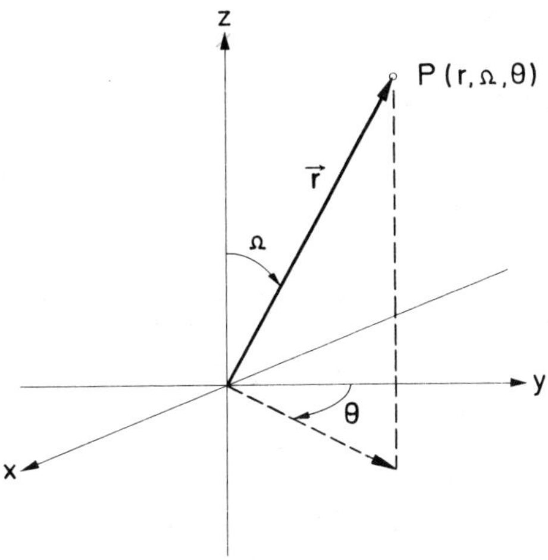

Fig. VI-10. Definitions of the geometrical factors used in the description of the pseudocontact shifts. The carthesian coordinates indicate the orientation in space of the principal axes of the g-tensor, and the origin coincides with the location of the electron spin. The position P of the nucleus considered is determined by the distance r from the electron spin, and the angles Ω and θ.

$g_{\parallel} > g_{\perp}$ then shows that all the nuclei in the porphyrin ring plane experience upfield, i.e. negative pseudocontact shifts, while positive pseudocontact shifts would be expected for nuclei lying within a cone with an apex angle of 54.7° about the z-axis. In $Fe^{3+}(Porphin)(CN)_2$, $\Omega = 90°$ for all the nuclei of porphin, and hence from Equation VI-10 the relative pseudocontact shifts for any two nuclei i and j are related by

$$\frac{\delta_{pc}^{(i)}}{\delta_{pc}^{(j)}} = \frac{r_j^3}{r_i^3} \qquad \text{(VI-11)}$$

With $r = 3.2$ Å, 3.5 Å, 4.5 Å, 4.3 Å, and 5.3 Å for the quaternary carbon, meso-carbon, meso-proton (b in Fig. VI-3), β-carbon (1–8 in Fig. VI-3), and β-proton (a in Fig. VI-3) positions, respectively[25], we find that the relative extent of the pseudocontact shifts for these nuclei is 1:0.76:0.36:0.41:0.22. The corresponding ratios of the experimental hyperfine shifts (Table VI-4) are 1:0.49:0.09:0.40:0.24. The discrepancies between the calculated pseudocontact shift ratios and those observed for the hyperfine shifts indicate that the latter must also contain contributions from interactions other than dipole–dipole coupling. This is even more strikingly manifested in the hyperfine shifts of $Fe^{3+}(PROTO)(CN)_2$ (Table VI-5), since pseudocontact interactions could not possibly cause hyperfine shifts of opposite sign for nuclei in identical geometrical locations, e.g. the ring methyl carbon and hydrogen atoms (a in Table VI-5), or the ^{1}H and ^{13}C resonances c of the propionic acid side chains.

Contact Shifts

Finite amounts of unpaired electron spin density can be transferred to the locations of individual nuclei in paramagnetic molecules by electron spin delocalization through the chemical bonds. Fermi type contact shifts[14] may then arise from interactions of the nuclear spins with fractions of the unpaired electrons located on the nuclei considered.

Electron spin delocalization pathways in low spin ferric hemes, where in the absence of spin delocalization the single unpaired electron would be localized essentially entirely in the d_{xz} and d_{yz} orbitals of the iron, are illustrated in Fig. VI-11. The primary delocalization is from the d_{xz} and d_{yz} iron orbitals to the p_z-orbitals of the directly bonded ligand nitrogen atoms, as indicated by the arrows in Fig. VI-11. Since the π-orbital system extends further over the entire conjugated porphyrin ring system, and in $Fe^{3+}(PROTO)(CN)_2$ also to the two vinyl side chains, spin delocalization in the paramagnetic heme groups can produce finite spin densities on each individual aromatic carbon atom. From the p_z-orbitals of the peripheral ring carbons 1–8 (Fig. VI-4), the unpaired electron spin is then further transferred to the nuclei of the porphyrin side chains by "spin polarization"[26–28], or by "hyperconjugation"[29, 30]. As we shall see in the following, the NMR contact shifts δ_c are proportional to the fractional spin densities localized on the

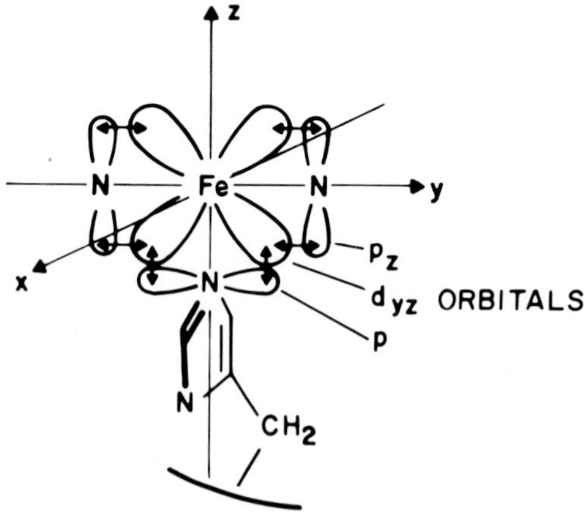

Fig. VI-11. Schematic representation of electron spin delocalization pathways in iron porphyrin complexes. The drawing includes the orbitals of π-symmetry in the yz plane of the immediate environment of the iron atom. A histidyl side chain is attached to one of the axial coordination sites, which is a typical structural feature in hemoproteins[6]. The unpaired electron spin in the d_{xz} and d_{yz} iron atomic orbitals can be delocalized into the π orbitals of the ligand nitrogen atoms, and further to the π molecular orbitals of the aromatic carbon skeleton of the ligand molecules.

individual aromatic carbon atoms, and thus delineate the electronic molecular orbital wave functions in the paramagnetic species.

Quantitative expressions have been derived to relate the ^{1}H and ^{13}C-NMR contact shifts with the unpaired spin density localized on the aromatic carbon atoms[26–37]. Here, we shall restrict the discussion to the contact shifts for aromatic protons and protons in methyl groups attached to aromatic carbons, which will provide the basis for a semi-quantitative discussion of the spin density distribution in the iron porphyrin complexes, and in the heme groups of hemoproteins. A more thorough discussion, which includes also ^{13}C contact shifts, is presented in the Appendix VI.

The contact shifts for a nucleus I are usually expressed by Equation VI-12

$$\delta_c = (hA^I)\frac{g\beta}{\hbar\gamma_I}S(S+1)\frac{1}{3kT} \qquad \text{(VI-12)}$$

where A^I is the contact interaction constant in units of Hz, g the electronic g-factor, β the Bohr magneton, γ_I the gyromagnetic ratio for the nucleus, S the total electronic spin, k the Boltzmann constant, and T the absolute temperature. In writing Equation VI-12, it was assumed that the influence of the g-tensor anisotropy on δ_c can be omitted from consideration[23, 24], an approximation which seems to be adequate for many systems, including the low spin ferric hemes[6], and that only the electronic ground state is populated[23, 24].

The contact interaction constant A^H for aromatic protons is related to the integrated unpaired electron spin density centered on the p_z-orbital of the neighboring carbon atom, ρ_C^π, by the McConnell relation[27]

$$A^H = Q^H \cdot \rho_C^\pi. \qquad \text{(VI-13)}$$

The factor Q^H, which describes electron–nucleus interactions through spin polarization[27, 28], is of the order $Q^H \approx -63$ MHz, when ρ_C^π is given in % of one unpaired electron. A corresponding relation holds for the protons of a methyl group attached to an aromatic carbon atom[29, 30]. Q^{CH_3}, which describes electron–nucleus interactions by hyperconjugation, can be quite different depending on the radical[30–32]. $Q^{CH_3} \approx 75$ MHz was found for uncharged aromatic radicals[29, 30], and for low spin ferric hemes a value of $Q^{CH_3} \approx 50$ MHz has been proposed[6, 17, 33] (see Appendix VI).

Since Q^H and Q^{CH_3} have opposite sign, it is now readily apparent that the overall qualitative features of the ^{1}H-NMR spectra of low spin ferric hemes are due to contact shifts, e.g. the resonances of the 2,4-protons in $Fe^{3+}(DEUT)(CN)_2$ are shifted upfield by the positive spin density centered on the β-carbon atoms, whereas those of the ring methyl protons are shifted downfield. In the Appendix VI it is shown that the qualitative features of the ^{13}C-NMR spectra (Figs. VI-8 and VI-9) can be rationalized in an analogous way.

Temperature Dependence of the Hyperfine Shifts

According to Equations VI-9, VI-10, and VI-12, both pseudocontact and contact shifts are proportional to the reciprocal of temperature. Over the accessible temperature range, the chemical shifts of the ^{1}H-NMR lines in Fe^{3+}(PROTO)$(CN)_2$ (Fig. VI-12) are actually to a good approximation linearly dependent on $1/T$. This temperature dependence may e.g. also be used to identify the hyperfine shifted resonances in the spectrum of a protein[46].

The Equations VI-9 and VI-12 further imply that the hyperfine shifts vanish at $1/T \to 0$. Fig. VI-12 shows, however, that the chemical shifts attained by extrapolation of the experimental temperature dependence to $1/T \to 0$ deviate markedly from those measured in Zn^{2+}(PROTO) (Fig. VI-5). This implies that even though the application of Equations VI-9, VI-10, and VI-12, which are strictly valid only for systems with a single populated electronic state, may be quite adequate for the analysis of the data on low spin ferric hemes, a more elaborate treatment of the system would have to include thermal admixture and Zeeman admixture of excited electronic

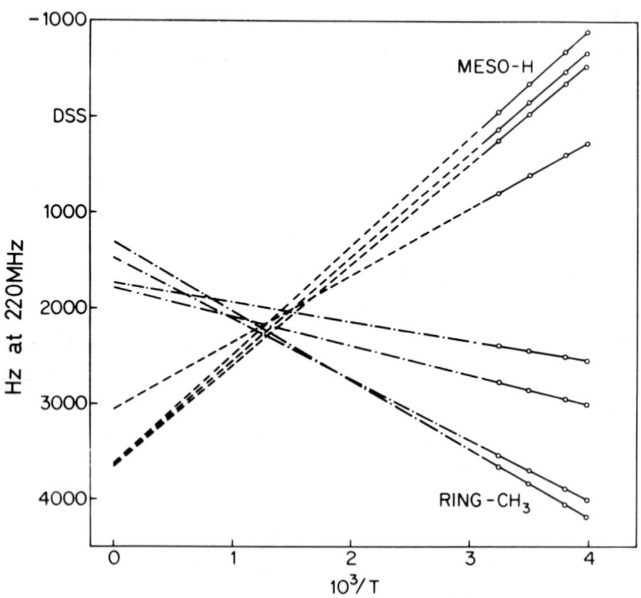

Fig. VI-12. Dependence on the reciprocal of temperature of the chemical shifts of the meso proton and ring methyl ^{1}H-NMR lines of Fe^{3+}(PROTO)$(CN)_2$ in a mixed solvent of d_5-pyridine and D_2O 4:1. The solid lines indicate the experimental temperature range, which extended from $-20°$ to $40°$, and the broken lines represent the extrapolation of the measured temperature dependence to $1/T = 0$ (from Wüthrich[6]).

states to the ground state[6, 23, 24]. This is a feature which is quite commonly encountered in paramagnetic transition metal complexes[3]. For extensions of the theory to include these additional temperature dependent phenomena, the reader is referred to the original papers[23, 24].

Separation of Contact and Pseudocontact Shifts

In general, the ^{1}H-NMR hyperfine shifts in paramagnetic molecules may be expressed as the sum of the pseudocontact and contact shifts

$$\delta_{hf} = \delta_{pc} + \delta_{c} \tag{VI-14}$$

In practice, the nuclei which experience extensive contact shifts will usually also be subject to sizeable pseudocontact interactions. However, since only the individual terms taken separately can yield information on the molecular geometry and the electron spin density distribution (Equations VI-9, VI-10, and VI-12), δ_{pc} and δ_{c} rather than δ_{hf} are the relevant data to be derived from studies of paramagnetic compounds. Therefore the separation of the contact and pseudocontact shifts in Equation VI-14 is a critical step in the analysis of δ_{hf}. No generally applicable procedure lends itself to this task. Rather, the most promising approach will depend in each case on many different aspects. The following are some typical situations which may be encountered in practice with biological systems.

(i) If the electronic g-tensor is known from single crystal EPR experiments*, and the molecular geometry has also independently been determined, e.g. by X-ray methods, δ_{pc} can readily be computed for all the nuclei with Equation VI-9 or VI-10, and the contact shifts are then obtained from Equation VI-14.

(ii) A molecule may contain a considerable number of atoms which are located near the metal ion without being directly involved in the binding of the latter. As a result, $\delta_{hf} = \delta_{pc}$ for these nuclei. If the molecular geometry has independently been determined, these pseudocontact shifts may be used with equations of the type VI-11 to estimate δ_{pc} for the nuclei of the metal bound ligands, and δ_{c} for the latter can then be obtained via Equation VI-14. This situation may typically be encountered in paramagnetic metalloproteins, where numerous nuclei of the polypeptide chain may be subject to dipolar coupling with the electron spin (see Section VI.4).

(iii) If in the situation (ii) one further knows the principal g-values from EPR studies in a polycrystalline sample or a frozen solution*, which is often

---

*Note that here the implicit meaning is that these EPR data are also valid under the conditions of the NMR experiments, i.e. in solution and at ambient temperature. This is in practice quite frequently not the case.

the case for biological materials, the spatial orientation of the g-tensor may be obtained from the NMR pseudocontact shifts. One can thus arrive at the situation (i) for the analysis of the hyperfine shifts in the metal bound ligand atoms. Again this constellation may quite likely be encountered in metalloproteins, and will be illustrated in Section VI-4.

Spin Density Distribution in Low Spin Ferric Hemes

From the foregoing discussion it is quite obvious that the first step in the interpretation of the NMR spectra of the iron porphyrin complexes is the separate evaluation of the pseudocontact and contact shifts. From the ^{1}H-NMR data alone, the spin densities centered on the peripheral ring carbon atoms, i.e. the meso- and β-carbons, can then be computed from δ_c (Equation VI-13). As is shown in the Appendix VI, a more complete description of the electronic structure may be attained when the ^{1}H-NMR data are combined with ^{13}C-NMR studies.

For the separation of δ_{pc} and δ_c in the heme compounds, one has that the molecular geometries are known from X-ray studies[25], and the principal g-values from low temperature EPR studies in the frozen solutions[36]. Two different methods have mainly been chosen, and the results checked for internal consistency with the available NMR data. In the first approach[17], the hyperfine shifts of the methylester protons of Fe^{3+}(DEUT)$(CN)_2$ dimethylester (Fig. VI-5e) were measured. Since no electron delocalization is to be expected to this location, one has that for these protons $\delta_{hf} = \delta_{pc}$. With Equation VI-11, the corresponding values of δ_{pc} for the other nuclei were then computed. This experiment seemed to imply that the pseudocontact shifts were rather small, with $\delta_{pc} \approx -2.5$ ppm for the meso-protons[17]. The second approach made use of the low temperature EPR data[6, 15, 20, 33]. For Fe^{3+}(Porphin)$(CN)_2$, $g_1 = 3.64$, $g_2 \approx 2.29$, and $g_3 \approx 1.0$ had been obtained in a frozen solution at 1.4°, and in analogy to data on other heme compounds, the largest g-value is in the direction perpendicular to the heme plane[6, 36]. Since the β-protons 1–8, and the four meso-protons in Fe^{3+}(Porphin)$(CN)_2$ are equivalent (Fig. VI-3), one has that, contrary to what would have been anticipated from Equation VI-9, the anisotropy in g_2 and g_3 observed in the frozen solution is not manifested in the NMR spectra. This indicates that under the conditions of the NMR experiment, the magnetic anisotropies in the heme plane are averaged out on the NMR time scale. Using $g_\parallel = 3.6$ along the z-axis (Fig. VI-10), and $g_\perp = 1.6$ in Equation VI-10, $\delta_{pc} \approx -10$ ppm is found for the meso-protons. Hence, according to this criterion, the hyperfine shifts of the meso-protons in Fe^{3+}(Porphin)$(CN)_2$ would come mainly from pseudocontact coupling. A similar situation is encountered in Fe^{3+}(PROTO)$(CN)_2$[15, 33]. The two different approaches appear to provide an upper and a lower limit for δ_{pc}, and spin density distributions for various heme compounds have been derived for both limiting cases[15, 17, 20, 33].

As an illustration, let us compute the spin densities in Fe^{3+}(Porphin)$(CN)_2$. First, with $\delta_{pc} \approx -2.5$ ppm for the meso-protons, one has from Equation VI-14 that $\delta_c \approx -6.8$ ppm (Table VI-4). From Equation VI-12, using $g = 2.4$, and from Equation VI-13, $\rho_C^\pi \approx 0.3\%$ of one unpaired electron is then obtained for the meso carbon atoms. With $r = 4.5$ Å and 5.3 Å for the meso-protons and the β-protons, respectively, δ_{pc} for the latter follows from Equation VI-11 to be ≈ -1.5 ppm. Thus for the β-protons, $\delta_c \approx -23.1$ ppm (Table VI-4), and with Equations VI-12 and VI-13 one computes that $\rho_C^\pi \approx 1.1\%$ of one electron for the β-carbon atoms. Second, with $\delta_{pc} \approx -10$ ppm for the meso-protons, one has that $\delta_{hf} \approx \delta_{pc}$, and hence $\rho_C^\pi \approx 0$ for the meso carbon atoms. The corresponding spin density on the β-carbons is $\rho_C^\pi \approx 0.8\%$ of one electron.

3. NUCLEAR SPIN RELAXATION IN PARAMAGNETIC MOLECULES

In addition to the nuclear spin relaxation arising from the different types of nucleus–nucleus interactions in Equation IV-24, dipole–dipole coupling and scalar coupling between nuclear and electronic spins can greatly influence the nuclear relaxation times T_1 and T_2 in paramagnetic systems. The formalism describing the relaxation through electron–nuclear dipolar interactions is quite analogous to the expressions V-5 and V-6 for dipole–dipole coupling between different nuclear spins, and for chemically inert systems, scalar electron–nucleus relaxation is of the "second kind" (Equations IV-35 and IV-36). If one uses that $\omega_X \gg \omega_S$, the contribution of the electron spin to the spin relaxation of nuclei located near a paramagnetic center is thus given by[38–40]

$$\frac{1}{T_{1M}} = \frac{2}{15}\frac{\gamma_I^2 g^2 \beta^2 S(S+1)}{r^6}\left(\frac{3\tau_{c1}}{1+\omega_I^2\tau_{c1}^2} + \frac{7\tau_{c2}}{1+\omega_S^2\tau_{c2}^2}\right) + \frac{2S(S+1)A^2}{3}\frac{\tau_2}{1+\omega_S^2\tau_2^2} \tag{VI-15}$$

$$\frac{1}{T_{2M}} = \frac{1}{15}\frac{\gamma_I^2 g^2 \beta^2 S(S+1)}{r^6}\left(4\tau_{c1} + \frac{3\tau_{c1}}{1+\omega_I^2\tau_{c1}^2} + \frac{13\tau_{c2}}{1+\omega_S^2\tau_{c2}^2}\right) + \frac{S(S+1)A^2}{3}\left(\tau_1 + \frac{\tau_2}{1+\omega_S^2\tau_2^2}\right) \tag{VI-16}$$

where the first term describes dipolar relaxation, and the second term scalar relaxation. T_{1M} and T_{2M} are the longitudinal and transverse relaxation times of the nuclear spin I which arise from the interactions with the electron spin,

S is the total electronic spin, r is the electron–nucleus distance, τ_{c1} and τ_{c2} are the effective correlation times for the dipolar interactions, A is the scalar coupling constant in radians $\sec^{-1}$, ω_I and ω_S are the Larmor frequencies for the nucleus and the electron, and τ_1 and τ_2 are the correlation times for scalar coupling, which are in the absence of chemical exchange given by the longitudinal and transverse electron spin relaxation times, i.e. $\tau_1 = T_{1e}$ and $\tau_2 = T_{2e}$. The correlation times τ_{c1} and τ_{c2} for chemically inert systems are determined by the combination of the rotational correlation time τ_c, which in rigid molecules is equal to τ_R (Equation IV-31), and the electron spin relaxation times according to

$$\frac{1}{\tau_{c1}} = \frac{1}{\tau_c} + \frac{1}{T_{1e}}; \quad \frac{1}{\tau_{c2}} = \frac{1}{\tau_c} + \frac{1}{T_{2e}}. \qquad \text{(VI-17)}$$

Let us first consider the impact of electron–nucleus dipolar coupling on the nuclear relaxation. Several different situations can be distinguished: (i) For long electronic relaxation times T_{1e} and T_{2e}, the correlation times τ_{c1} and τ_{c2} are equal to τ_c, which is also the effective correlation time for nucleus–nucleus dipolar coupling. Because of the larger gyromagnetic ratio of the electron (Table VI-1), the electron–nucleus interactions will be the dominant relaxation mechanism for nuclei located near the paramagnetic center. (ii) For very short electronic relaxation times (Table VI-3), $\tau_{c1} = T_{1e}$ and $\tau_{c2} = T_{2e}$, and despite the dominant electron magnetic moment, the contributions from electron–nucleus dipolar coupling may be small compared to the overall nuclear relaxation (Equation IV-24) even for nuclei located near the unpaired electrons. (iii) In the extreme motional narrowing limit with $\omega_I\tau_{c1} \ll 1$ and $\omega_S\tau_{c2} \ll 1$, one has in analogy to the interactions between nuclear spins that the dipolar contributions to T_{1M} and T_{2M} are identical. In practice, since at the magnetic field strength typically used for biological high resolution NMR experiments ω_S is of the order $4 \cdot 10^{11}$ radians $\cdot \sec^{-1}$ or larger, a situation with $\omega_S\tau_{c2} > 1$ and $\omega_S\tau_{c1} \ll 1$ is more likely to be encountered. Again, nearly identical dipolar contributions to T_{1M} and T_{2M} result in this case, i.e. $T_{1M} : T_{2M} = 7:6$. For all the different limiting cases we have that once the effective correlation times τ_{c1} and τ_{c2} are known, the distance r between the electron and the nucleus considered can be obtained from Equations VI-15 and VI-16. In much the same way as in the interpretation of the pseudocontact shifts, the determination of this geometrical factor may contribute to the resonance assignments, and the elucidation of the molecular structures in the paramagnetic compounds. In this context it should be added that the Equations VI-15 and VI-16 are strictly valid only for systems with isotropic electronic g-tensor[41], and for the isotropic rotational motions of the dipolar relaxation vector r_{IS} (Fig. IV-11)[42].

In the chemically inert systems considered here, the nuclear relaxation through electron–nucleus scalar coupling depends on the contact interaction constant A, and the electronic relaxation times T_{1e} and T_{2e}. As mentioned in the preceding section, scalar coupling has to be considered only for nuclei in ligand molecules which are directly bonded to the metal ion. For these nuclei values of $|A| = 6 \cdot 10^5$ to $6 \cdot 10^6$ radians $\cdot$ sec^{-1} are typically encountered, with $|A| = 4 \cdot 10^6$ radians $\cdot$ sec^{-1} corresponding to a proton bonded to an aromatic carbon atom carrying a spin density of $\rho_C^\pi \approx 1$ percent of one electron. Appreciable contributions to the overall relaxation rates are then to be expected mainly for values of τ_2 such that $\omega_S \tau_2 \gtrsim 1$. Typically, scalar coupling between electron and nuclear spin will thus be manifested essentially only in the transverse relaxation time T_{2M}, and may thus cause extensive line broadening in the NMR spectra. The resulting difference between T_{1M} and T_{2M} may be used to separate the dipolar and scalar coupling terms in Equations VI-15 and VI-16.

In conclusion, we have that some of the structural parameters which can be obtained from the analysis of the NMR hyperfine shifts of paramagnetic molecules may in principle also be derived from studies of the nuclear spin relaxation times. In close analogy to the analysis of the chemical shifts, one has to determine the individual terms in Equations VI-18 and V-19, where $1/T_{1d}$ and $1/T_{2d}$ would describe the relaxation in the absence of the unpaired electron spin (Equation IV-24), and $1/T_{1M}^{DD}$, $1/T_{2M}^{DD}$, $1/T_{1M}^{SC}$, and $1/T_{2M}^{SC}$ the relaxation through dipolar and scalar electron–nucleus coupling, respectively.

$$\frac{1}{T_1} = \frac{1}{T_{1d}} + \frac{1}{T_{1M}^{DD}} + \frac{1}{T_{1M}^{SC}}, \qquad \text{(VI-18)}$$

$$\frac{1}{T_2} = \frac{1}{T_{2d}} + \frac{1}{T_{2M}^{DD}} + \frac{1}{T_{2M}^{SC}} \qquad \text{(VI-19)}$$

On the one hand, if the effective correlation times in Equations VI-15 and VI-16 can independently be obtained, e.g. from the frequency dependence of T_{1M} and T_{2M}, the electron–nucleus distance r and the absolute value of the scalar coupling constant, $|A|$, can be derived from the relaxation studies. On the other hand, if r and $|A|$ are known, e.g. from the analysis of the NMR hyperfine shifts, the correlation times τ_{c1}, τ_{c2}, τ_1, and τ_2, and hence in favorable situations the electronic relaxation times T_{1e} and T_{2e} and the correlation time τ_R, may be computed from the NMR relaxation data. On the basis of Equations VI-15 and VI-16 we also acquire a more fundamental insight into the previously introduced classification of paramagnetic centers as NMR shift reagents and NMR line broadening reagents (Table VI-3). In the systems with very short electron relaxation times T_{1e} and T_{2e}, the unpaired electrons cause only little NMR line broadening so that the hyperfine shifts can readily be measured. Systems with long T_{1e} and T_{2e}, on the other hand, will be characterized primarily by excessive broadening of the NMR lines.

While the above classification of paramagnetic centers with short T_{1e} as NMR shift reagents is very useful in practice, it has recently been pointed out[43] that the line broadening arising from electron–nucleus interactions in these systems may exceed that expected from Equation VI-16. This additional transverse nuclear relaxation is due to modulation by the rotational molecular motions of the chemical shifts produced by dipole–dipole coupling between the nuclear spin moment and the thermal equilibrium magnetization of the electronic spin system, and is proportional to H_0^2. Sizeable contributions from this mechanism to the overall line widths would be expected in NMR studies at high magnetic fields H_0 of large paramagnetic molecules with a large total electronic spin S. The magnitude of the resultant line broadening is such that it will usually be noticeable primarily in systems where the other relaxation mechanisms (Equation VI-16) are ineffective because of very short electronic relaxation times. Typical examples are the high spin ferrous hemoproteins (Fe^{2+}, $S = 2$)[43].

4. HIGH RESOLUTION NMR STUDIES OF HEMOPROTEINS

Among the natural paramagnetic metalloproteins, those containing iron have been most extensively investigated by high resolution NMR techniques, e.g. hemoproteins[6] and iron–sulfur proteins[44]. These investigations were on the one hand concerned with the polypeptide conformations, comparison of the molecular conformations in solution and in single crystals, and studies of the relations between the biological roles and the molecular conformations of these proteins. In this respect this section presents additional examples to the theme of Section III.4. In addition, the NMR studies bear also on the electronic structures of the paramagnetic centers, and their relations with the biological function of the iron containing proteins. This will in the following be illustrated by some experiments with hemoproteins.

Structural Features and Biological Roles of Hemoproteins

A hemoprotein molecule consists of one or several polypeptide chains, each of which is typically made up of 100 to 200 amino acid residues, and one or several heme groups (Section VI.2, Fig. VI-4). The heme groups can be combined with the polypeptide moiety of the molecule through one or several covalent bonds, one or two coordinative bonds with axial ligands of the heme iron, and a multitude of weaker interactions. The heme groups appear to be an integral part of the active centers in all the hemoproteins known to-date.

Four different electronic configurations of the heme iron are commonly encountered (Table VI-6). Three of these are paramagnetic, i.e. the high spin ferric (Fe^{3+}, $S = 5/2$), low spin ferric (Fe^{3+}, $S = 1/2$), and high spin ferrous (Fe^{2+}, $S = 2$) forms, whereas low spin ferrous iron (Fe^{2+}, $S = 0$) is diamagnetic. According to the previously established criteria, the electronic relaxation times qualify the low spin ferric hemes as NMR shift reagents, whereas high spin ferric hemes have an intermediate position and may cause both sizeable line broadening and hyperfine shifts. At high fields H_0, line broadening may also be a prominent feature in the spectra of high spin ferrous hemoproteins[43]. It is important to note in Table VI-6 that different electron configurations may occur in a given hemoprotein.

Hemoproteins are involved in many vital processes in living organisms. Prominent among their biological functions are those of hemoglobin, which is the oxygen transporting protein in the blood, myoglobin, which binds and

TABLE VI-6

Electron configurations of the heme iron

Oxidation state		Fe^{2+}	Fe^{2+}	Fe^{3+}	Fe^{3+}
Spin state		$S = 2$	$S = 0$	$S = 5/2$	$S = 1/2$
Electronic structure	$d_{x^2-y^2}$	↑	—	↑	—
	d_{z^2}	↑	—	↑	—
	$d_{xy}d_{xz}d_{yz}$	↑ ↑ ↑↓	↑↓ ↑↓ ↑↓	↑ ↑ ↑	↑ ↑↓ ↑↓
T_{1e}		....	—	2×10^{-10} sec	2×10^{-12} sec
Selected proteins		Mb^{II} Hb^{II}	$Mb^{II}O_2$ $Hb^{II}(O_2)_4$ Cyt c^{II} Cyt b_5^{II} Cyt b_2^{II}	$Mb^{III}(H_2O)$ $Hb^{III}(H_2O)_4$	$Mb^{III}(CN)$ $Hb^{III}(CN)_4$ Cyt c^{III} Cyt b_5^{III} Cyt b_2^{III}

Abbreviations used:
Mb^{II} = deoxymyoglobin, Hb^{II} = deoxyhemoglobin;
$Mb^{II}O_2$ = oxymyoglobin, $Hb^{II}(O_2)_4$ = oxyhemoglobin;
Cyt c^{II} = reduced cytochrome *c* or ferrocytochrome *c*, Cyt b_5^{II} and Cyt b_2^{II} are the reduced cytochromes b_5 and b_2;
$Mb^{III}(H_2O)$ = metmyoglobin, $Hb^{III}(H_2O)_4$ = methemoglobin;
$Mb^{III}(CN)$ = cyanometmyoglobin, $Hb^{III}(CN)_4$ = cyanomethemoglobin;
Cyt c^{III} = oxidized cytochrome *c* or ferricytochrome *c*, Cyt b_5^{III} and Cyt b_2^{III} are the oxidized cytochromes b_5 and b_2.

stores oxygen in the muscles, various cytochromes, which act as electron transferring oxidation–reduction carriers, and various enzymatically active hemoproteins. Many biochemical reactions of hemoproteins involve changes of the oxidation and spin states of the heme iron, e.g. the oxygenation of myoglobin and hemoglobin

$$\mathrm{Mb^{II}}(\mathrm{Fe^{2+}}, S = 2) + \mathrm{O_2} \rightleftarrows \mathrm{Mb^{II}O_2}(\mathrm{Fe^{2+}}, S = 0) \qquad \text{(VI-20)}$$

and the electron transfer in cytochromes, e.g.

$$\mathrm{Cyt}\ b_5^{\mathrm{III}}(\mathrm{Fe^{3+}}, S = \tfrac{1}{2}) + \mathrm{e}^{\ominus} \rightleftarrows \mathrm{Cyt}\ b_5^{\mathrm{II}}(\mathrm{Fe^{2+}}, S = 0) \qquad \text{(VI-21)}$$

In the investigations of structure–function relationships in this class of proteins, prominent questions bear e.g. on the molecular conformations in the different states of a reaction cycle, and on the pathways through which electrons are transferred to and from the heme iron during biological oxidation–reduction reactions.

NMR Spectral Features of Hemoproteins

In addition to the NMR properties of polypeptide chains discussed in Chapter III, the interactions with the heme groups may cause extensive conformation dependent chemical shifts for certain nuclei in hemoproteins, which makes these molecules particularly attractive for high resolution NMR studies. In diamagnetic hemoproteins (Fe^{2+}, $S = 0$), the most outstanding chemical shifts result from the local ring current field of the heme groups. Qualitatively the ring current field of the porphyrin ring (Fig. VI-13) is quite similar to that of the aromatic amino acids (Figs. II-2 and II-3). The extent of the conformation dependent ring current shifts in proteins, however, may be up to approximately three-fold those caused by a phenylalanine ring. In the paramagnetic hemoproteins, the nuclei of the polypeptide chain which are located near the heme groups may experience conformation dependent pseudocontact shifts, broadening of the NMR lines, or both, depending on the electronic structure of the heme iron. In addition to these hyperfine shifted resonances of the polypeptide chains, the NMR lines of the heme group and the axial ligands of the heme iron may also be in quite unique positions due to the scalar and dipolar interactions with the electron spin.

As will be illustrated below, the above mentioned interactions with the heme groups produce numerous well resolved resonance lines even in the spectra of rather large hemoproteins, e.g. hemoglobin which has a molecular weight of 64000. The well resolved NMR spectral features have been used extensively for investigations of the molecular conformations and the electronic structures in a variety of hemoproteins. These include the oxygen

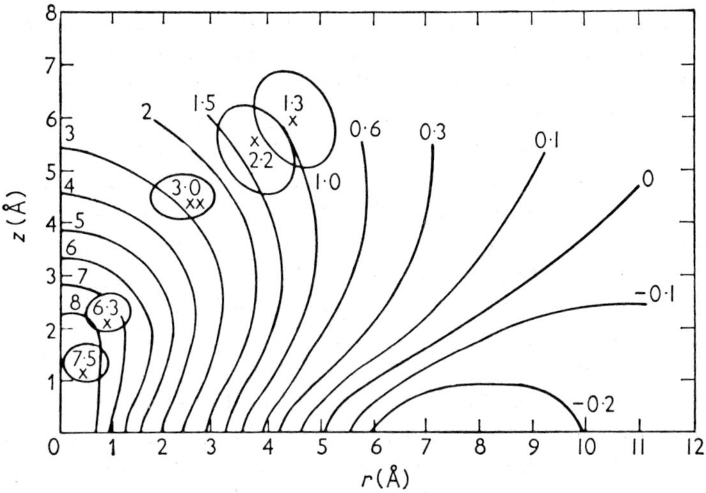

Fig. VI-13. The shielding zone about a porphyrin ring. The plot represents one quadrant of a plane through the center of the ring and normal to the ring plane. The origin of the plot is at the ring center, and the r-axis is in the ring plane. r and z are in Å-units. The isoshielding lines are in ppm, where positive numbers indicate shifts to higher field. The numbers given in the ellipses are ring current shifts measured in porphyrin and phtalocyanine complexes, where the atom positions could be deduced from the chemical formulae. The size of the ellipses indicates the uncertainty in the determination of the atom positions (from Shulman et al.[45]).

binding proteins myoglobin[45–52] and hemoglobin[53–66]. In tetrameric hemoglobins, the chemical shifts of individual hyperfine and ring current shifted resonances had been used to monitor separately changes of the tertiary and quaternary molecular structures during the oxygen uptake, and to distinguish between the ligand binding to the individual subunits[55–59, 62, 65, 66]. Among the cytochromes, many investigations have dealt with structure–function relations in cytochrome c[67–77], cytochrome c_3[78, 79], cytochrome b_5[80], cytochrome b_2[81], and cytochrome P-450.[82] In the following, some experiments with cytochrome b_5 and cytochrome c will be described in more detail to provide an illustration of some practical aspects of NMR studies of hemoproteins.

^{1}H-NMR Studies of Cytochrome b_5

Cytochrome b_5 is an electron transporting hemoprotein (Equation VI-21) found in the microsomal fraction of liver tissue homogenates. The rabbit liver cytochrome b_5 preparation used for the NMR studies has a molecular weight of ca. 12000, and consists of one polypeptide chain with 93 amino acid residues and one protoheme IX group (Fig. VI-4). The axial positions of the heme iron are occupied by two histidyl residues of the polypeptide chain. At the outset of the NMR studies, the amino acid sequence and the molecular conformation in single crystals of the protein were known, as well as the principal values of the electronic g-tensor obtained from low temperature EPR studies of the frozen protein solution. Cytochrome b_5 is included here to illustrate the application of ring current and pseudocontact shifts for comparative studies of the molecular conformations in single crystals and in solution, and to provide a basis for the separate evaluation of the contact and pseudocontact shifts (Equation VI-14) of the NMR lines of the heme group and the axial iron ligands. Similar studies have also been described for cyanometmyoglobin[51].

The principal NMR spectral information employed in the following discussion are the ^{1}H-NMR spectra of the reduced and oxidized forms of cytochrome b_5, and the temperature dependence of the latter. The proton NMR spectrum of the diamagnetic reduced cytochrome b_5 is shown in Fig. VI-14. Outside the spectral region from 0 to 10 ppm, which contains the bulk of the resonances, there are 5 high field methyl resonances at -1.3, -1.1, -0.7, -0.7, and -0.5 ppm. Following the discussion in Section III.4, it appears quite likely that the high field positions of these resonances result from interactions with the ring current fields of the aromatic rings in the protein. Comparison with a spectrum computed with the single crystal atomic coordinates and with empirical expressions for the ring current fields in the different aromatic rings (Figs. II-3 and VI-13)[45] then showed that four of these resonances correspond to methyl groups of aliphatic amino acid side chains located near the heme group. The line at -1.1 ppm corresponds to

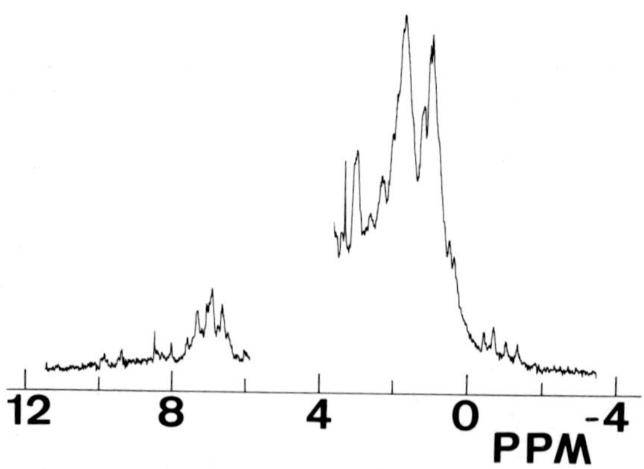

Fig. VI-14. ^{1}H-NMR spectrum at 220 MHz of a ca. 0.008-M solution of ferrocytochrome b_5 in deuterated 0.2-M phosphate buffer, pD = 7.0 at 29°. The spectral region from 4 to 6 ppm, which contains the resonance of HDO and its spinning side bands, has been omitted (from Keller and Wüthrich[80]).

an aliphatic methyl group located near the only tryptophanyl residue in the molecule. The close similarity of the experimental chemical shifts with those computed from the X-ray data implied also that the molecular conformations of ferrocytochrome b_5 in single crystals and in solution are quite similar. On this basis the high field methyl resonances in the spectrum of Fig. IV-14 could be assigned to specific aliphatic amino acid residues near the heme group, and to isoleucine-76 which is located near the tryptophanyl residue in cytochrome b_5, as indicated in Fig. VI-17.

The spectrum of the oxidized cytochrome b_5 in Fig. VI-15 contains more

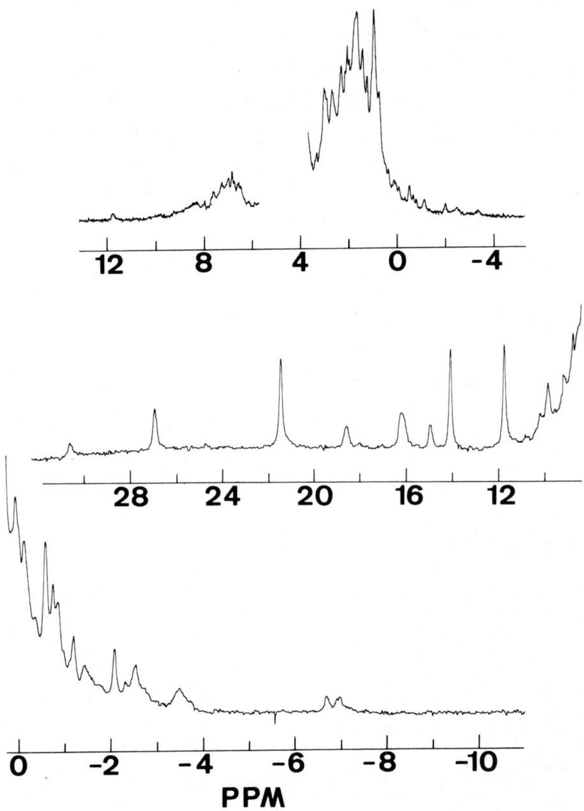

Fig. VI-15. 220 MHz ^{1}H-NMR spectrum of 0.008-M ferricytochrome b_5 in 0.2 M deuterated phosphate buffer, pD 6.3, at 29°C. The three spectral regions are represented with different horizontal and vertical scales (the resonance intensities of the resolved lines in the regions 10 to 30 and 0 to − 10 ppm correspond to 1 to 3 protons, whereas the spectral region from 0 to 10 ppm contains the resonances of approx. 500 protons). The spectral region from 4 to 6 ppm, which contains the resonance of HDO and its spinning side bands, has been omitted (from Keller and Wüthrich[80]).

than 20 resolved resonance lines in the spectral regions from 10 to 30 and 0 to − 10 ppm. Most of these unusual resonance positions come from interactions with the unpaired electron of the low spin ferric heme iron (Table VI-6). This is also supported by the temperature dependence of the chemical shifts of the resolved resonances at high (Fig. VI-16) and low fields. From the spectra of the isolated low spin ferric hemes (Section VI.2, Fig. VI-4) it is quite obvious that the resonances of the heme group may be located outside the spectral region from 0 to 10 ppm. The same holds for the protons of the two imidazole rings of the axial histidyl residues, which also experience both contact and pseudocontact shifts. However, the number of protons corresponding to the hyperfine shifted resonances in ferricytochrome b_5 exceeds the number of protons of the iron bound ligands, and hence the spectrum must also contain sizeably pseudocontact shifted resonances of protons of the polypeptide chain. In the following we will concentrate on the methyl resonances between 0 and − 3 ppm (Figs. VI-15 and VI-16), which all have to

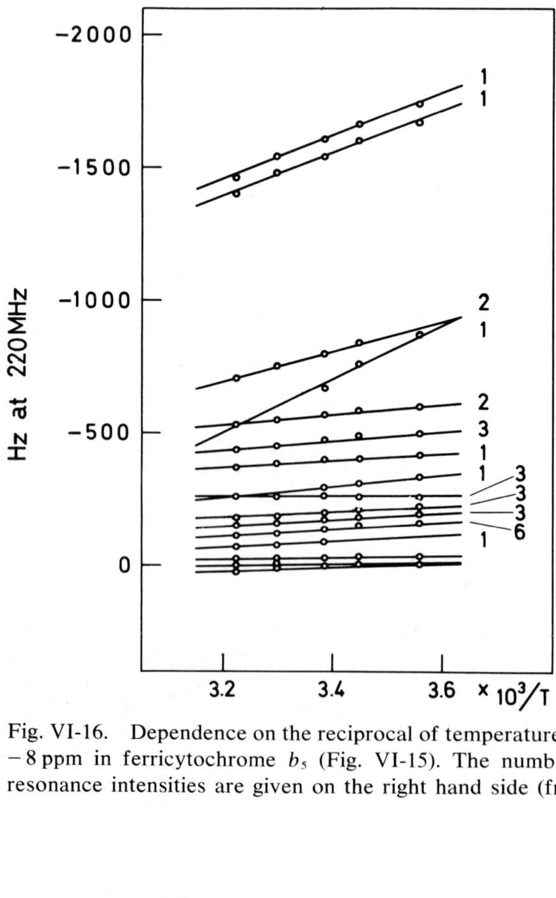

Fig. VI-16. Dependence on the reciprocal of temperature of the ^{1}H-NMR lines between 0 and − 8 ppm in ferricytochrome b_5 (Fig. VI-15). The number of protons corresponding to the resonance intensities are given on the right hand side (from Keller and Wüthrich[80]).

come from aliphatic amino acid side chains, since the ring methyl resonances of the heme are known to be at low fields (Fig. VI-4).

In the present example cytochrome b_5, the investigation of the pseudocontact shifted lines was based on the atomic coordinates in single crystals, and the principal values of the electronic g-tensor. In analogy to the reduced cytochrome b_5, it was assumed that the conformation in single crystals and in solution are identical, and the values $g_1 = 3.03$, $g_2 = 2.23$, and $g_3 = 1.43$ obtained in the frozen solution were used. Chemical shifts for the different protons in the molecule were computed with Equation VI-6, where the chemical shifts of ferrocytochrome b_5 described above were used as the diamagnetic chemical shifts δ_d, and $\delta_{hf} = \delta_{pc}$ was obtained with Equation VI-9. Two subtle corrections were included in this computation. First, since the atomic coordinates obtained from X-ray studies are for carbon atoms, appropriate modifications were made in the calculation of ^{1}H-NMR shifts. Thus in the methyl groups, which were of central importance in this study, it was assumed that the average position for each of the 3 protons is 0.37 Å away from the methyl carbon in the direction of the elongated C—C bond. Second, Equation VI-9 was multiplied with an empirical correction factor $F = 0.77$[51,80].

Since only the principal g-values were known, Equation VI-9 could not be applied for a straightforward calculation of the pseudocontact shifts. Instead a computer program was used which started with an arbitrary orientation of the g-tensor, then varied the directions of the principal g-tensor axes in small steps, and computed the corresponding pseudocontact shifts for the protons in the protein in search for agreement with the observed chemical shifts. In all, 9 methyl resonances between 0 and -3 ppm were considered in this analysis, and it was found that only a narrow range of g-tensor orientations fulfilled the criteria of a good fit with the experiment[80]. At this point one had a complete description of the electronic g-tensor, with the principal g-values determined by EPR, and the direction of the principal axes from the above fit with the NMR data. The pseudocontact shifts for all the protons in the protein could now be calculated with Equation VI-9, and some resonance lines assigned to specific protons on the basis of their unique chemical shifts in both the experimental and the computed spectrum. The results for the high field region of the spectrum are shown in Fig. VI-17.

In Fig. VI-17, the observed temperature dependences of the resonance positions in the oxidized protein have been extrapolated to $1/T = 0$ (solid lines). The resonance positions observed in the diamagnetic reduced protein are plotted at $1/T = 0$, and the approximate dependence on temperature that these protons would have in the oxidized protein (dashed lines) was evaluated from the pseudocontact shifts computed with Equation VI-9, using the g-factor obtained as described above. It is seen that the ring current field and

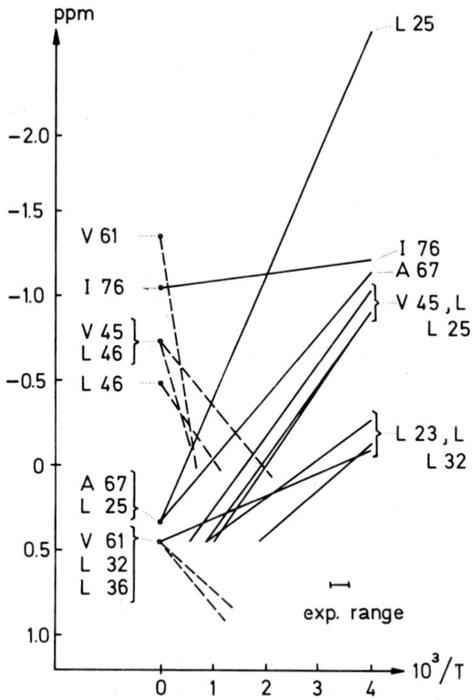

Fig. VI-17. Dependence on the reciprocal of temperature and assignments of the high field methyl resonances in ferric and ferrous cytochrome b_5 (V = valine, I = isoleucine, L = leucine, A = alanine, the numbers indicate the positions in the amino acid sequence). The circles on the left indicate the observed resonance positions in the reduced protein (Fig. VI-14). The broken lines indicate the calculated dependence on the reciprocal of temperature for these resonances in the ferric protein. The solid lines show the extrapolation to $1/T = 0$ of the measured temperature dependence (Fig. VI-16) of the high field methyl resonances in the oxidized protein (from Keller and Wüthrich[80]).

the pseudocontact field of the heme have opposite sign for the protons considered, and hence the resonances of different amino acid residues are shifted to high field positions in the two oxidation states of the protein. The resonance of isoleucine-76, which is located near tryptophan-22, and is approximately 17 Å away from the heme iron, is essentially not affected by the paramagnetic center. Qualitatively, the situation of Fig. VI-17 is quite representative for low spin ferric hemoproteins: The protons above and below the heme plane experience sizeable upfield shifts by the ring current field (Fig. VI-13), which produces a number of well resolved high field lines in the spectrum of the diamagnetic form of the protein. In the spectrum of the paramagnetic protein, the dipolar field of the unpaired electron is more prominently manifested than the ring current field of the heme. The discussion of Fe^{3+}(Porphin)$(CN)_2$ in Section VI.2 applies in as far to the paramagnetic natural hemes as the direction of the largest g-value is always nearly perpendicular to the heme plane. As a consequence, the protons in the heme plane experience upfield pseudocontact shifts, those above and below the heme plane downfield pseudocontact shifts. One can say that the low spin ferric hemes function like typical "paramagnetic NMR shift reagents" (Section VI.5).

A spectral analysis of the type outlined here for ferricytochrome b_5 is in principle generally applicable for paramagnetic metalloproteins, provided the NMR spectrum contains an adequate number of resolved NMR lines. The procedure becomes considerably more straightforward for proteins where the g-tensor is known from single crystal EPR studies[51].

The investigation of the pseudocontact interactions in ferricytochrome b_5 solutions in D_2O revealed also that the resonance lines at lower field than 10 ppm, and at higher field than − 5 ppm had to come from protons of the heme and the axial histidines, which also experience contact shifts[80]. Since at this stage δ_{pc} for these protons can be calculated with Equation VI-9, δ_c may be obtained from Equation VI-14. The general features, and in particular the symmetry of the spin density distribution in the heme group can thus be derived from the ^{1}H-NMR spectra of the hemoprotein by following the procedures outlined in Section VI-2 for the isolated hemes. In one example, cyanometmyoglobin (Table VI-1), the individual heme ring methyl resonances were identified by the introduction of specifically deuterated protoheme IX into the protein[83]. From this the measured spin densities could then be assigned to specific ones of the β-carbon atoms of the porphyrin ring.

^{1}H-NMR Studies of Cytochrome c

Cytochrome *c* has a molecular weight of 12500. The molecule consists of one polypeptide chain with 103 to 109 amino acid residues, depending on the species, and one group of heme *c* (Fig. VI-18). Analogous to cytochrome b_5, the biological function of cytochrome *c* as an electron transfer protein is intimately linked with the interconversion between the ferric and ferrous oxidation states (Equation VI-21). Cytochrome *c* is included here to illustrate that even though the combination of X-ray and NMR data is in many cases the most desirable approach for studies of protein structures, the interpretation of the NMR spectra does not necessarily have to rely on the X-ray data. The experiments with cytochrome *c* give also an indication of the versatility of the NMR methods, which enable one to extend the investigations with reasonable expense of working time and finances to entire classes of related protein molecules.

The ^{1}H-NMR spectra of reduced and oxidized cytochrome *c* (Fig. VI-19)

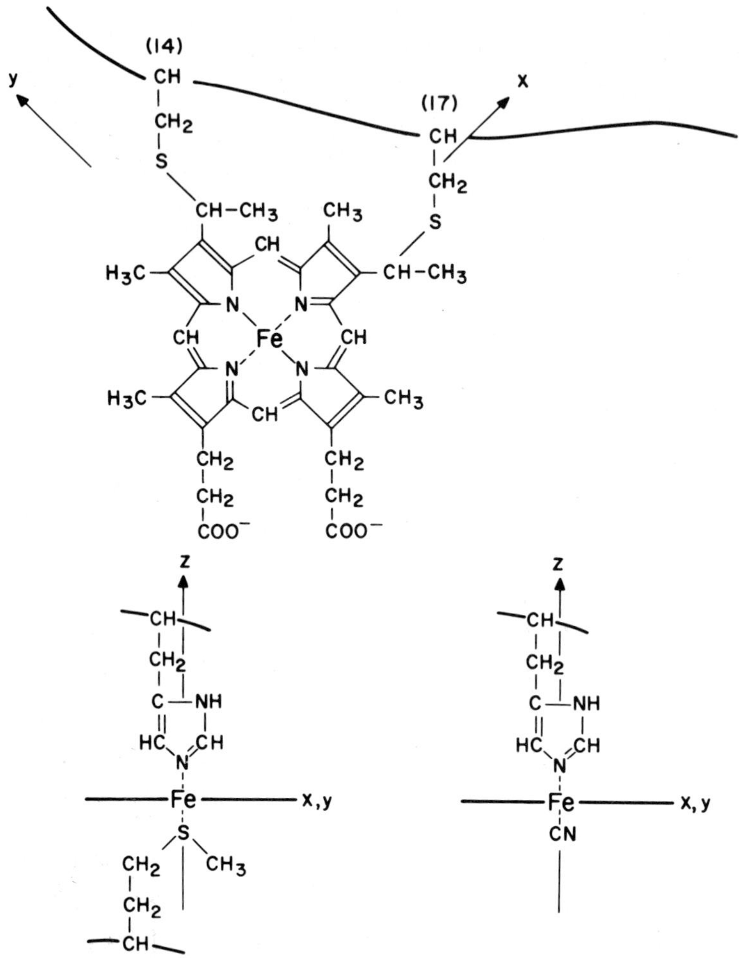

Fig. VI-18. Structure of "heme *c*", the heme group of cytochrome *c*. The axial ligands of the heme iron in cytochrome *c*, i.e. a histidyl and a methionyl residue of the polypeptide chain, and its complex with cyanide ion are also indicated. Heme *c* is covalently linked with the cysteinyl residues in positions 14 and 17 of the polypeptide chain.

are in their overall traits quite similar to those of cytochrome b_5. In the diamagnetic ferrocytochrome *c*, one has the strongly overlapping resonances of ca. 650 protons in the spectral region from 0.5 to 10 ppm, and some resolved lines at high fields from DSS. The ring current shifted high field resonances include lines of intensity three protons at -3.3, -0.7, -0.6 and -0.6 ppm, a resonance of two protons at 0.1 ppm, and one-proton resonances at -3.7, -2.7, -1.9, -0.2, and 0.2 ppm[68, 69]. The spectrum of ferricytochrome *c* contains a number of well resolved hyperfine shifted lines in the spectral regions from 10 to 35 ppm, and from 0 to − 25 ppm. In particular, at 35° there are resonances of intensity three protons at 34.0, 31.4, 10.3, − 2.1, − 2.5, and − 23.2 ppm, and one-proton lines at 18.1, 13.8, 11.8, 11.5, − 4.0, and − 6.0 ppm (Fig. VI-19). In various different NMR experiments, these spectral features were used to identify the axial ligands of the heme iron (Fig. VI-18)[68,70], to study the reaction of cytochrome *c* with cyanide ion[68,70], to investigate the electron exchange between oxidized and reduced cytochrome *c*[72–74], and to investigate the relations between the electronic heme structure and the biological activity[70,74].

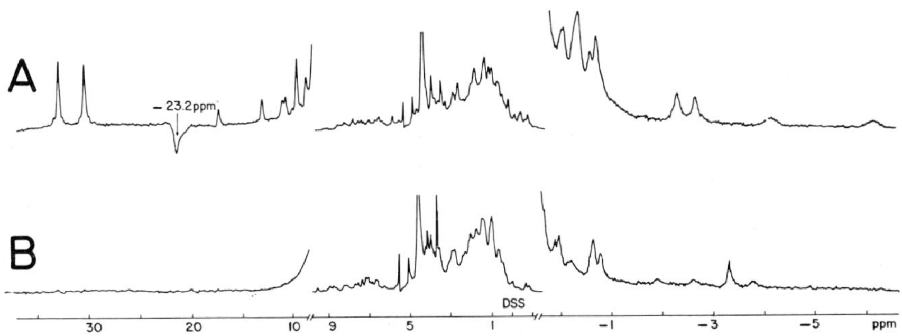

Fig. VI-19. ^{1}H-NMR spectra at 220 MHz of cytochrome *c* from guanaco in D_2O solution, pD = 7.0, T = 35°. A. Ferricytochrome *c*. B. Ferrocytochrome *c*. The sharp lines between 4 and 6 ppm correspond to the HDO resonance and its first and second spinning side bands. The vertical and horizontal scales are different for the three parts of the spectrum. The intensities of the individual lines in the spectral regions from 10 to 35 ppm, and from 0 to − 25 ppm correspond to 1 to 3 protons, whereas the spectrum between 0 and 10 ppm comes from ca. 650 protons. The high-field line at − 23.2 ppm is observed as an inversed resonance of the center band of the spectrum. (The spectrometer operated with a 10 kHz field modulation. Usually one observes the first upfield side band. If large hyperfine shifts occur, parts of the center band and the different side bands of the spectrum may overlap.)

Let us first look at the reactions between cytochrome *c* and cyanide ion, which provided the clue for the identification of methionine as an axial ligand (Fig. VI-18). The hyperfine-shifted resonances of ferricytochrome *c* and its complex with cyanide ion are markedly different (Fig. VI-20). At 35°, the spectrum of cyanoferricytochrome *c* contains no lines at lower field than 25 ppm, and the highest field resonance is a one-proton line at approximately − 2.8 ppm. Because the exchange of cyanide ions between different cytochrome *c* molecules is slow on the NMR time scale, the complex formation can readily be monitored on the basis of the spectral differences in Fig. VI-20. The experiment represented in Fig. VI-21 showed that in neutral aqueous solution a 1 : 1 complex Cyt c^{III}(CN) is formed. This complex can be reduced with dithionite to the cyanoferrocytochrome *c* complex, which is diamagnetic

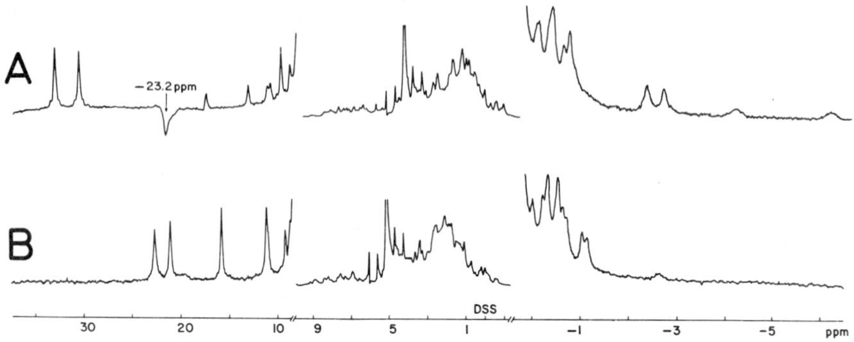

Fig. VI-20. ^{1}H-NMR spectra at 220 MHz of cytochrome *c* from guanaco. A. Ferricytochrome *c*. B. Cyanoferricytochrome *c*. The experimental conditions were the same as in Fig. VI-19 (from Wüthrich[68]).

(Fe^{2+}, $S = 0$). Fig. VI-22 shows that Cyt c^{II}(CN) contains no resonance lines at higher field than − 1.0 ppm and, in analogy to ferrocytochrome *c* (Fig. VI-19), there were no lines at lower field than 11 ppm. With time elapsing after the reduction, the typical ferrocytochrome *c* (Fig. VI-19) resonances between 0 and − 4 ppm appeared. This shows that under the conditions of the experiment in Fig. VI-22, the rapid reduction of cyanoferricytochrome *c* to Cyt c^{II}(CN) is followed by slow dissociation of the unstable cyanoferrocytochrome *c*.

$$\text{Cyt}\, c^{II}(\text{CN}) \rightarrow \text{Cyt}\, c^{II} + \text{CN}^- \qquad \text{(VI-22)}$$

In the course of this dissociation, cyanide ion is replaced by the axial ligand of native ferrocytochrome *c*, which by this becomes again subjected to the ring current field of the heme group (Figs. VI-13 and VI-18). From the experiment of Fig. VI-22, the methyl resonance at − 3.3 ppm, and the one-proton lines at − 3.7, − 2.7, and − 1.9 ppm could be assigned to the axial iron ligand which replaced CN^- in the reduced cytochrome *c*. As indicated in Fig. VI-18, this ligand then has to be methionine, which is the only iron binding amino acid which contains a methyl group. The other axial position had previously been shown by various lines of evidence to be occupied by histidine in both the reduced and the oxidized state of the protein.

It remained to be seen whether methionine is also bound to the heme iron in ferricytochrome *c*. Evidence in support of this comes from the observation of the methyl resonance at − 23.3 ppm in Cyt c^{III} (Fig. VI-19), which from considerations of the molecular orbitals in heme *c* has to come from one of the axial ligands[68, 70], and also from the disappearance of this high field line after complexation with cyanide ion (Fig. VI-21). In the present context, it is instructive that the heme *c* could be employed as an NMR line broadening

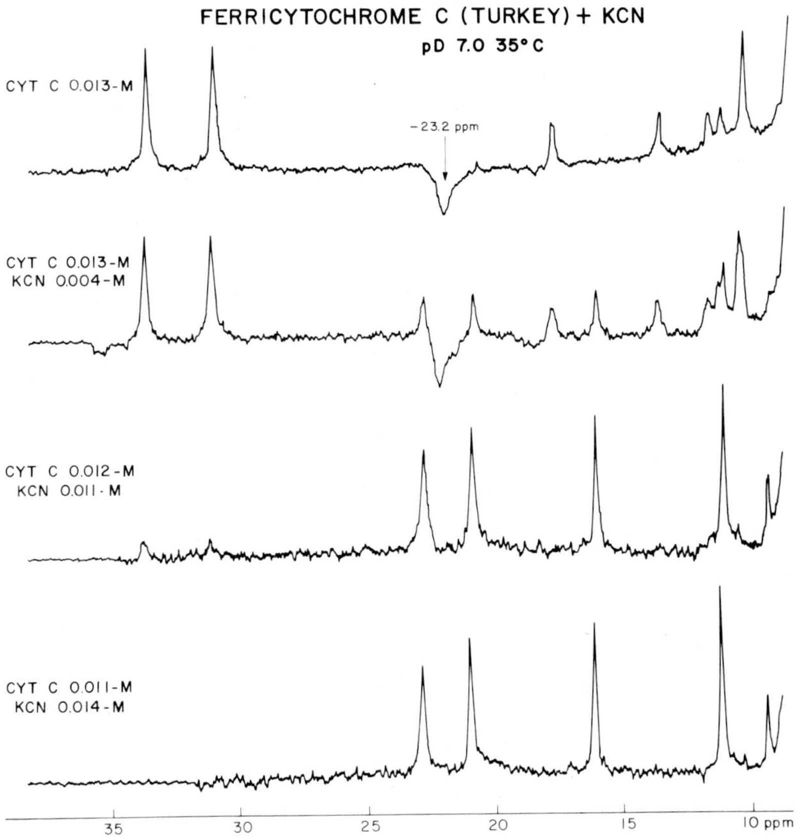

Fig. VI-21. ^{1}H-NMR spectral changes following the addition of variable amounts of KCN to a solution of ferricytochrome *c* in neutral deuterated 0.2-M phosphate buffer solution. No further spectral changes were observed upon addition of an excess of KCN (from Wüthrich[70]).

reagent to corroborate this resonance identification[6]: The overall line width of the ^{1}H-NMR lines in ferricytochrome *c* is given by $1/T_2$ in Equation VI-19. $1/T_{2d}$ comes essentially entirely from proton–proton dipolar coupling, and should be quite similar for the ring methyl resonances of the heme group, and the methyl resonances of an axial methionyl residue. $1/T_{2M}^{SC}$, the line-broadening from electron–proton scalar coupling, can be estimated from the second term of Equation VI-16. With $T_{1e} = T_{2e} = 2 \cdot 10^{-12}$ sec, one finds that for a ring methyl resonance which is at 33 ppm at 35° (Fig. VI-19), $1/\pi T_{2M}^{SC}$ should be ca. 8.5 Hz, and for a methyl resonance of methionine at − 23.2 ppm, ca. 3.8 Hz. A difference of this order would hardly be noticeable in the spectra. On the other hand appreciably different line-widths result from dipole–dipole coupling between the electron and the protons. The average

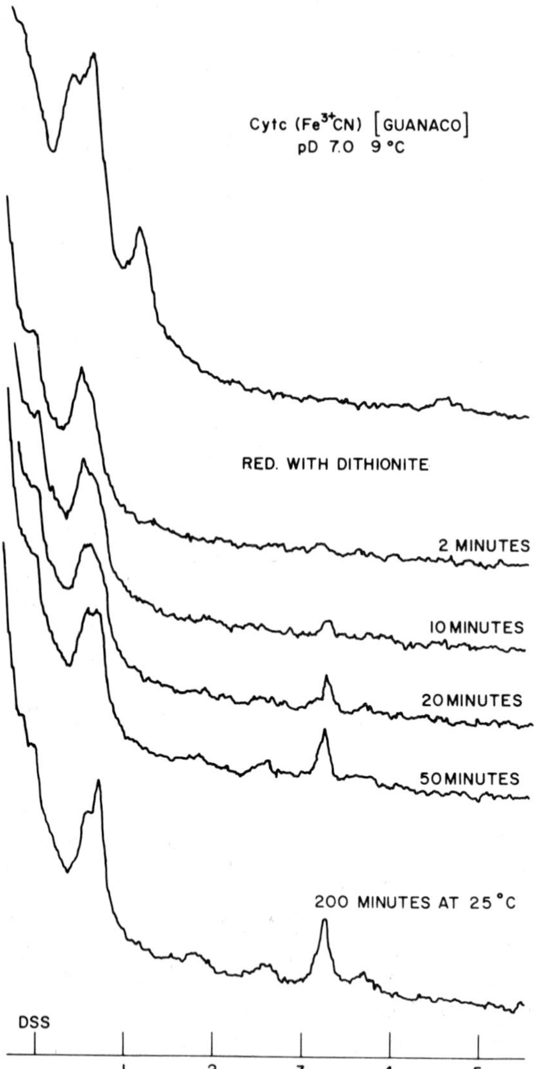

Fig. VI-22. ^{1}H-NMR spectra between 0 and −5 ppm of cyanoferricytochrome *c* (top trace), and of the reaction products observed at variable times after reduction of cyanoferricytochrome *c* with dithionite. The second trace from the top corresponds to cyanoferrocytochrome *c* (from Wüthrich[68]).

distance from the heme iron is ca. 6.4 Å for the ring methyl protons of the heme group, and ca. 3.6 Å for the methyl protons of an axial methionine. Inserting these values for r, and $\tau_{c1} = \tau_{c2} = 2 \cdot 10^{-12}$ sec into the first term of Equation VI-16, one finds that at 220 MHz the line-broadening from electron–proton dipolar coupling, $1/\pi T_{2M}^{DD}$, should be ca. 1.1 Hz for the ring methyls, and ca. 36 Hz for the methyl resonance of methionine. The spectrum of Fig. VI-19 indicates indeed that the resonance at − 23.2 ppm is appreciably broader than the other resolved methyl resonances. Equation VI-16 would further predict that the line-broadening from electron–proton dipolar coupling should increase at lower fields, whereas in the range considered the line broadening through the two other relaxation mechanisms is comparatively little dependent of the field strength. At 100 MHz the dipolar line-broadening should be ca. 1.7 Hz for the ring methyls, and ca. 55 Hz for the methyl resonance of methionine. In agreement with these calculations, measurements at 100 MHz gave a full width at half height of the resonance at − 23.2 ppm of ca. 100 Hz, as compared to ca. 30 Hz for the resonances at 31 and 34 ppm[6].

In addition to their implications on structural features of the heme cavity in cytochrome *c*, these NMR studies provided some insight into the electronic structure of the heme group. Fig. VI-19A shows that two of the four heme ring methyl resonances are largely shifted by contact interactions, whereas the other two ring methyls have to be between 11 and − 3 ppm, and are thus only little affected by contact coupling. The spin density distribution in the heme *c* of the native ferricytochrome *c* is thus highly anisotropic, with spin densities of ca. 2% of one electron on two of the β-carbons 1, 3, 5, and 8 (see Fig. VI-4 for the numerotation, which applies also to heme *c*), and less than 0.4% of one electron on the other two of these four β-carbon atoms. NMR investigations of cytochromes *c* from different species, and of chemically modified cytochromes *c* implied that this anisotropic spin density distribution is an essential structural feature of biologically active ferricytochrome *c*[70,71]. In the following we will see that heme *c* and its molecular environment have also been investigated by a very different NMR approach.

Resonance Identification by Comparison of Mutant Proteins

When working with synthetic peptides, comparison with suitable analogue molecules can often be employed to identify individual resonance lines. At present, synthesis of proteins for this purpose is not practicable. However, in the form of homologous proteins from different species, which contain different amino acids in certain positions of the polypeptide chain, nature provides us in many cases with suitable sets of analogue proteins. As an illustration, Fig. VI-23 shows a comparison of the ^{1}H-NMR spectra of ferricytochrome *c* from turkey, and ferricytochrome *c*-557 from *Crithidia*

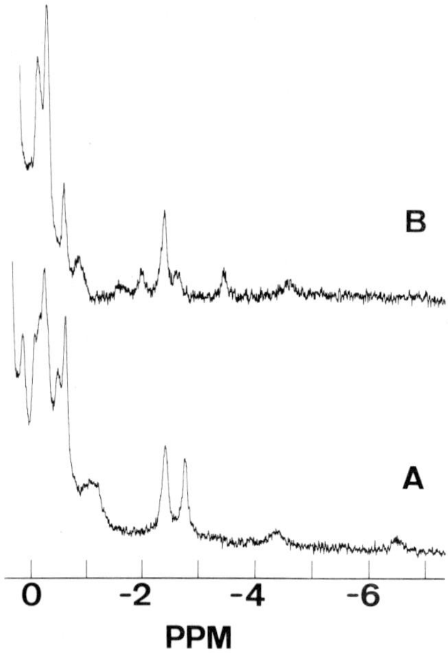

Fig. VI-23. Spectral region between 0 and −7 ppm of the 220 MHz ^{1}H-NMR spectra at 25° of ferricytochrome *c* from turkey (A), and ferricytochrome *c*-557 from *Crithidia oncopelti*; (B). The spectrum of ferricytochrome *c* from turkey is representative for the NMR spectra of all the vertebrate ferricytochromes *c* studied so far (from Keller et al.[76]).

oncopelti[76]. The most outstanding difference between the amino acid sequences of these two proteins concerns the replacement of Cys-14 (Fig. VI-18) by Ala in cytochrome *c*-557. The heme group in cytochrome *c*-557 is still covalently linked with Cys-17, but instead of the second thioether linkage with the polypeptide chain (Fig. VI-18) the heme contains a vinyl group. Overall the ^{1}H-NMR spectrum of ferricytochrome *c*-557 contains very similar contact shifted lines as the mammalian cytochromes *c* (Fig. VI-19), indicating that the electronic heme structure of the mammalian ferricytochromes *c* is essentially maintained in this species[76]. On the other hand, Fig. VI-23 shows that one of the two methyl resonances which are at around −2 to −3 ppm in mammalian ferricytochromes *c*, is missing in cytochrome *c*-557. This methyl resonance could thus be assigned to a thioether methyl group of cytochrome *c* (Fig. VI-18), and it was then very probable that the methyl line at 2.4 ppm in ferricytochrome *c*-557 and in ferricytochrome *c* comes also from a thioether methyl group. In this particular example, the resonance assignment through comparison of mutant proteins opened a new avenue for assessing the pseudocontact shifts of the ^{1}H-NMR of the heme group, since the thioether methyl protons experience at most very small contact shifts.

Comparison of homologous proteins played an import role in the investigations of human hemoglobin. A large number of "abnormal" hemoglobins are available, which all differ from "normal" hemoglobin A by a single amino acid substitution in either the α- or β-subunits. Despite the high molecular weight of 64000, several resonance lines could be identified from such comparative investigations[55, 56, 58, 63].

Pulsed Fourier Transform NMR Cross Saturation Studies of Cytochrome c

With a novel technique developed by Redfield and Gupta[72–74], the correspondence of the previously established methionine methyl resonance positions in reduced and oxidized cytochrome *c* could be confirmed in a very elegant way. Furthermore, corresponding resonance positions in the two oxidation states of the protein could also be established for a number of additional protons.

In these experiments a pulse of radiofrequency radiation was applied at the resonance frequency of one of the hyperfine-shifted lines of ferricytochrome *c* (Fig. VI-19). The intensity of the pulse was fixed at a level just sufficient to saturate the resonance to which it was applied. The rest of the spectrum was then scanned by an observation pulse just after the saturating pulse to search for the changes induced by the latter. Since these changes are small they were recorded by taking a difference spectrum with and without the saturating pulse. In this way changes could also be observed in the spectral region of the strongly overlapping polypeptide resonances between 0 and 10 ppm. In mixed solutions of ferro- and ferri-cytochrome *c*, "cross saturation" effects could thus be observed (Fig. VI-24), e.g. between the methionine methyl resonances at -23.2 ppm in Cyt c^{III}, and -3.3 ppm in Cyt c^{II} (Fig. VI-19). The cross saturation depends on the following two facts: (i) The nuclear spin relaxation time T_1 is sufficiently long so that a given nucleus "remembers" its past history for about half a second (in the reduced state), e.g. after being "labelled" by saturation with a resonant radio frequency pulse. (ii) At practical concentrations of half-reduced protein, i.e. ca. 0.005-M, a given cytochrome *c* molecule stays in its reduced state for about 50 msec, on the average, before its electron transfers to another cytochrome *c* molecule. Thus, a given proton near the heme has its resonance shifted back and forth randomly about 20 times per sec between its two well-defined positions in the oxidized and reduced states. Since the rate of the electron exchange in a half-reduced cytochrome *c* solution thus exceeds the proton relaxation rate $1/T_1$, the saturating effect of the radiofrequency field at -23.2 ppm is transmitted to the line at -3.3 ppm when, as a consequence of the electron exchange, the methionine methyl group changes its resonance frequency between these two positions.

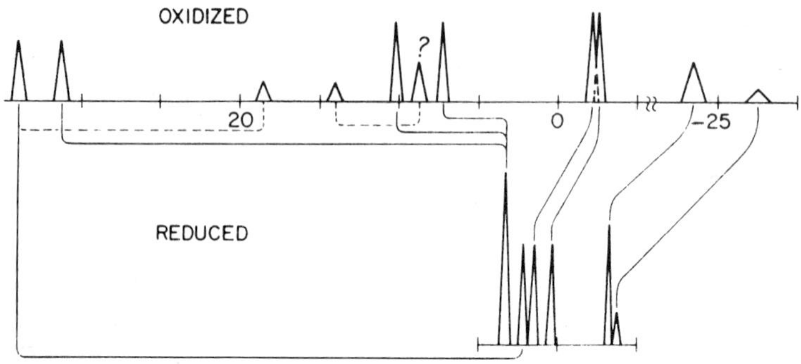

Fig. VI-24. Schematic representation of all the resonances in a mixed solution of Cyt c^{III} and Cyt c^{II} for which cross-saturation has been observed. The resonances at 18.5, 14.0, −27.4 in the oxidized, and −3.8 ppm in the reduced state have intensities characteristic of a single proton. The intensity of the resonance at 8.5 ppm is not known, and all the others are methyl resonances. Lines are drawn between resonances which are found experimentally to be connected by cross-saturation. Solid lines indicate cross-saturation occurring via electron exchange, broken lines indicate that occurring via interspin cross-relaxation (from Redfield and Gupta[74]).

Fig. VI-24 shows that in addition to the methionine methyl resonance, the corresponding resonance positions in Cyt c^{II} and Cyt c^{III} could be determined for a number of additional resonance lines, including the four ring methyls of heme *c*. On this basis a more definite assessment of the spin density distribution in the heme group could be derived, since the chemical shifts of the heme ring methyls in ferrocytochrome *c* represent probably a better approximation for δ_d (Equation VI-6) in Cyt c^{III} than the chemical shifts of the Zn^{2+}-porphyrin complexes (Fig. VI-5).

In addition to the cross relaxation mediated by the electron exchange, a second kind of cross relaxation was observed in ferricytochrome *c* (dotted lines in Fig. VI-24), i.e. "interspin cross relaxation"[74]. Here the cross relaxation is mediated by dipolar interactions between protons which are located in close proximity of each other in the molecule.

^{13}*C-NMR Spectra of Hemoproteins*

Polypeptide ^{13}C resonances of hemoproteins can readily be obtained in analogous experiments to those described in Section V.5 for various other proteins. For example, a detailed analysis of the aromatic ring carbon resonances of cytochrome *c* corresponding to the experiments with lysozyme in Fig. V-22 was recently described[84, 85], where the spectral differences between Cyt c^{II} and Cyt c^{III} were also used for the resonance assignments. It would of course be of particular interest for a refined analysis of the electronic heme structure, to investigate the contact shifted ^{13}C-NMR lines, as described in the foregoing for the isolated heme groups. Yet, even though the resonances of certain carbon atoms of the heme could be observed in the diamagnetic ferrocytochrome *c* both with a home built instrument using 20 mm sample tubes[84], and with a commercial spectrometer (Fig. VI-25, resonances b)[86], the contact shifted ^{13}C-NMR lines in ferricytochrome *c* have so far not been observed.

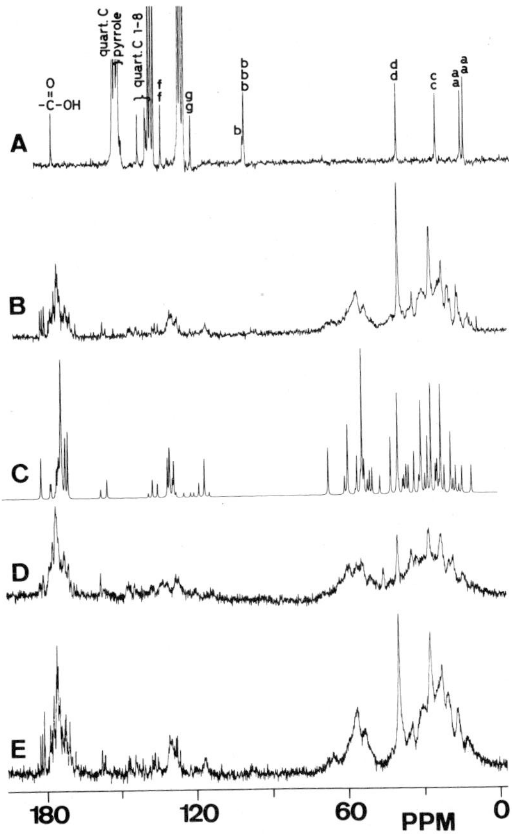

Fig. VI-25. FT ^{13}C-NMR spectra at 25.16 MHz of reduced horse cytochrome *c* and some of its constituent components, $T = 25°$. A. Zn^{2+}(PROTO) in d_5-pyridine, proton noise-decoupled. For the resonance assignments see Fig. VI-5. The corresponding resonances in diamagnetic heme *c* should be in very nearly identical positions. The intense lines at 124, 137, and 150 ppm come from the solvent. B. Cyt c^{II}, ca. 20% in D_2O, pD = 7.0, proton noise-decoupled. C. Hypothetical spectrum for a random coil form of the polypeptide chain of cytochrome *c* computed as the sum of the amino acid resonances from the known amino acid composition of horse cytochrome *c* and the spectra for individual amino acid residues (Table V-8). D. Same as B, without proton double resonance irradiation. E. Same as B and D, but with selective proton double resonance irradiation of the meso protons at 10.6 ppm (Fig. VI-5b). Note that the resonances of the meso carbon atoms of the heme (b) can be observed as very weak lines in the spectra B and E of ferrocytochrome *c* (from Wüthrich[86]).

5. EXTRINSIC PARAMAGNETIC NMR SHIFT AND RELAXATION PROBES

In the foregoing Section VI.4 we have seen that hemoproteins are endowed with two special natural "NMR probes", i.e. the local ring current field of the heme groups, and the local magnetic fields produced by the unpaired electrons in the paramagnetic states of the heme groups. The improved spectral resolution which results from the presence of these local fields is readily apparent in the Figs. VI-14 to VI-24. An analogous favorable situation may be encountered in other conjugated proteins, e.g. in flavoproteins, and quite generally in paramagnetic metalloproteins. Most peptides and proteins, however, do not fall into this class, and the NMR studies are then restricted to the spectral features inherent in the polypeptide chains (Chapter III). The improved spectral resolution in certain paramagnetic compounds suggested the use of extrinsic paramagnetic species as a technical aid to improve the spectral resolution in diamagnetic molecules. Experience has then shown that in certain favorable cases the pseudocontact shifts produced by the extrinsic paramagnetic probes (Equations VI-8 to VI-11, Fig. VI-10), combined with their influence on the nuclear spin relaxation (Equations VI-15 to VI-19), can be of much help for the elucidation of the molecular geometry of diamagnetic compounds.

The concept of paramagnetic probes as a technical aid in NMR spectroscopy is not restricted to biological applications, but is equally attractive in organic and inorganic chemistry. Thus, studies of the impact of paramagnetic metal ions on the NMR of the solvent featured prominently among the very first NMR experiments[87], and the use of "free" hydrated metal ions as shift and relaxation probes in aqueous media was also among the early developments[88–93]. A real breakthrough came in 1969, when it was discovered that the addition of rare earth chelate complexes to non-aqueous solutions of organic materials could cause sizeable selective resonance shifts in the spectrum of the solute[94]. The use of lanthanide shift reagents, e.g. the bis-pyridine adduct of tris-(dipivaloyl-methanato)europium(III), "Eu(dpm)$_3$", has since become an important technique for NMR applications in organic chemistry[95–97], and these chelate complexes may also be employed for studies of peptides in non-polar solvents. Most biological applications of paramagnetic probes, however, are in aqueous media, and are more directly related to the early NMR studies of paramagnetic metal ions in aqueous solution[88–93].

The theory describing the interaction of nuclear spins with paramagnetic centers was introduced in the Sections VI.2 and VI.3. However, compared to the previously described situations encountered in chemically inert compounds, an additional aspect must now be considered, i.e. the chemical exchange reactions in a solution containing the diamagnetic molecules of interest, the paramagnetic metal ions, and the solvent. This will first be illustrated with simple systems containing Co^{2+} as a NMR shift probe, or Mn^{2+} as a relaxation probe. Thereafter, the discussion will focus on the rare earth metal ions, which are at present the most widely used paramagnetic NMR probes, and on some special applications of paramagnetic metal ions in high resolution NMR studies.

Co^{2+} as a NMR Shift Probe

First, we consider a solution of a Co^{2+}-salt in water. The relevant species for the present discussion are the hydrated cobaltous ion, $Co^{2+}(H_2O)_6$, and free water. It has been shown that the chemical exchange of water molecules between the bulk of the solution and the first coordination sphere of Co^{2+} is very rapid, the average life time in the binding sites of Co^{2+}, τ_M, being of the order $1 \cdot 10^{-6}$ sec at 25°[91].

$$Co^{2+}(H_2O)_6 \underset{(\tau_M = 1 \cdot 10^{-6}\,\text{sec})}{\overset{H_2O}{\rightleftharpoons}} \text{bulk water.} \qquad \text{(VI-23)}$$

For the protons of the metal bound water molecules, the chemical shift with respect to that of the free water in ppm, $\Delta\delta_M \equiv \delta_M - \delta_{H_2O}$, is ≈ 70 ppm[93]. From the discussion of rate processes in Section II.5 (Fig. II-10), it thus follows that the proton resonances of the free and the metal bound water molecules in aqueous solutions of $Co^{2+}(H_2O)_6$ cannot be observed as separate lines. Rather, a single resonance line is observed which corresponds to the average of the resonances for metal-bound and free water. The chemical shift of the average resonance with respect to that of pure water, $\Delta\delta_p = \delta_p - \delta_{H_2O}$, corresponds to that for the metal bound protons, $\Delta\delta_M$, weighted with the ratio of metal-bound to free water.

$$\Delta\delta_p = P_M \cdot q \cdot \Delta\delta_M. \qquad \text{(VI-24)}$$

$P_M = [Me^{n+}]/[H_2O]$ is the ratio of the concentrations in moles liter^{-1} of the paramagnetic metal ion, and water. q is the coordination number of the metal ion with respect to water, i.e. it indicates the number of water molecules bound to Co^{2+} at any given instant. As a numerical example, let us compute $\Delta\delta_p$ in a 0.1-M Co^{2+}-solution. With $[H_2O] = 55$, $q = 6$ and $\Delta\delta_M = 70$ ppm, one has that the proton NMR line of the water is shifted by 0.76 ppm. Because of the very short electronic relaxation time T_{1e} (Table VI-3), the nuclear spin relaxation of H_2O is very little affected by Co^{2+}, and hence this chemical shift is readily observable. Note that $\Delta\delta_p$ increases proportional to the metal ion concentration $[Me^{n+}]$.

Equation VI-24 is valid only for the limiting situation of "very rapid" exchange. A more general description of the chemical shift of the solvent resonance in a solution of one kind of metal ions with one type of coordination sites was given by Swift and Connick[91]

$$\Delta\delta_p = \frac{P_M \cdot q \cdot \Delta\delta_M}{\tau_M^2\left[\left(\frac{1}{T_{2M}} + \frac{1}{\tau_M}\right)^2 + \Delta\omega_M^2\right]} \qquad \text{(VI-25)}$$

T_{2M} is the transverse relaxation time of the metal bound nuclei, and $\Delta\omega_M$ is the chemical shift $\Delta\delta_M$ expressed in radians sec^{-1}. It is readily seen that Equation VI-24 corresponds to the limiting case of VI-25 where $1/\tau_M \gg 1/T_{2M}, \Delta\omega_M$. In the limit of very slow exchange, the solvent resonance will not be affected by the presence of the metal ions, and one observes two separate lines with chemical shifts δ_{H_2O} and δ_M.

It may be instructive at this point to recount the application of Co^{2+} as a shift reagent in the determination of the hydration number of aluminium ions[92]. The solution studied was 1.5-M in Al^{3+}, 0.23-M in Co^{2+}, and 55.5-M in H_2O. It is known that the water exchange in and out of the first coordination sphere of Al^{3+} is very slow, so that separate NMR lines may in principle be observed for free water, and water bound to Al^{3+}. From the relative intensity of the NMR lines, which are in the above solution given by $(55.5\text{–}1.5\,q):1.5\,q$, the number of water molecules bound to Al^{3+} should thus be accessible for direct measurement by NMR. In practice, however, the relative chemical shifts of free water and water bound to the diamagnetic Al^{3+}-ions are too small for the two lines to be resolved. After addition of Co^{2+}, one has the situation described by Equation VI-26

$$Al^{3+}(H_2O)_q \xrightleftharpoons{\text{Slow Exchange}} \text{Bulk Water} \xrightleftharpoons{\text{Rapid Exchange}} Co^{2+}(H_2O)_6. \qquad \text{(VI-26)}$$

On the NMR time scale, $Al^{3+}(H_2O)_6$ is kinetically inert. Therefore only the resonance of the bulk water is shifted by Co^{2+}, and as a result the resonance lines of $Al^{3+}(H_2O)_q$ and the bulk water become well separated. In analogy to this example, shift reagents may discriminate between individual nuclei in a protein, depending on whether or not they are accessible as binding sites for the paramagnetic metal ion. (By the way, from observations of the ^{17}O NMR of H_2O in the system of Equation VI-26, $q = 6$ was found for $Al^{3+}(H_2O)_q$[92].)

In experiments with peptides and proteins, Co^{2+} was e.g. used as a paramagnetic probe to improve the resolution in the high field region of the ^1H-NMR spectrum of lysozyme (Fig. III-20)[98]. Some representative spectra obtained after the addition of increasing amounts of $CoCl_2$ to an aqueous solution of lysozyme are shown in Fig. VI-26. It is seen that some resonances are shifted to higher field, some to lower field, the magnitudes of the shifts are quite different for different resonances, and, as one would

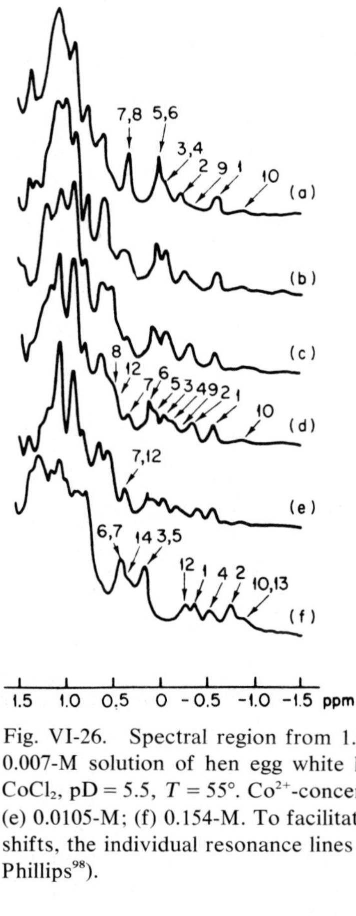

Fig. VI-26. Spectral region from 1.5 to − 1.5 ppm in the ^{1}H-NMR spectra at 220 MHz of a 0.007-M solution of hen egg white lysozyme in D_2O after addition of variable amounts of $CoCl_2$, pD = 5.5, $T = 55°$. Co^{2+}-concentrations: (a) 0; (b) 0.0017-M; (c) 0.0035-M; (d) 0.0070-M; (e) 0.0105-M; (f) 0.154-M. To facilitate the observation of the gradual Co^{2+}-induced resonance shifts, the individual resonance lines are identified with roman numbers (from McDonald and Phillips[98]).

expect from the above discussion, the Co^{2+}-induced shifts increase with increasing Co^{2+}-concentration (Fig. VI-27). Thus we have obviously a situation where Co^{2+} is exchanging rapidly in and out of the binding sites of lysozyme. The NMR spectra which were observed correspond to the average of the spectra of metal-free lysozyme, and lysozyme with one or several Co^{2+}-ions bound. The gradual change of the relative positions of individual resonances with increasing Co^{2+}-concentration helped to dissect out the components of spectral regions which are incompletely resolved in the absence of Co^{2+}, and to determine the relative intensities of the individual resonance lines. In this, the application of the shift reagent is to some extent an alternative to the application of higher magnetic fields H_0. In the spectrum of lysozyme (Fig. VI-26a), only the resonances 1, which corresponds to a methyl group, and 10, which corresponds to a single proton,

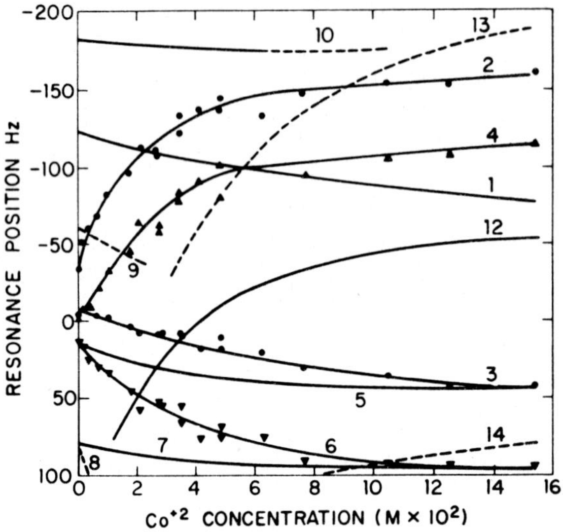

Fig. VI-27. Plot of the ^{1}H-NMR shifts (in Hz) in the high field region of the 220 MHz spectrum of lysozyme (Fig. VI-26) vs. the concentration of Co^{2+} added to the lysozyme solution (from McDonald and Phillips[98]).

and the one-proton line 11 at − 2.0 ppm, which is not shown in the figure, are completely resolved in the absence of Co^{2+}. At some stage of the stepwise addition of Co^{2+} (Fig. VI-26, b–f), each of the resonance lines numbered 2 to 9 is quite well resolved. In addition, three lines, numbered 12 to 14, are shifted out of the bulk of the spectrum into high field positions between 0.5 and − 1.0 ppm by the Co^{2+}. It could thus be determined that all the resonance lines 1 to 7 have an intensity of 3 protons, and thus correspond to methyl groups of aliphatic amino acid side chains.

In this experiment, the principal result of the use of Co^{2+} was the improved spectral resolution. Yet it is instructive to further investigate the origin of the Co^{2+}-induced shifts in the spectrum of lysozyme, since this illustrates some of the potential ambiguities which may arise in the application of paramagnetic probes for studies of the molecular geometry. The Co^{2+}-induced shifts could be caused via different mechanisms, of which the following three, or combinations thereof, are in principle quite probable. (i) The addition of Co^{2+} might induce changes of the protein conformation which would be manifested in the variations of the NMR chemical shifts. (ii) Co^{2+} might bind to one or several specific binding sites of the protein, and then cause hyperfine shifts of the resonances of nuclei located near these binding sites. (iii) Unspecific binding of metal ions on the surface of protein molecules in solution is a quite general phenomenon, and NMR shifts might arise from the hyperfine interactions with these loosely bound metal ions. The explanation (i) was excluded on the basis of optical rotatory dispersion (ORD) experiments, which implied that the lysozyme conformation was maintained essentially unchanged after the addition of Co^{2+}[98], and the hypothesis (iii) was dismissed on the grounds that the Co^{2+}-induced shifts in lysozyme are much larger than those in other proteins, where the influence of unspecifically bound Co^{2+} should be comparable. On the other hand, evidence for specific Co^{2+}-binding was obtained from the observation, that for most of the lines the shifts are almost maximum at a cobaltous ion concentration of 0.15-M (Fig. VI-27), indicating that these shifts are a consequence of the formation of thermodynamically rather stable Co^{2+}-lysozyme complexes. From various lines of evidence, in particular the pH dependence of the Co^{2+}-induced shifts, McDonald and Phillips[98] concluded that for $[Co^{2+}] \leqslant 0.1$-M, the observed shifts came essentially entirely from the formation of a 1:1 complex of Co^{2+} and lysozyme, with an association constant of 57 liters · mole^{-1} at pD = 5.5, and $T = 55°$. The unique binding site was tentatively assigned to a location near the active site of the lysozyme molecule. Assuming that the binding site of Co^{2+} is correctly assigned, the situation is at this point very much alike that encountered in cytochrome b_5 (Section VI.4)[80]. One would have to evaluate separately the contact and pseudocontact shifts, and the latter could then be used to characterize the molecular geometry near the Co^{2+} binding site, and to assign the hyperfine shifted resonance lines to individual amino acid residues (see Fig. VI-17). Such a study of the structure of the metal environment is naturally of particular interest when Co^{2+}, or a different shift probe, can be introduced into a metal binding site which is essential for the biological function of the protein.

Mn^{2+} as a NMR Relaxation Probe

In aqueous solutions of manganous ions, one has an equilibrium situation between metal-bound and free water, and the exchange of water molecules in and out of the first coordination sphere of Mn^{2+} is very rapid, with $\tau_M = 3 \cdot 10^{-8}$ sec at 25°[91].

$$Mn^{2+}(H_2O)_6 \underset{(\tau_M = 3\cdot 10^{-8}\,\text{sec})}{\overset{H_2O}{\rightleftharpoons}} \text{bulk water.} \qquad \text{(VI-27)}$$

The dominant influence of Mn^{2+} on the NMR of the nuclei in H_2O is on the relaxation times, as one would anticipate from the relatively long electronic relaxation time T_{1e} (Table VI-3).

Let us first examine the relaxation rates for the metal-bound water molecules, T_{1M} and T_{2M}. The equations VI-15 and VI-16 can be applied with the following modification: Compared to chemically inert systems, the chemical exchange of nuclei in and out of the coordination sphere of the paramagnetic metal ion is an additional mechanism for the modulation of both dipolar and scalar electron–nucleus interactions. Therefore, for systems with chemical exchange, the effective correlation times in Equations VI-15 and VI-16 become

$$\frac{1}{\tau_{c1}} = \frac{1}{\tau_c} + \frac{1}{T_{1e}} + \frac{1}{\tau_M}; \quad \frac{1}{\tau_{c2}} = \frac{1}{\tau_c} + \frac{1}{T_{2e}} + \frac{1}{\tau_M} \tag{VI-28}$$

$$\frac{1}{\tau_1} = \frac{1}{T_{1e}} + \frac{1}{\tau_M}; \quad \frac{1}{\tau_2} = \frac{1}{T_{2e}} + \frac{1}{\tau_M}. \tag{VI-29}$$

For typical relaxation probes, T_{1M} and T_{2M} cannot usually be measured by the observation of the metal bound ligands, since these NMR lines are broadened beyond detection by the electron–nucleus interactions.

Next, we consider the relaxation rates for the free water molecules in a solution of Mn^{2+}. If there were no chemical exchange of water molecules in the system of Equation VI-27, the relaxation rates of the free water would be unaffected by the paramagnetic species. The relaxation rates would then be identical to those in pure water, $1/T_{1L}$ and $1/T_{2L}$, where the subscript L stands for "ligand". With exchange of ligand molecules between the two different sites M and L, one has[91, 99]

$$\frac{1}{T_{1L}} + \frac{1}{T_{1p}} = \frac{1}{T_{1L}} + \frac{P_M \cdot q}{T_{1M} + \tau_M} \tag{VI-30}$$

$$\frac{1}{T_{2obs}} = \frac{1}{T_{2L}} + \frac{1}{T_{2p}} = \frac{1}{T_{2L}} + \frac{P_M \cdot q}{\tau_M} \left[\frac{\dfrac{1}{T_{2M_2}} + \dfrac{1}{T_{2M}\tau_M} + \Delta\omega_M^2}{\left(\dfrac{1}{T_{2M}} + \dfrac{1}{\tau_M}\right)^2 + \Delta\omega_M^2} \right]. \tag{VI-31}$$

T_{1obs} and T_{2obs} are the relaxation times observed for the NMR lines of the bulk of the ligands in the presence of the paramagnetic species (Equation VI-27), T_{1L} and T_{2L} the relaxation times of the ligands in the absence of the paramagnetic species, $T_{1p}^{-1} = T_{1obs}^{-1} - T_{1L}^{-1}$, $T_{2p}^{-1} = T_{2obs}^{-1} - T_{2L}^{-1}$, $P_M = [Me^{2+}]/[H_2O]$, q is the number of binding sites on the metal ion which are available for the molecules L, T_{1M} and T_{2M} are the relaxation times of the metal bound nuclei, τ_M is the average life time with respect to chemical exchange of the molecules L in the binding sites of the metal ion, and $\Delta\omega_M^2$ is the relative chemical shift between corresponding nuclei in the free and the metal-bound molecules L in radians sec^{-1}. For typical relaxation probes, the T_{2p} relaxation arising from the "$\Delta\omega_M$-mechanism"[91] may be omitted from consideration, and the Swift-Connick Equation VI-31 then has an analogous form to that of the expression VI-30 for T_1

$$\frac{1}{T_{2\mathrm{obs}}} = \frac{1}{T_{2\mathrm{L}}} + \frac{1}{T_{2p}} = \frac{1}{T_{2\mathrm{L}}} + \frac{P_{\mathrm{M}} \cdot q}{T_{2\mathrm{M}} + \tau_{\mathrm{M}}}. \qquad \text{(VI-32)}$$

T_{1p} and T_{2p} are thus governed by $T_{1\mathrm{M}}$ or $T_{2\mathrm{M}}$, τ_{M}, or a combination of $T_{1\mathrm{M}}$ and τ_{M}, or $T_{2\mathrm{M}}$ and τ_{M}, respectively. In the limit of "slow exchange"

$$\frac{1}{T_{1p}} = \frac{P_{\mathrm{M}}q}{\tau_{\mathrm{M}}}; \quad \tau_{\mathrm{M}} \gg T_{1\mathrm{M}}$$

$$\frac{1}{T_{2p}} = \frac{P_{\mathrm{M}}q}{\tau_{\mathrm{M}}}; \quad \tau_{\mathrm{M}} \gg T_{2\mathrm{M}} \qquad \text{(VI-33)}$$

and for the case of "fast exchange"

$$\frac{1}{T_{1p}} = \frac{P_{\mathrm{M}}q}{T_{1\mathrm{M}}}; \quad \tau_{\mathrm{M}} \ll T_{1\mathrm{M}}$$

$$\frac{1}{T_{2p}} = \frac{P_{\mathrm{M}}q}{T_{2\mathrm{M}}}; \quad \tau_{\mathrm{M}} \ll T_{2\mathrm{M}}. \qquad \text{(VI-34)}$$

Note that the terms "slow" and "fast" exchange may have somewhat different meanings depending on whether one considers the chemical shifts (Section II.5), or the relaxation times.

Equations VI-30 to VI-34 show that the relaxation times of the bulk water contain information on the electron–nucleus interactions in $Mn^{2+}(H_2O)_6$, and the kinetics of the water exchange. However, to extract this information, one has to determine the mechanisms governing T_{1p} and T_{2p}. The different limiting situations may be distinguished on the basis of the different temperature dependences of τ_{M} and $T_{1\mathrm{M}}$, or $T_{2\mathrm{M}}$, respectively. The variation of τ_{M} with temperature is given by the usual expression for the temperature dependence of the rate of chemical reactions

$$\tau_{\mathrm{M}} = \frac{h}{kT}\, \mathrm{e}^{(\Delta H^{\ddagger}/RT - \Delta S^{\ddagger}/R)} \qquad \text{(VI-35)}$$

where T is the absolute temperature, $\Delta H^{\ddagger}$ and $\Delta S^{\ddagger}$ are the enthalpy and entropy of activation for the ligand exchange reaction, and R is the gas constant. It is convenient to rewrite Equation VI-35 in the form

$$\ln \tau_{\mathrm{M}} = \frac{\Delta H}{R} \frac{1}{T} + \text{const.} \qquad \text{(VI-36)}$$

where const. is to a good approximation invariant over the temperature range covered with aqueous solutions. The activation enthalpies for ligand exchange reactions of the type considered here are typically of the order of $10\ \mathrm{kcal} \cdot \mathrm{mol}^{-1}$.[91] The temperature variation of $T_{1\mathrm{M}}$ or $T_{2\mathrm{M}}$ depends on the mechanism which dominates the modulation of the electron–nucleus interactions (Equations VI-28 and VI-29), where τ_c, T_{1e}, and T_{2e} vary comparatively little with temperature. As an illustration, the temperature dependence of T_{2p} of ^{17}O in an aqueous solution of Mn^{2+} is shown in Fig. VI-28. At temperatures around 20°, $\tau_M \gg T_{2M}$, and hence according to Equations VI-33 and VI-36, ln T_{2p} increases proportional to $1/T$. With increasing temperature, τ_M becomes shorter, and at higher temperatures than ca. 50°, T_{2p} is governed essentially entirely by T_{2M} (Equation VI-34). To

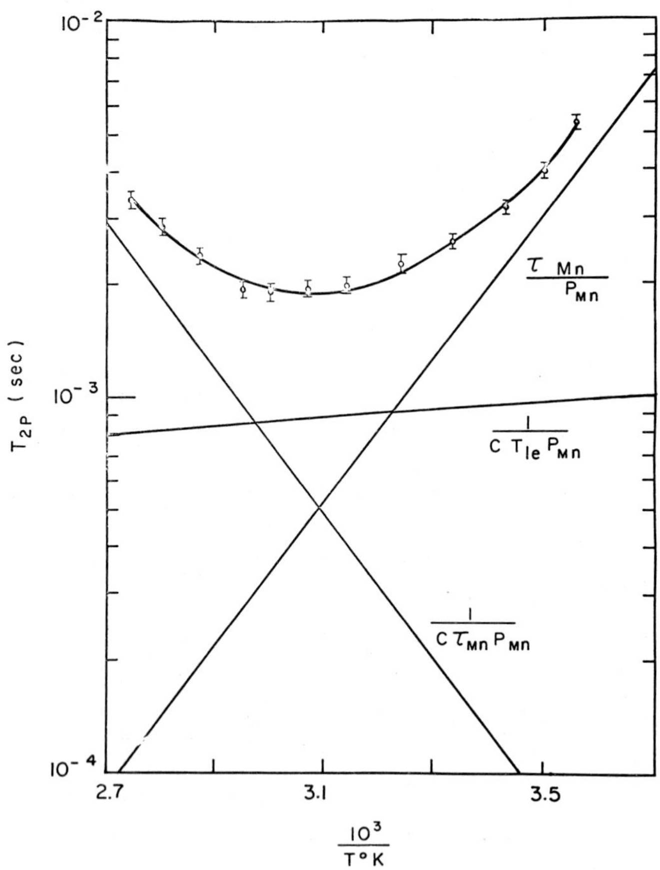

Fig. VI-28. Dependence on the reciprocal of temperature of log T_{2p} of ^{17}O in a $1.8 \cdot 10^{-4}$-M solution of $MnSO_4$ in ^{17}O enriched water. The temperature range covered is 5° ($10^3/T = 3.6$) to 95° ($10^3/T = 2.75$). The straight lines represent the functions for the slow and fast exchange limits (Equations VI-33 and VI-34) which gave the best fit with the experiment. The following parameters were obtained from this fit: $\tau_M = 3 \cdot 10^{-8}$ sec at 25°, $\Delta H^{\ddagger} = 8.1$ Kcal · mole^{-1}, $\Delta S^{\ddagger} = 2.9$ e.u., $A = 9.2 \times 10^6$ Hz, $T_{1e} = 3.8 \times 10^{-9}$ sec at 25° (from Swift and Connick[91]).

analyse the temperature dependence of T_{2p} in this region, the T_2 studies were complemented by measurements of T_1, which showed that T_{1obs} was only little affected by the presence of Mn^{2+}. This implies that the T_{2p} relaxation in Fig. VI-28 had to come predominantly from scalar coupling, with $\omega_S\tau_2 \gg 1$ in Equations VI-15 and VI-16, and the effective correlation time for the T_{2M}-relaxation is thus τ_1 (Equation VI-29). At around 50°, we have that $\tau_1 \approx T_{1e}$, which is almost independent of temperature. Accordingly, $T_{2p} = 1/CT_{1e}P_Mq$ varies little with temperature (Fig. VI-28), where $C = \frac{1}{3}S(S+1)A^2$ (Equation VI-16). If the temperature is increased beyond 50°, τ_M finally becomes short compared even to T_{1e}. We then have that $T_{2p} = 1/C\tau_MP_Mq$, and the temperature dependence is the inverse of that in the chemical exchange controlled region at low temperature (Fig. VI-28).

As was mentioned above, the data of Fig. VI-28 were obtained for the ^{17}O NMR of the water in Mn^{2+} solutions. The same qualitative features were found for the corresponding ^1H-NMR data[100]. In particular, τ_M is identical for ^1H and ^{17}O, which shows that the proton exchange rate is governed by the exchange of the whole water molecules, and not by a hydrolysis mechanism. Overall, if P_M and q are known, τ_M, T_{1e}, and the scalar coupling constant A may be obtained from a set of data corresponding to those of Fig. VI-28. If dipolar coupling is the dominant interaction, the electron–nucleus distance r (Equation VI-16) may be obtained instead of A.

In analogy to the use of Co^{2+} as a shift probe, Mn^{2+} and other relaxation probes may be applied to selectively broaden part of the resonance lines in the NMR spectrum of a protein. As will be illustrated below, the spectral resolution can thus be improved dramatically. However, more important perhaps than these potential applications in high resolution NMR, Mn^{2+} has been widely applied in NMR relaxation studies of enzymes[101]. This will be followed up in Chapter IX. At this point we conclude the discussion with a brief comment on a principal aspect of extending the NMR relaxation studies from aqueous Mn^{2+}-solutions to systems containing also protein molecules, and possibly other metal-binding components: Usually, the temperature range which may be covered with solutions of biological materials is too narrow to allow an unambiguous differentiation of the limiting situations VI-33 and VI-34 solely on the basis of the temperature dependence of T_{1p} and T_{2p}. Therefore, such studies have to include an investigation of the dependence of T_{1p} and T_{2p} on the resonance frequency $\omega_I = \gamma_I H_0$, which requires equipment for studies at variable field strength H_0. It is readily apparent that T_{1p} and T_{2p} are independent of H_0 in the limit of slow chemical exchange (Equation VI-33), whereas the field dependence of T_{1M} and T_{2M} (Equations VI-15, VI-16, VI-28, and VI-29) is usually manifested in T_{1p} and T_{2p} in the limit of rapid chemical exchange (Equation VI-34). Some illustrative examples will be presented in Chapter IX.

The Lanthanide Ions as NMR Probes

The lanthanides represent a group of 15 trivalent ions, including two diamagnetic species, La^{3+} and Lu^{3+}, a typical NMR relaxation probe with a long electronic relaxation time T_{1e}, Gd^{3+}, and 12 potential shift probes, Ce^{3+}, Pr^{3+}, Nd^{3+}, Pm^{3+}, Sm^{3+}, Eu^{3+}, Tb^{3+}, Dy^{3+}, Ho^{3+}, Er^{3+}, Tm^{3+}, and Yb^{3+}. One of the principal advantages of the lanthanides as NMR probes is the uniform chemical behaviour throughout the series, e.g. isomorphous replacement of different lanthanide ions in chelate complexes is quite common. Furthermore, with the possible exception of the nuclei which are directly bonded to the lanthanide ion, the hyperfine shifts come essentially entirely from pseudocontact interactions. The lanthanide-induced ^{1}H-NMR shifts can thus be related to the geometry of the surroundings of the lanthanide ion in the molecule to which it is bonded, and the differences in the hyperfine shifts and relaxation effects observed after isomorphous replacement of different lanthanide ions may greatly reduce the ambiguities usually inherent in the application of a single perturbing probe.

The magnetic properties of the lanthanides were investigated in detail[102, 103]. In contrast to the transition metal ions, spin-orbit coupling in the lanthanides is dominant over the crystal field effects, and hence $J = (L + S)$ rather than S is a good quantum number. Since the electronic g-tensor in the lanthanides is isotropic at ambient temperature, pseudocontact shifts are observed only to a second approximation. One thus has the expression VI-37 for the pseudocontact shifts in ppm in an axially symmetric ligand field [103]

$$\delta_{pc} = \frac{g^2\beta^2 J(J+1)(2J-1)(2J+3)}{60(kT)^2 r^3} D_z\,(3\cos^2\Omega - 1) \tag{VI-37}$$

g is the electronic g-factor, β the Bohr magneton, k the Boltzmann constant, T the absolute temperature, D_z a characteristic constant for a given lanthanide ion[103], r the distance from the metal ion to the nucleus, and Ω the angle between the principal symmetry axis and the vector $\boldsymbol{r}$ linking the metal ion with the nucleus. For axial symmetry, which appears to be quite common for lanthanide ions[102, 103], the ratio of the pseudocontact shifts of two different nuclei i and j is then

$$\frac{\delta^i_{pc}}{\delta^j_{pc}} = \frac{\left\langle \dfrac{3\cos^2\Omega_i - 1}{r_i^3} \right\rangle}{\left\langle \dfrac{3\cos^2\Omega_j - 1}{r_j^3} \right\rangle} \tag{VI-38}$$

The brackets indicate that Ω and r are averaged over rate processes, e.g. intramolecular rotational and vibrational motions, which are rapid on the NMR time scale. The contact shifts are given by Equation VI-12, where S has to be replaced by J. However, as mentioned above, the scalar coupling constant A are usually small for protons in lanthanide complexes, and sizeable contact shifts result only for directly bonded nuclei, e.g. ^{31}P. Since this applies also for Gd^{3+}, ratios in terms of the electron–nucleus distance r can also be formulated for the longitudinal or transverse relaxation times induced by Gd^{3+} in different protons i and j (see Equations VI-15 and VI-16).

$$\frac{\left(\frac{1}{T_1^i}\right)}{\left(\frac{1}{T_1^j}\right)} = \frac{\left(\frac{1}{T_2^i}\right)}{\left(\frac{1}{T_2^j}\right)} = \frac{\langle r_j^6 \rangle}{\langle r_i^6 \rangle} \qquad \text{(VI-39)}$$

In writing Equation VI-39, it is implicitly assumed that the same effective correlation time applies for the nuclei i and j, which is in general valid for protons in the same ligand molecule, if τ_{c1} and τ_{c2} are governed by T_{1e}, T_{2e}, and τ_M (Equation VI-28).

A particularly attractive feature of the lanthanides as a group of NMR probes is that they not only allow isomorphous substitution of diamagnetic ions, relaxation probes, and shift inducing ions, but in addition the different lanthanide shift probes produce markedly different pseudocontact shifts. Theoretical and observed relative pseudocontact shifts of nuclei in ligands bound to lanthanide ions are given in Table VI-7. The wide range of shifts obtained with different lanthanides, which may further be modified by gradual changes of the lanthanide concentration (see Fig. VI-27), allows very often to find suitable conditions for improvement of the spectral resolution. They can also provide the experimental data needed to quantitatively relate the geometry of the lanthanide environment to the observed NMR shifts.

Some General Comments on the Use of Paramagnetic NMR Probes

In the foregoing, we have encountered two different uses of paramagnetic probes in high resolution NMR. These are the applications as a technical aid to simplify the spectra of complex molecules, and for the determination of molecular structures. If only a simplification of the spectrum is desired, be it e.g. by selective broadening of part of the resonance lines, spreading out of the spectrum by selective shifts of individual resonances, or quenching of the nuclear Overhauser enhancement[104], any reagent may in practice be used in any desired manner[95–97]. These uses of NMR probes may then be combined with other NMR techniques, e.g. double resonance studies in the well-resolved spectra[105–108]. For structural conclusions to be put on a firm basis, the origin of the observed shifts has to be established, as was briefly discussed for the Co^{2+}-induced shifts in lysozyme (Figs. VI-26 and VI-27). The following procedures have been suggested for structural studies with

TABLE VI-7

Relative NMR shifts of ligand nuclei in lanthanide complexes at room temperature scaled to $Dy^{3+} = 100$*.

Lanthanide Ion	Theoretical pseudo-contact shift (ppm)	Experimental hyperfine shift
La^{3+}	0	0
Ce^{3+}	6.3	
Pr^{3+}	11.0	13
Nd^{3+}	4.2	7.6
Pm^{3+}	−2.0	
Sm^{3+}	0.7	1.3
Eu^{3+}	−4.0	−12
Gd^{3+}	0	not observed
Tb^{3+}	86	92
Dy^{3+}	100	100
Ho^{3+}	39	51
Er^{3+}	−33	−31
Tm^{3+}	−53	−64
Yb^{3+}	−22	−23
Lu^{3+}	0	

*Negative shifts are to higher field at constant radio frequency. The experimental hyperfine shifts are for the tris-complexes of the rigid ligand 2,6-dipicolinate (from Ref. 102).

lanthanides[102], and may also be applied for transition metal ions, if suitable isomorphous substitutions can be found.

(i) Only pseudocontact shifts may readily be quantitatively related with the coordinates in space of the nucleus understudy (Equation VI-37). Lanthanide ions which produce pseudocontact shifts both to high and low field must be used (Table VI-7).

(ii) To eliminate effects which might arise from changes of the protein conformation upon binding of the metal ion, the observed shifts must be related to those in the diamagnetic lanthanide complexes with La^{3+} or Lu^{3+}.

(iii) To eliminate possible contributions from contact coupling, the ratios of the shifts for different proton resonances (Equation VI-38) should be compared for the different lanthanides. If these ratios are independent of the lanthanide cation, the observed shifts have their origin in dipolar coupling. Furthermore, the constancy of the shift ratios for the different lanthanides is a strong indication for axial symmetry of the dipolar field[102].

(iv) The stoichiometry of the protein-lanthanide complexes must be determined by following the dependence of the lanthanide induced shifts, and the shift ratios for different protons, on the lanthanide concentration (see Fig. VI-27).

(v) A search for the best fit of the NMR shift data to a particular structure is best done with a computer. The search should include positioning of the metal ion, unless its location could independently be determined, and the orientation of the magnetic axis. These procedures are quite similar to those described in Section VI.4 for cytochrome b_5 (Fig. VI-17).

(vi) The chances for a successful computer search (v) are improved if relaxation data obtained with Gd^{3+} as a relaxation probe (Equation VI-39) are also included.

(vii) If the procedures (v) and (vi) fail to find a solution, one may want to examine the temperature dependence of the shifts, since a possible reason for the failure might be that the observed shifts correspond to the average of several molecular conformations present in a thermal equilibrium.

A complete investigation following these lines was described for mononucleotides in solution[106].

In addition to these technical considerations, the applicability of paramagnetic metal ions for structure determinations depends on the molecules to be studied. The technique will quite generally be suitable for systems which contain metal ions in the native state, e.g. metalloproteins, or proteins which require metal ions as cofactors for biological activity. For example, Mn^{2+} can in enzymatic reactions be substituted for Ca^{2+} or Mg^{2+},[101] and it has been suggested that the lanthanide ions might be substituted for Ca^{2+} and Mg^{2+} in biological ATP reactions. In metal free systems, on the other hand, the biological relevance of the metal binding site of a protein may often not be obvious. In addition, the metal ion may induce conformation changes which are possibly not very likely to occur under physiological conditions. To cite a particular example, the complexation with a transition metal ion, or a lanthanide ion, will almost certainly impose a new, and possibly quite rigid conformation on a flexible linear oligopeptide in solution.

Simplification of Protein ^{1}H-NMR Spectra by Difference Spectroscopy

In Section IV.3 (Fig. IV-9) we have seen that in Fourier transform NMR experiments, the line shape in the frequency domain spectrum may be altered by manipulations of the time domain response, e.g. by multiplication of the latter with suitable exponential functions, the signal : noise ratio may be improved at the expense of some line broadening ("convolution"), or the resolution may be enhanced at the expense of a lesser signal : noise ratio ("deconvolution"). Here we shall see that a quite dramatic resolution enhancement in protein ^{1}H-NMR spectra may be obtained by using the difference between a broadened and a normal spectrum. The broadened spectrum can either be obtained by convolution in the computer, or by the addition of a line broadening probe to the protein solution, e.g. Gd^{3+} or Mn^{2+}.

The convolution difference method is a straightforward mathematical manipulation of the free induction decay. Assume that the latter is of the form

$$f(t) = e^{-t/T_2^*} \cos[(\omega' - \omega_0)t] \qquad \text{(VI-40)}$$

where the decay constant T_2^* was defined in Equation IV-15, ω' is the frequency of the applied radio, frequency pulse, and ω_0 is the resonance frequency. The Fourier transform of this decay describes a Lorentzian line

$$F(\omega) = \mathrm{L}(T_2^*) = \frac{T_2^*}{1 + \Delta\omega^2 T_2^{*2}} \qquad \text{(VI-41)}$$

where $\Delta\omega = \omega - \omega_0$. If we multiply the decay of Equation VI-40 with e^{-t/τ_A}, the resulting frequency domain function is also Lorentzian, with T_2^* replaced by T_A, where $1/T_A = 1/T_2^* + 1/\tau_A$. Similarly, a Lorentzian $\mathrm{L}(T_B)$ may be obtained, with $1/T_B = 1/T_2^* + 1/\tau_B$. The convolution difference spectrum is obtained by multiplying $\mathrm{L}(T_B)$ with a constant K, and subtracting $K \cdot \mathrm{L}(T_B)$ from $\mathrm{L}(T_A)$. The following new line shape is thus generated

$$\mathrm{L(CD)} \propto \frac{T_A - KT_B + \Delta\omega^2(T_A T_B^2 - KT_B T_A^2)}{1 + \Delta\omega^2(T_A^2 + T_B^2) + \Delta\omega^4 T_A^2 T_B^2}. \qquad \text{(VI-42)}$$

The constant K may be adjusted empirically to give a convenient line shape. The two cases $K = T_B/T_A$, and $K = 1$ are discussed in detail in the original literature[107].

Fig. VI-29 shows a convolution difference ^{1}H-NMR spectrum of lysozyme. It is seen that for a considerable number of lines, the fine structure from spin–spin coupling is resolved. Thus, spin decoupling can then be applied even in regions where individual resonances were hardly resolved in the original spectrum. (See also illustrations in III.4.)

In an alternative approach, the spectrum $\mathrm{L}(T_B)$ to be used in Equation VI-42 may be recorded experimentally by adding a line-broadening reagent to the protein solution. In the situation of fast chemical exchange of the metal ion in and out of the binding site on the protein (Equation VI-34), T_{2p} (Equation VI-32) is then equivalent to τ_B in the digital convolution of the spectrum. However, in contrast to the digital convolution, the paramagnetic probe which may be a cation, e.g. Gd^{3+} or Mn^{2+}, an anion, e.g. $Cr(CN)_6^{3-}$, or an uncharged particle, e.g. a nitroxide spin label, selectively broadens only the resonances of the protons located near the binding site. With $K = 1$ (equation VI-42), only these resonances will appear in the difference spectrum. Since T_{2p} decreases proportional to the metal ion concentration (Equation VI-32), and in the fast exchange situation increases with r^6

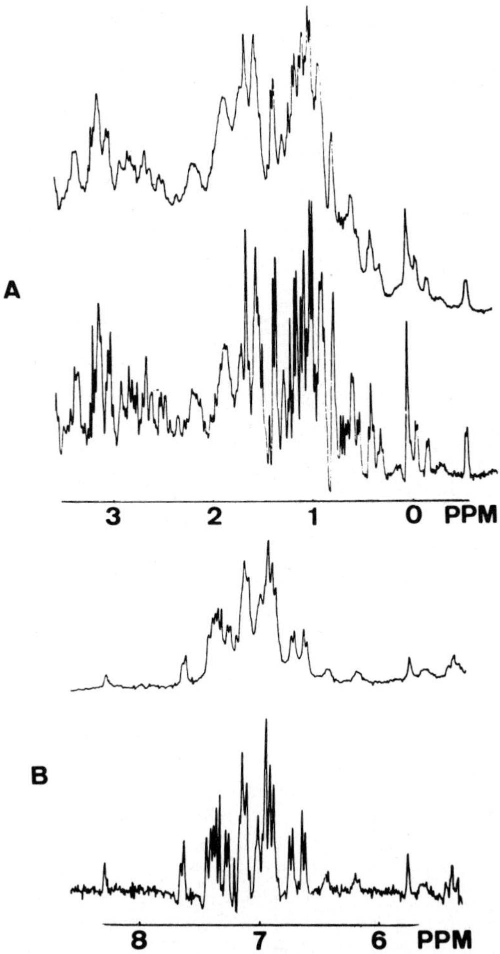

Fig. VI-29. Normal 270 MHz ^{1}H-NMR spectrum and convolution difference ^{1}H-NMR spectrum of hen egg white lysozyme. A. aliphatic region from − 1 to 3 ppm. B. aromatic region from 6 to 8 ppm. Upper trace, normal spectrum; lower trace, convolution difference spectrum. The following parameters were used in Equation VI-42: $\tau_A = 0.25$ sec, $\tau_B = 0.08$ sec, $K = 0.94$ (from Campbell et al.[107]).

(Equation VI-16), the radius of the sphere around the metal binding site within which the proton resonances are sufficiently broadened to appear in the paramagnetic difference spectrum, can be adjusted by variation of the concentration of the line broadening reagent (Fig. VI-30). In Fig. VI-31 a conventional paramagnetic difference spectrum of lysozyme is compared with a deconvoluted paramagnetic difference spectrum[108]. The advantage of the latter in providing an improved spectral resolution is readily apparent.

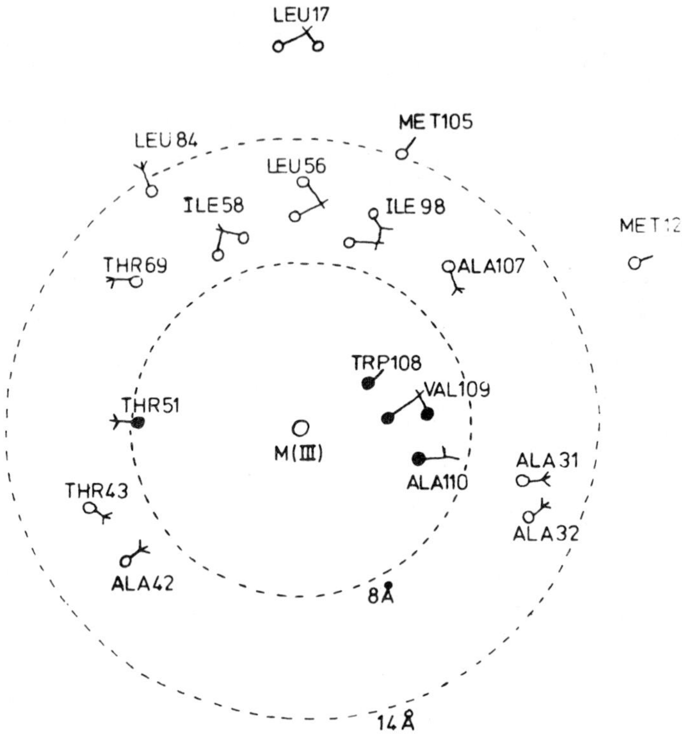

Fig. VI-30. Schematic two-dimensional representation of amino acid residues around the lanthanide metal binding site in single crystals of lysozyme, which is thought to be a binding site also in solution. The numbers indicate the position of the amino acid residues in the amino acid sequence. In the situation of fast exchange of the paramagnetic relaxation probe in and out of the binding site, the radius of the sphere within which the NMR lines are extensively broadened (two different spheres with radius 8 Å, and 14 Å, respectively, are indicated by the broken lines) is different depending on the concentration of the paramagnetic species added. Only the resonances of nuclei located inside the sphere will appear in a paramagnetic difference spectrum (see Fig. VI-31) (from Campbell et al.[108]).

Nitroxide Spin Labels as NMR Probes

Because of the long electronic relaxation time T_{1e} (Table VI-3), nitroxide radicals may be used as NMR relaxation probes[109, 110]. The principles are the same as for the use of paramagnetic metal ions, e.g. Gd^{3+} or Mn^{2+}. The spin labels represent an interesting addition to the metal probes, in that they are neutral molecules, and can be covalently attached to a macromolecular structure.

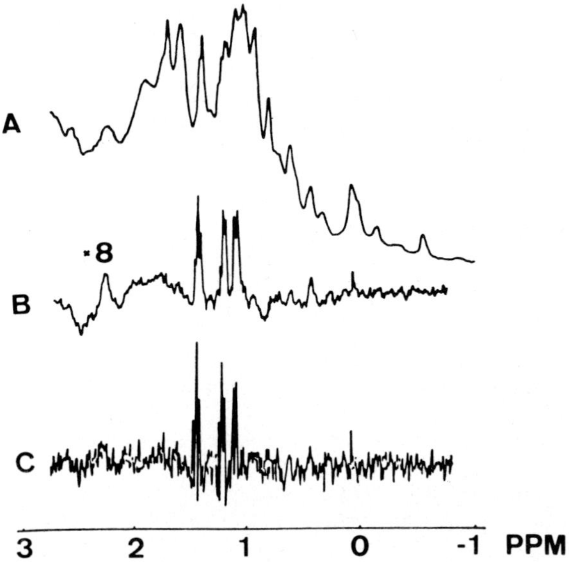

Fig. VI-31. Paramagnetic ^{1}H-NMR difference spectra of hen egg white lysozyme at 270 MHz. A. Normal spectrum of a D_2O solution of 0.005-M lysozyme and 0.05-M La^{3+}. B. Conventional paramagnetic difference spectrum obtained as the difference between spectrum A, and the spectrum obtained after adding 5×10^{-5}-M Gd^{3+} to the solution A. C. Deconvoluted paramagnetic difference spectrum. Only this spectrum contains well resolved multiplet structures, which could e.g. be used for spin decoupling studies (from Campbell et al.[108]).

Nitroxide spin labels are primarily EPR probes[4,5]. From the EPR measurements, one obtains information on the effective rotational correlation time for the spin label, which is manifested in the more or less complete averaging of the *g*-tensor and hyperfine coupling tensor anisotropies (Table VI-2)[4,5]. Additional data may be obtained from studies of the interactions between the spin label and other paramagnetic centers in the same system[8]. The present discussion shall be limited to a "NMR aspect" of such experiments. The EPR experiments will usually provide information on the mobility of the spin label, but not on its location in the system of interest. Because of the line broadening effect of the spin label on the NMR lines of nearby nuclei (see Fig. VI-30), its location may in certain cases be determined from a comparison of the NMR spectra of the labelled and the unlabelled system. This is of particular interest when the system studied is labelled with a chemically unreactive nitroxide radical, which may be introduced e.g. by diffusion into the system. Typical examples are found among the applications of nitroxide spin labels for studies of membranes[4,5].

6. BIBLIOGRAPHY

1 Eaton D. R. and Phillips W. D., Adv. Magnetic Res. *1*, 103 (1965)
2 De Boer E. and van Willigen H., Progress in NMR Spectroscopy *2*, 111 (1965)
3 La Mar G. N., Horrocks W. D. and Holm R. H. (Eds.), NMR of Paramagnetic Molecules. Academic Press, New York, 1973.
4 McConnell H. M. and McFarland B. G., Quart. Rev. Biophys *3*, 91 (1970)
5 Griffith O. H. and Waggoner A. S., Acc. Chem. Res. *2*, 17 (1968)
6 Wüthrich K., Structure and Bonding *8*, 53 (1970)
7 Phillips W. D. and Poe M., in Iron–Sulfur Proteins, W. Lovenberg (Ed.). Academic Press, New York, 1974
8 Dwek R. A., Nuclear Magnetic Resonance in Biochemistry: Applications to Enzyme Systems. Clarendon Press, Oxford, 1973
9 Ayscough P. B., Electron Spin Resonance in Chemistry. Methuen, London, 1967
10 Abragam A. and Bleaney B., Electron Paramagnetic Resonance of Transition Ions. Clarendon Press, Oxford, 1970
11 Feher G., Electron Paramagnetic Resonance with Applications to Selected Problems in Biology. Gordon and Breach, New York, 1970
12 Swift T. J. and Connick, R. E., J. Chem. Phys. *37*, 307 (1962)
13 McConnell H. M. and Robertson R. E., J. Chem. Phys. *29*, 1361 (1958)
14 Fermi E., Z. Physik *60*, 320 (1930)
15 Wüthrich K. and Baumann R., Helv. Chim. Acta *57*, 336 (1974)
16 Caughey W. S. and Koski W. S., Biochemistry *1*, 923 (1962)
17 Wüthrich K., Shulman R. G., Wyluda B. J. and Caughey W. S., Proc. Nat. Acad. Sci. *62*, 636 (1969)
18 Doddrell D. and Caughey W. S., J. Amer. Chem. Soc. *94*, 2510 (1972)
19 Wüthrich K., Keller R. M. and Baumann R., in Dynamic Aspects of Conformation Changes in Biological Macromolecules, C. Sadron (Ed.). Reidel, Dordrecht-Holland, 1973, p. 151
20 Wüthrich K. and Baumann R., Helv. Chim. Acta *56*, 585 (1973)
21 Kreilick R. W., in Ref. 3, p. 595
22 McConnell H. M. and Robertson R. E., J. Chem. Phys. *29*, 1361 (1958)
23 Jesson J. P., J. Chem. Phys. *47*, 579 (1967)
24 Kurland R. J. and McGarvey B. R., J. Magn. Res. *2*, 286 (1970)
25 Hoard J. L., in Structural Chemistry and Molecular Biology, A. Rich and N. Davidson (Eds.) Freeman, London, 1968, p. 573
26 Bloembergen N., J. Chem. Phys. *27*, 595 (1957)
27 McConnell H. M., J. Chem. Phys. *24*, 764 (1956)
28 McConnell H. M. and Chesnut D. B., J. Chem. Phys. *28*, 107 (1958)
29 Bersohn R., J. Chem. Phys. *24*, 1066 (1956)
30 Bolton J. R., Carrington A. and McLachlan A. D., Mol. Phys. *5*, 31 (1962)
31 Eaton D. R., Josey A. D., Phillips W. D. and Benson R. E., J. Chem. Phys. *37*, 347 (1962)
32 Eaton D. R., Josey A. D., Benson R. E., Phillips W. D. and Cairna T. L., J. Am. Chem. Soc. *84*, 4100 (1962)
33 Shulman R. G., Glarum S. H. and Karplus M., J. Mol. Biol. *57*, 93 (1971)
34 Karplus M. and Fraenkel G. K., J. Chem. Phys. *35*, 1312 (1961)
35 Horrocks W. D., in Ref. 3, p. 127
36 Blumberg W. E., private communication
37 Carrington A. and McLachlan A. D., Introduction to Magnetic Resonance. Harper and Row, New York, 1967, p. 84
38 Solomon I., Phys. Rev. *99*, 559 (1955)

39 Connick R. E. and Fiat D., J. Chem. Phys. *44*, 4103 (1966)
40 Reuben J., Reed G. H. and Cohn M., J. Chem. Phys. *52*, 1617 (1970)
41 Sternlicht H., J. Chem. Phys. *42*, 2250 (1965)
42 Woessner D., J. Chem. Phys. *36*, 1 (1962)
43 Guéron M., J. Magn. Res. (in press)
44 Phillips W. D. and Poe M., in Iron–Sulfur Proteins, W. Lovenberg (Ed.). Academic Press, New York, 1974
45 Shulman R. G., Wüthrich K., Yamane T., Patel D. and Blumberg W. E., J. Mol. Biol. *53*, 143 (1970)
46 Wüthrich K., Shulman R. G. and Peisach J., Proc. Nat. Acad. Sci. *60*, 373 (1968)
47 Kurland R. J., Davis D. G. and Ho C., J. Amer. Chem. Soc. *90*, 2700 (1968)
48 Shulman R. G., Wüthrich K., Yamane T., Antonini E. and Brunori M., Proc. Nat. Acad. Sci. *63*, 623 (1969)
49 Wüthrich K., Shulman R. G., Yamane T., Wyluda B. J., Hügli T. E. and Gurd F. R. N., J. Biol. Chem. *245*, 1947 (1970)
50 Patel D. J., Kampa L., Shulman R. G., Yamane T. and Wyluda B. J., Proc. Nat. Acad. Sci. *67*, 1109 (1970)
51 Sheard B., Yamane T. and Shulman R. G., J. Mol. Biol. *53*, 35 (1970)
52 Morishima I. and Iizuka T., J. Amer. Chem. Soc. *96*, 5279 (1974)
53 Wüthrich K., Shulman R. G. and Yamane T., Proc. Nat. Acad. Sci. *61*, 1199 (1968)
54 Shulman R. G., Ogawa S., Wüthrich K., Yamane T., Peisach J. and Blumberg W. E., Science *165*, 251 (1969)
55 Davis D. G., Lindstrom T. R., Mock N. H., Baldassare J. J., Charache S., Jones R. T. and Ho C., J. Mol. Biol. *60*, 101 (1971)
56 Ho C., Davis D. G., Mock N. H., Lindstrom T. and Charache S., Biochem. Biophys. Res. Comm. *38*, 779 (1970)
57 Lindstrom T. R. and Ho C., Proc. Nat. Acad. Sci. *69*, 1707 (1972)
58 Lindstrom T. R., Norén I. B. E., Charache S., Lehmann H. and Ho C., Biochemistry *11*, 1677 (1972)
59 Ogawa S. and Shulman R. G., Biochem. Biophys. Res. Comm. *42*, 9 (1971)
60 Ogawa S., Shulman R. G. and Yamane T., J. Mol. Biol. *70*, 291 (1972)
61 Ogawa S., Shulman R. G., Fujiwara M. and Yamane T., J. Mol. Biol. *70*, 301 (1972)
62 Ogawa S. and Shulman R. G., J. Mol. Biol. *70*, 315 (1972)
63 Winterhalter K. and Wüthrich K., J. Mol. Biol. *63*, 477 (1972)
64 Wüthrich K., Keller R. M., Brunori M., Giacometti G., Huber R. and Formanek H., FEBS Letters *21*, 63 (1972)
65 Ogawa S., Mayer A. and Shulman R. G., Biochem. Biophys. Res. Comm. *49*, 1485 (1972)
66 Mayer A., Ogawa S., Shulman R. G. and Gersonde K., J. Mol. Biol. *81*, 187 (1973)
67 Kowalsky A., Biochemistry *4*, 2382 (1965)
68 Wüthrich K., Proc. Nat. Acad. Sci. *63*, 1071 (1969)
69 McDonald C. C., Phillips W. D. and Vinogradov S. N., Biochem. Biophys. Res. Comm. *36*, 442 (1969)
70 Wüthrich K., in Probes of Structure and Function of Macromolecules and Membranes, Vol. II: Probes of Enzymes and Hemoproteins, B. Chance, T. Yonetani and A. S. Mildvan (Eds.). Academic Press, New York, 1971, 465
71 Wüthrich K., Aviram I. and Schejter A., Biochim. Biophys. Acta *253*, 98 (1971)
72 Gupta R. K. and Redfield A. G., Science *169*, 1204 (1970)
73 Gupta R. K., Koenig S. H. and Redfield A. G., J. Magn. Res. *7*, 66 (1972)
74 Redfield A. G. and Gupta R. K., Cold Spring Harbor Symposia on Quantitative Biology *36*, 405 (1971)
75 Stellwagen E. and Shulman R. G., J. Mol. Biol. *80*, 559 (1973)

76 Keller R. M., Pettigrew G. W. and Wüthrich K., FEBS Letters *36*, 151 (1973)
77 Smith G. M. and Kamen M. D., Proc. Nat. Acad. Sci. *71*, 4303 (1974)
78 McDonald C. C., Phillips W. D. and Le Gall J., Biochemistry *13*, 1952 (1974)
79 Dobson C. M., Hoyle N. J., Geraldes C. F., Wright P. E., Williams R. J. P., Bruschi M. and Le Gall J., Nature *249*, 425 (1974)
80 Keller R. M. and Wüthrich K., Biochim. Biophys. Acta *285*, 326 (1972)
81 Keller R. M., Groudinsky O. and Wüthrich K., Biochim. Biophys. Acta *328*, 233 (1973)
82 Keller R. M., Wüthrich K. and Debrunner P., Proc. Nat. Acad. Sci. *69*, 2073 (1972)
83 Mayer A., Ogawa S., Shulman R. G., Yamane T., Cavaleiro J. A. S., D'A Rocha Gonsalves A. M., Kenner G. W. and Smith K. M., J. Mol. Biol. *86*, 749 (1974)
84 Oldfield E. and Allerhand A., Proc. Nat. Acad. Sci. *70*, 3531 (1973)
85 Oldfield E. and Allerhand A., J. Amer. Chem. Soc. *97*, 221 (1975)
86 Wüthrich K., Pure Appl. Chem. 1975 *40*, 127 (1974)
87 Bloembergen N., Purcell E. M. and Pound R. V., Phys. Rev. *73*, 679 (1948)
88 Bloembergen N., J. Chem. Phys. *27*, 572 (1957)
89 Phillips W. D., Looney C. E. and Ikeda C. K., J. Chem. Phys. *27*, 1435 (1957)
90 Jackson J. A., Lemons J. and Taube H., J. Chem. Phys. *33*, 553 (1960)
91 Swift T. J. and Connick R. E., J. Chem. Phys. *37*, 307 (1962)
92 Connick R. E. and Fiat D. N., J. Chem. Phys. *39*, 1349 (1963)
93 Luz Z. and Shulman R. G., J. Chem. Phys. *43*, 3750 (1965)
94 Hinckley C. C., J. Amer. Chem. Soc. *91*, 5160 (1969)
95 Sanders J. K. M. and Williams D. H., Nature *240*, 385 (1972)
96 von Ammon R. and Fischer R. D., Angew. Chemie *84*, 737 (1972)
97 Horrocks W. DeW., in Ref. 3, p. 479
98 McDonald C. C. and Phillips W. D., Biochem. Biophys. Res. Comm. *35*, 43 (1969)
99 Luz Z. and Meiboom S., J. Chem. Phys. *40*, 2686 (1964)
100 Bloembergen N. and Morgan L. O., J. Chem. Phys. *34*, 842 (1961)
101 Cohn M., Quart. Rev. of Biophysics *3*, 61 (1970)
102 Bleaney B., Dobson C. M., Levine B. A., Martin R. B., Williams R. J. P. and Xavier A. V., Chem. Comm. *1972*, 791
103 Bleaney B., J. Magn. Res. *8*, 91 (1972)
104 Gutowsky H. S. and Natusch D. F. S., J. Chem. Phys. *57*, 1203 (1972)
105 Birdsall B., Feeney J., Glasel J. A., Williams R. J. P. and Xavier A. V., Chem. Comm. *1971*, 1473
106 Barry C. D., North A. C. T., Glasel J. A., Williams R. J. P. and Xavier A. V., Nature *232*, 236 (1971)
107 Campbell I. D., Dobson C. M., Williams R. J. P. and Xavier A. V., J. Magn. Res. *11*, 172 (1973)
108 Campbell I. D., Dobson C. M., Williams R. J. P. and Xavier A. V., Ann. New York Acad. Sci. *222*, 163 (1973)
109 Sternlicht H. and Wheeler E., in Magnetic Resonance in Biological Systems, A. Ehrenberg, B. G. Malmström and T. Vänngard (Eds.). Pergamon Press, New York, 1967, p. 325.
110 Kopple K. D. and Schamper T. J., J. Amer. Chem. Soc. *94*, 3644 (1972)

APPENDIX VI:

INTERPRETATION OF ^{1}H AND ^{13}C-NMR HYPERFINE SHIFTS IN PARAMAGNETIC MOLECULES

In Section VI.2 the interpretation of ^{1}H-NMR hyperfine shifts in paramagnetic molecules was discussed, and illustrated with the spectra of low spin ferric porphyrin complexes. It was shown that the spin densities centered on the peripheral aromatic carbon atoms of the porphyrin ring could on the basis of the Equations VI-6 to VI-14 be evaluated both in the isolated heme groups, and in hemoproteins (Section VI.4). The extent of the electron delocalization, and the symmetry of the electronic wave functions in the heme groups could thus be assessed. More detailed information may be obtained through the combination of ^{1}H and ^{13}C-NMR data. It is the purpose of this appendix to introduce some of the concepts used for the interpretation of the ^{13}C hyperfine shifts.

Contact Shifts

In analogy to the situation encountered with protons, the contact interaction constants A^{C} may be derived from the measured contact shifts with the use of Equation VI-12. In practice, one deals quite often with the molecular fragments I and II.

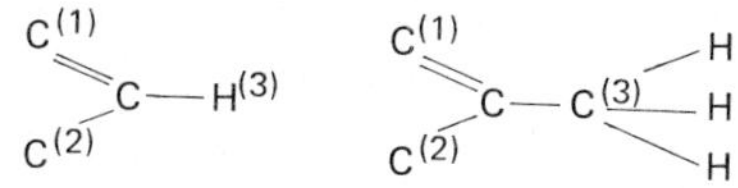

I II (VI-43)

In these fragments, $C^{(1)}$, C, and $C^{(2)}$ are part of an aromatic ring system and may contain finite spin densities ρ_{C}^{π} in their p_z-orbitals. $H^{(3)}$ (I) and $C^{(3)}H_3$(II) are the substituents attached to the central aromatic carbon atom. In contrast to A^{H}, A^{C} for the carbon atoms in these fragments is in general related to the electron spin densities centered on more than one aromatic atom. To describe the relations between the contact interaction constants A^{I} and the spin densities ρ_{C}^{π}, we shall therefore use the following notation for the factors Q (Equation VI-13): $Q_{C'H}^{H}$ describes the interactions between an aromatic proton and the spin density centered on the carbon atom to which it is attached, $Q^{C}_{C'C}$ the interactions between a ^{13}C nucleus and the spin density on the carbon atom to which it is attached, Q^{C}_{CX} the interactions of a ^{13}C nucleus attached to an atom X with the spin density centered on its own p_z-orbital, etc.

Using the above notation, the McConnell relation VI-13 becomes

$$A^{H} = Q^{H}_{C'H} \cdot \rho^{\pi}_{C'} \qquad \text{(VI-44)}$$

where, as discussed previously, the factor $Q^{H}_{C'H}$, which describes electron–nucleus interactions through spin polarization[27,28], is of the order $Q^{H}_{C'H} \approx -63$ MHz, and $\rho^{\pi}_{C'}$ is given in percent of one unpaired electron. The corresponding relation for the methyl protons in the molecular fragment II is

$$A^{H} = Q^{H}_{C'CH_3} \cdot \rho^{\pi}_{C'} \qquad \text{(VI-45)}$$

where the electron–proton interactions are mainly due to hyperconjugation[29,30], and $Q^{H}_{C'CH_3}$ can vary significantly among different molecules[6,15,30–33]. As mentioned in Section VI.2, $Q^{H}_{C'CH_3}$ is positive, and a value of ≈ 75 MHz is quite typical for uncharged aromatic radicals[29,30].

For an aromatic carbon atom, the contact interaction constant A^{C} is related to the spin densities ρ^{π}_{C} and $\rho^{\pi}_{X_i}$ centered on its π-orbital, and on the π-orbitals of the three atoms X_i to which it is bonded[34]. For the central carbon atoms in the molecular fragments I and II (Equation VI-43), the electron-^{13}C interactions are mainly by spin polarization, and one has[34]

$$A^{C} = \left(S^{C} + \sum_{i=1}^{3} Q^{C}_{CX_i}\right) \rho^{\pi}_{C} + \sum_{i=1}^{3} Q^{C}_{X_iC} \cdot \rho^{\pi}_{X_i} \qquad \text{(VI-46)}$$

S^{C} accounts for the polarization of the 1s-electrons, the factors $Q^{C}_{CX_i}$ for the polarization of the 2s-electrons by the π-electron spin density on the observed carbon atom, and the factors $Q^{C}_{X_iC}$ for the polarization of the 2s-electrons by the π-electron spin densities on the neighboring atoms. The following numerical values have been derived from theoretical considerations[34], and were found to be in good agreement with the experimental data in a variety of organic radicals: $S^{C} \approx -35$ MHz, $Q^{C}_{CH} \approx 54$ MHz, $Q^{C}_{CC'} \approx 40$ MHz, and $Q^{C}_{C'C} \approx -39$ MHz. It is seen that the factors $Q^{C}_{CX_i}$ and $Q^{C}_{X_iC}$ in Equation VI-46 depend on the atoms X.

For the methyl carbon atom in the molecular fragment II, one obtains from Equation VI-46 an analogous expression to the McConnell relation for aromatic protons

$$A^{C} = Q^{C}_{C'C} \cdot \rho^{\pi}_{C'} \qquad \text{(VI-47)}$$

where C′ denotes the aromatic carbon to which the methyl group is bound, and $Q^{C}_{C'C} \approx -39$ MHz.

Returning our attention briefly to the low spin ferric hemes, we see that the qualitative features of the hyperfine shifts (Tables VI-4 and VI-5) can be interpreted in terms of contact shifts resulting from π-delocalization of the electron. For example, Equations VI-45 and VI-47 show that the sign of δ_c is different for the ^{13}C-NMR and the ^{1}H-NMR of the ring methyls, which coincides with the experimental observations (Table VI-5). Similarly, Equations VI-44 and VI-45 imply that with positive spin densities on all the β-carbons, the sign of δ_c is different for the 2,4-protons and the ring methyl protons in Fe^{3+}(DEUT)$(CN)_2$. This is again compatible with the experimental data.

In addition to the spin delocalization through the π-bonding system discussed so far, delocalization through σ-bonds might in principle also contribute to the observed hyperfine shifts. Very few examples have so far been reported where σ spin delocalization appeared to be of importance, and as a general trend it was found that the interactions through σ-bonds are rapidly attenuated with increasing distance from the metal ion[35].

Metal Localized and Ligand Localized Pseudocontact Shifts

In the foregoing discussion of pseudocontact shifts, we assumed implicitly that the electron spin interacting with the nuclear spins through dipole–dipole coupling was localized on the metal ion. For nuclei outside the first coordination sphere of the paramagnetic center, this is in most cases a reasonable assumption. However, if finite amounts of spin density ρ_C^π are localized on the nuclei of the metal-bound ligands, their dipolar interactions might also contribute to the observed hyperfine shifts. One would then have ligand localized pseudocontact shifts δ_{pc}^{L} in addition to the metal localized pseudocontact shifts δ_{pc}^{M}.

While the spin density at the ligand nucleus is usually much less than at the metal ion, $1/r^3$ can be much larger for the ligand localized fraction of the electron spin, so that δ_{pc}^{M} and δ_{pc}^{L} may be of comparable magnitude. It has been shown that ligand localized pseudocontact shifts are to be expected when unpaired electron density is localized in a *p*-orbital of the observed nucleus, whereas δ_{pc}^{L} should be negligibly small if the electron density on the nucleus in question is predominantly in *s*-orbitals[24]. For the heme groups, one might thus expect ligand localized pseudocontact shifts for the aromatic ring carbon atoms, and possibly also for the pyrrole nitrogens, while on the other hand δ_{pc}^{L} should be absent for all the ^{1}H-NMR lines, and for the ^{13}C-NMR of the aliphatic side chains in positions 1, 3, and 5–8 in Fe^{3+}(PROTO)$(CN)_2$ (Fig. VI-4).

Separation of Contact and Pseudocontact Shifts

Summarizing the discussion up to this point, we have that in general the following contributions to the hyperfine shifts observed in paramagnetic transition metal compounds may have to be taken into consideration

$$\delta_{hf} = \delta_{pc}^{M} + \delta_{pc}^{L\sigma} + \delta_{pc}^{L\pi} + \delta_{c}^{\sigma} + \delta_{c}^{\pi} \qquad \text{(VI-48)}$$

d_{pc}^{M} is the pseudocontact shift arising from dipole–dipole coupling with the electron spin localized on the central metal ion, $\delta_{pc}^{L\sigma}$ and $\delta_{pc}^{L\pi}$ are the pseudocontact shifts from interactions with the electron spin delocalized to the ligands through the σ- and π-bonds, respectively, δ_{c}^{σ} the contact shift arising from Fermi contact coupling with the electron spin delocalized to the ligand atoms through the σ- bond system, and δ_{c}^{π} the contact shifts from spin delocalization in the π-bond system. The foregoing discussions have shown that σ-delocalization of the electron spin can in most cases be neglected, and that $\delta_{pc}^{L\pi}$ vanishes for protons and for aliphatic carbon atoms attached to an aromatic ring. Thus for the aromatic proton $H^{(3)}$ and the ^{1}H and ^{13}C nuclei of the methyl group in the molecular fragments I and II (Equation VI-43), Equation VI-48 reduces to

$$\delta_{hf} = \delta_{pc}^{M} + \delta_{c}^{\pi} \qquad \text{(VI-49)}$$

which is identical to the previously used Equation VI-14. For the aromatic carbons, one has

$$\delta_{hf} = \delta_{pc}^{M} + \delta_{pc}^{L\pi} + \delta_{c}^{\pi} \qquad \text{(VI-50)}$$

The separate evaluation of the two terms in Equation VI-49 was discussed in Section VI.2. Similarly δ_{pc}^{M} in equation VI-50 may separately be evaluated. As the following discussion of low spin ferric hemes shows, the occurrence of ligand localized pseudocontact shifts may be demonstrated by comparison of the spin densities ρ_{C}^{π} obtained from the ^{1}H and ^{13}C-NMR data.

Spin Density Distribution and Ligand Localized Pseudocontact Shifts of ^{13}C in Low Spin Iron(III) Porphyrin Complexes

In the following, the relations VI-6 to VI-12 and VI-44 to VI-50 are applied to the NMR data on the symmetrical molecule Fe^{3+}(Porphin)$(CN)_2$ (Fig. VI-3 and Table VI-4), and the natural hemes (Figs. VI-4 to VI-9 and Table VI-5).

The separate evaluation of the contact and pseudocontact shifts of the ^{1}H resonances in low spin ferric hemes was discussed in Section VI.2. For Fe^{3+}(Porphin)$(CN)_2$, the upper limit for δ_{pc}^{M} of the meso-protons was estimated at ≈ -10 ppm, and the corresponding spin densities on the meso- and β-carbons were then found to be $\approx 0\%$, and $\approx 0.8\%$ of one unpaired electron, respectively. In this calculation, $g = 2.4$ had been used in Equation VI-12. As an illustration, the ^{13}C hyperfine shifts of Fe^{3+}(Porphin)$(CN)_2$ will now be discussed with the same assumptions for δ_{pc}^{M}, i.e. $\delta_{pc}^{M} \approx -10$ ppm for the meso-protons. First, from Equation VI-11, δ_{pc}^{M} for the β-carbon, meso-carbon, and quaternary carbon positions are -11.5, -21.3, and -27.8 ppm, respectively, and from Table VI-4 the corresponding values for $(\delta_c^{\pi} + \delta_{pc}^{L\pi})$ are -29.1, -28.4 and -73.6 ppm. Using $\rho_C^{\pi} = 0$ and 0.8% for the meso- and β-carbon atoms, respectively, in Equation VI-6 and checking for internal consistency, $\rho_C^{\pi} \approx 0.3$ and $\rho_N^{\pi} \approx 0.8$ were obtained for the quaternary carbons and the pyrrole nitrogens, where the latter had been substituted in the computation by aromatic carbon atoms. For the ligand localized pseudocontact shifts $\delta_{pc}^{L\pi}$ of the aromatic carbon atoms it was found that a good fit with the experiments could be obtained with the assumption that $\rho_{pc}^{L\pi}$ is proportional to the unpaired spin density ρ_C^{π} localized on the carbon atom considered[20]

$$\delta_{pc}^{L\pi} = D \cdot \rho_C^{\pi} \tag{VI-51}$$

with the factor $D \approx -74$ ppm, and ρ_C^{π} given in percent of one unpaired electron. $\delta_{pc}^{L\pi}$ thus becomes -22, 0, and -60 ppm for the quaternary, meso- and β-carbon atoms, respectively. In conclusion, the interpretation of the NMR hyperfine shifts has shown that the spin density distribution in $Fe^{3+}(Porphin)(CN)_2$ is characterized by relatively high spin densities on the peripheral aromatic carbons 1–8, whereas only small amounts of the electron are on the quaternary and the meso carbon atoms. Overall, approximately 15% of the unpaired electron would be localized on the porphin ligand, with 85% on the iron and the axial cyanide ligands.

Compared to $Fe^{3+}(Porphin)(CN)_2$, a new element enters into the treatment of $Fe^{3+}(PROTO)(CN)_2$ in that we have also electron–nucleus interactions via hyperconjugation (Equation VI-45). As was mentioned before, $Q_{C'CH_3}^{H}$ varies sizeably in different molecules, and should therefore in each case be checked upon. With $\delta_{pc}^{M} \approx -10$ ppm for the meso-protons, Equation VI-11 yields $\delta_{pc}^{M} \approx -5$ ppm and -4 ppm for the carbon atom and the protons of the ring methyl groups. For the average contact shift of the 4 ring methyl carbons, we then have from Table VI-5 $(\delta_c^{\pi})_{av} \approx -42$ ppm, and from Equations VI-12 and VI-47 the average of the spin densities on the β-carbons to which the ring methyls are bound is $(\rho_C^{\pi})_{av} \approx 0.8\%$. Using this value for ρ_C^{π} in Equation VI-45, and with $(A^H)_{av} \approx 4.3 \cdot 10^5$ Hz obtained from the ^{1}H-NMR of the four ring methyls (Table VI-5), one finds $Q_{C'CH_3}^{H} \approx 53$ MHz[15]. Values around 50 MHz for $Q_{C'CH_3}^{H}$ had also been obtained from a similar calibration experiment using the 2,4-proton lines of $Fe^{3+}(DEUT)(CN)_2$, where δ_C^{π} is also mainly caused by spin polarization, in the place of the ^{13}C-NMR of the methyl carbons[33]. Somewhat smaller values, i.e. $Q_{C'CH_3}^{H} \approx 40$ MHz were obtained with the assumption that δ_{pc}^{M} for the meso-protons is less than -10 ppm[6, 15, 17].

The sign of δ_{hf} for the methylene protons *c* and the carbon atoms *d* (Table VI-5) implies that these nuclei are also subject to hyperconjugation with the electron spin. While δ_{hf} for the carbon nuclei *c* shows that the spin density on the β-carbons 6 and 7 is very nearly identical to that on the methyl substituted ring carbon atoms (Table VI-5), much smaller contact shifts are observed for the methylene protons *c* than for the methyl protons *a*. For the protons *c* one thus obtains an effective *Q* factor in Equation VI-45 of $Q^{H}_{C'CH} \approx 20$ MHz. At this point one must recall that the methyl groups are normally free to rotate, and the contact shift observed for the methyl protons is, in fact, an average contact shift. In cases where the substituent is not rotating freely, it was found experimentally that for each individual proton *Q* becomes[37]

$$Q^{H}_{C'CH} = B_0 + B_2 \cos \chi \qquad \text{(VI-52)}$$

where B_0 and B_2 are constant factors, and χ is the torsion angle between the p_z orbital of the aromatic carbon, and the methyl or methylene C—H bond. In the present example of the heme substituents, the different contact shifts of the protons *a* and *c* provide direct evidence that the rotational motions of the propionate side chains about the C—C bond linking them with the porphyrin ring are severely restricted. In general, Equation VI-52 may be used to investigate the rotamer populations in aliphatic substituents of paramagnetic molecules.

CHAPTER VII

NITROGEN-15 NMR OF AMINO ACIDS AND PEPTIDES

Among the seven NMR nuclei encountered in polypeptide chains, i.e. ^{1}H, D, ^{13}C, ^{14}N, ^{15}N, ^{17}O, and ^{33}S, ^{15}N is unique besides ^{1}H and ^{13}C to have a spin $I = 1/2$ (Table I-1), and thus to be suitable for high resolution NMR. In addition, like ^{1}H and ^{13}C, ^{15}N is omnipresent in peptide and protein molecules, and is thus the third fundamental NMR nucleus for studies of structure and conformation in this class of biopolymers. In practice, the development of ^{15}N-NMR for studies of molecular structures was in the past rather slow, mainly because of the experimental difficulties arising from the low NMR sensitivity and natural abundance of ^{15}N (Tables I-1 and I-2). As we will see in the following, these technical difficulties have been overcome by recent advances in instrumentation, and the results obtained so far indicate that the prospects of ^{15}N-NMR as a complementary technique to ^{1}H and ^{13}C-NMR studies of peptides and proteins are most encouraging.

The other nitrogen isotope, ^{14}N, might from a first glance at Table I-1 also appear a promising choice because of its high natural abundance. We will see in the following section that the chemical shifts and spin couplings of ^{14}N correspond to those of ^{15}N. However, the line broadening due to the nuclear quadrupole moment of ^{14}N severely limits the applications of ^{14}N-NMR. Thus, while the ^{14}N chemical shifts have been compiled for the free amino acids[1], the ^{14}N resonances in peptides and proteins appear in most cases to be broadened beyond detection.

1. NMR PROPERTIES OF NITROGEN-15

^{15}N *Chemical Shifts*

A general guide to typical ^{15}N chemical shifts for various functional groups in organic substances is given in Table VII-1. It is seen that ^{15}N chemical shifts of diamagnetic compounds cover a range of approximately 900 ppm[2–5]. ^{14}N chemical shifts are the same within experimental error when expressed in ppm[3,5].

In Table VII-1, the chemical shifts are given relative to $NH_4^{\oplus}$, which is one

TABLE VII-1

Nitrogen-15 chemical shifts in some diamagnetic structures*.

Molecular structure	Nitrogen-15 chemical shift δ in ppm from $^{15}NH_4^{\oplus}$

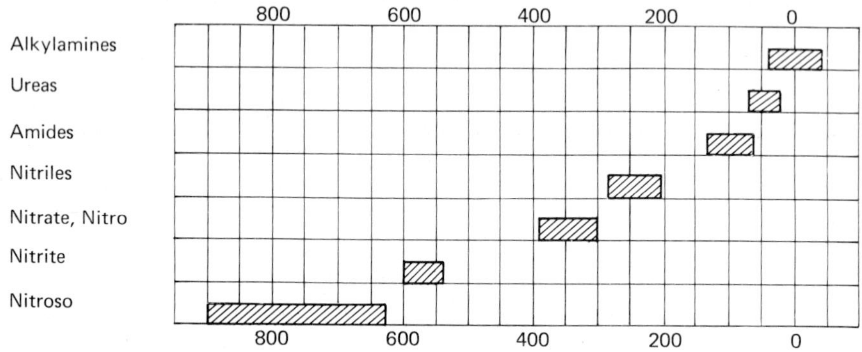

*The bars indicate the spectral regions where the resonances may occur under different conditions of solvent and temperature.

of the commonly used standards. A 5-M solution of isotope enriched $^{15}NH_4{}^{15}NO_3$ in 2-M aqueous HNO_3 is often used as an external standard[6], where the resonance of $NO_3^{\ominus}$ is 354 ppm to lower field from $NH_4^{\oplus}$.[7] In view of the large range of nitrogen chemical shifts, susceptibility corrections for the use of external standards are usually negligibly small, especially if one takes care to examine closely related substances under uniform conditions. Since heteronuclear field-frequency locking systems are a common attachment in modern spectrometers, direct measurements of ^{15}N chemical shifts relative to the ^1H-NMR frequency of TMS is also readily practicable[3,5,8].

Spin–Spin Couplings with ^{15}N

Table VII-2 contains some representative values for the $^{15}N—^1H$ and $^{15}N—^{13}C$ spin–spin coupling constants observed in simple amides. It is seen that one-bond couplings between directly bonded ^{15}N and ^1H are of the order of 90 Hz. Two-bond couplings via the amide link are between 10 and 25 Hz. Two-bond couplings via single bonds are of the order 1 Hz, and the three-bond $^{15}N—^1H$ coupling in acetamide is also very small. One-bond $^{15}N—^{13}C$ couplings through the peptide link are of the order 15 Hz, and the two-bond $^{15}N—^{13}C$ coupling in acetamide is 9.5 Hz. Spin–spin couplings with ^{14}N are equal to those with ^{15}N multiplied by $\gamma_{^{14}N}/\gamma_{^{15}N} \approx -0.71$, where the minus sign comes from the different signs of the magnetic moments of the two isotopes (Table I-1). As we will see below, the ^{15}N couplings appear not only in the ^{15}N-NMR spectra, but produce also well resolved multiplets of the ^{1}H and ^{13}C resonances. In contrast, the ^{14}N couplings are, as a result of the short relaxation times of ^{14}N, not usually manifested in the ^{1}H or ^{13}C-NMR spectra (see e.g. the ^{1}H-NMR spectra of peptides in Figs. III-13, III-16, and III-17).

TABLE VII-2

Some heteronuclear ^{15}N—^{1}H and ^{15}N—^{13}C spin–spin coupling constants in amides (from Ref.[5]).

Compound		Spin–spin couplings (Hz)	
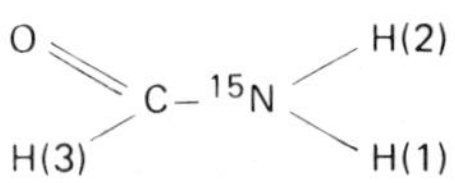	(in H_2O)	$^{1}J_{^{15}NH(1)}$	95.4
		$^{1}J_{^{15}NH(2)}$	91.9
		$^{2}J_{^{15}NH(3)}$	23.3
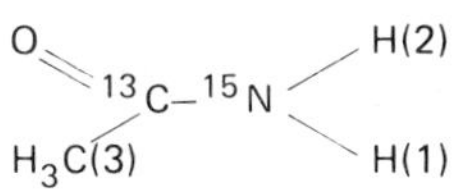	(in H_2O)	$^{1}J_{^{15}NH(1)}$	90.9
		$^{1}J_{^{15}NH(2)}$	88.4
		$^{3}J_{^{15}NH(3)}$	1.3
		$^{1}J_{^{15}N^{13}C}$	< 15
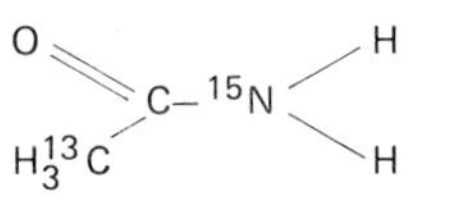		$^{2}J_{^{15}N^{13}C}$	9.5
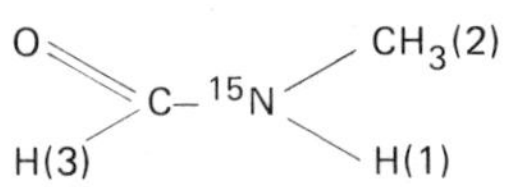	(neat)	$^{1}J_{^{15}NH(1)}$	92.6
		$^{2}J_{^{15}NH(3)}$	14.6
		$^{2}J_{^{15}NH(2)}$	0.8
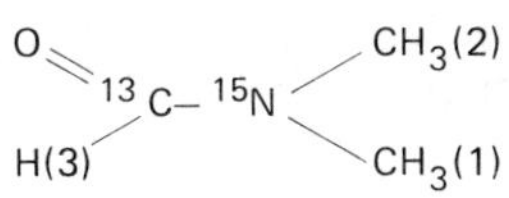		$^{2}J_{^{15}NH(3)}$	15.6
		$^{2}J_{^{15}NH(1)}$	1.2
		$^{2}J_{^{15}NH(2)}$	1.1
		$^{1}J_{^{15}N^{13}C}$	13.4

Spin Relaxation and Nuclear Overhauser Enhancement

Concerning the spin relaxation of ^{15}N, the overall situation is quite similar to that for ^{13}C (Chapter V). The relaxation times for ^{15}N are rather long[9], so that the natural line width should be very small. Fig. VII-1 represents the dependence of the relaxation times T_1 and T_2 on the correlation time τ_c for a ^{15}N spin relaxing by dipolar coupling with a covalently attached proton, which should be quite representative for ^{15}N in a peptide group. Larger line widths than those expected from Fig. VII-1 may obviously occur when there

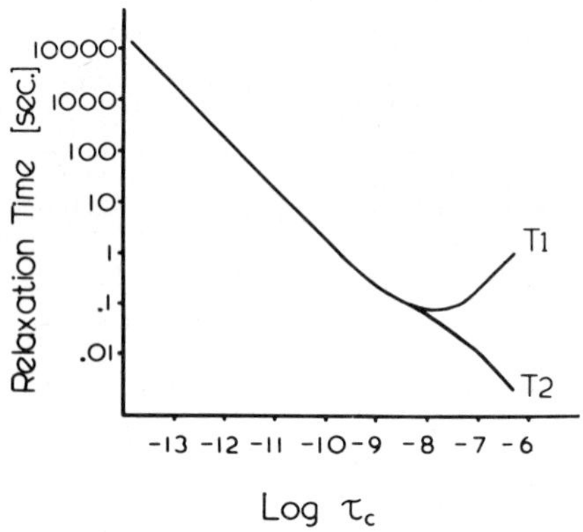

Fig. VII-1. Log–log plot of T_1 and T_2 vs. the effective correlation time for rotational motions, τ_c, for a ^{15}N spin relaxing by a dipolar interaction with a single covalently bonded proton ($r = 1.03$ A). The calculation is for a field H_0 of 21.1 KG ($\nu_{^{15}N} = 9.1$ MHz), and it was assumed that the rotational reorientation of the relaxation vector r_{IS} (Fig. IV-11) is isotropic, so that it can be described by one correlation time τ_c, and that there is complete proton decoupling (from Hawkes and Randall[10]).

is chemical exchange of protons coupled with ^{15}N (Ref. 9a).

The maximum possible nuclear Overhauser enhancement for a protonated ^{15}N nucleus relaxed entirely by ^1H—^{15}N dipolar coupling is in the extreme motional narrowing limit given by Equation VII-1.

$$NOE = 1 + \frac{\gamma_{^1H}}{2\gamma_{^{15}N}} = 1 - 4.94 \qquad \text{(VII-1)}$$

Because of the negative magnetic moment of ^{15}N, the NMR signals are inverted by the NOE, with the maximum signal intensity equal to -3.94 times that without proton irradiation[11]. The dependence of the NOE on the rotational correlation time τ_c is shown in Fig. VII-2. It is seen that at the resonance frequency 9.1 MHz, the ^{15}N signal vanishes for $\tau_c \approx 6 \cdot 10^{-9}$ sec. Note that in situations where the ^{15}N relaxation is not exclusively via ^1H—^{15}N dipolar coupling, vanishing intensity of the ^{15}N resonances may be obtained for shorter correlation times τ_c.[10, 11]

2. EXPERIMENTAL ASPECTS

Observation of the ^{15}N-NMR Spectra of Amino Acids and Peptides

As a result of the low inherent NMR sensitivity of ^{15}N (Tables I-1 and I-2), and the rather unfavorable situations which may be encountered in $^{15}N\{^1H\}$ double resonance experiments under conditions of less than maximum

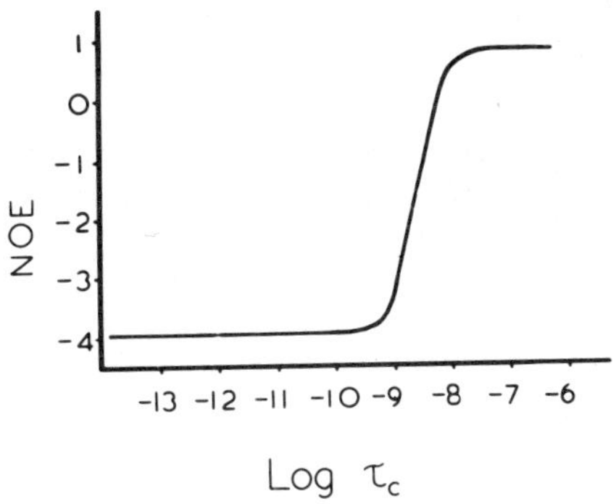

Fig. VII-2. Plot of the NOE vs. log τ_c for a ^{15}N spin relaxing by ^{15}N—^{1}H dipole–dipole coupling, $\nu_{^{15}N} = 9.1$ MHz. For details of the computation see Fig. VII-1 (from Hawkes and Randall[10]).

NOE, the observation of ^{15}N-NMR spectra in all but the simplest molecules was until very recently prohibitively time consuming. Today, the principal difficulties opposing the use of ^{15}N for studies of peptides have been overcome. Fourier transform NMR is now the approach of choice for the observation of ^{15}N, and spectra have been recorded for amino acids and peptides both at natural abundance of ^{15}N, and at various levels of isotope enrichment.

The Figs. VII-3 to VII-7 show some ^{15}N-NMR spectra of amino acids and peptides obtained at natural abundance of ^{15}N. Experimental details are indicated in the figure captions. It is seen that the negative NOE expected for small molecules (Fig. VII-2) is only manifested in Fig. VII-4, since the other figures represent power spectra*, which contain no phase information.

Two different Fourier transform NMR spectrometers were employed to record the spectra presented in this section. The "conventional" instrument operated at 21.1 KG, which corresponds to a ^{15}N resonance frequency of 9.12 MHz, and could be operated with sample tubes of 10, 13, or 15 mm outer diameter. The Figs. VII-3 to VII-5 show spectra obtained under these conditions. The other spectrometer, using a static field of 42.2 KG ($\nu_{^{15}N} =$ 18.24 MHz) produced by a wide bore superconducting solenoid which accomodates sample sizes of up to 25 mm diameter, is available specifically for work with ^{15}N and other rare nuclei. The gain in sensitivity over that of the spectrometers operating at resonance frequencies accessible with iron core magnets, and with the conventional sample tubes of 10 to 15 mm diameter, is quite strikingly illustrated by the spectra of viomycin in Fig. VII-6. On the

*The "power spectrum" corresponds to $(u^2 + v^2)^{1/2} \equiv (M_{x'}^2 + M_{y'}^2)^{1/2}$ (Equations IV-11 and IV-12).

other hand, the spectral resolution is already almost optimal at 9.12 MHz. The 13 nitrogen positions in viomycin (Fig. VII-6) give rise to 11 resonance lines, of which two correspond to two nitrogens each. At 18.24 MHz, 12 separate resonance lines can be distinguished, where the two resonances at

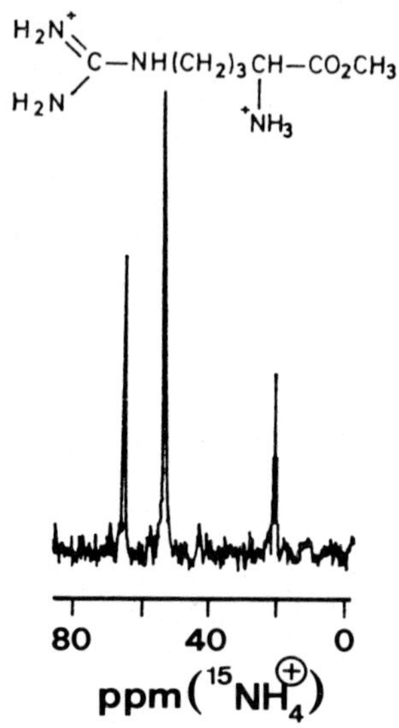

Fig. VII-3. Natural abundance ^1H-decoupled FT ^{15}N-NMR spectrum at 9.1 MHz (H_0 = 21.2 KG) of L-arginine methyl ester dihydrochloride. Note that the "power spectrum" (Section VII-2) has been recorded, with concomitant loss of the phase information. The chemical shift scale is in ppm relative to external $^{15}NH_4^{\oplus}$(5-M $^{15}NH_4NO_3$ in 2-M HNO_3), where positive numbers indicate shifts to lower field. The structure of the molecule studied is also indicated in the figure (from Pregosin et al.[6]).

lowest field are still essentially degenerate and give rise to a single line. Another spectrum obtained with the superconducting spectrometer at 18.24 MHz is that of gramicidin S in Fig. VII-7, which was recorded at natural abundance of ^{15}N in only 40 minutes.

Considering the data obtained at natural isotope abundance, the prospects for work with ^{15}N enriched peptides and even proteins are really encouraging. With ^{15}N enrichment from the natural abundance of 0.37% (Table I-1) to 95%, a 250 times smaller sample concentration will give the same signal:noise for the observation of amide nitrogens in corresponding molecules. For example, from Fig. VII-6B one estimates that the ^{15}N-NMR spectrum of a 10^{-2}-M solution of a peptide enriched to 95% in ^{15}N may be recorded with an acceptable signal:noise ratio after signal accumulation for approximately 10 hrs with a spectrometer operating at around 10 MHz, and

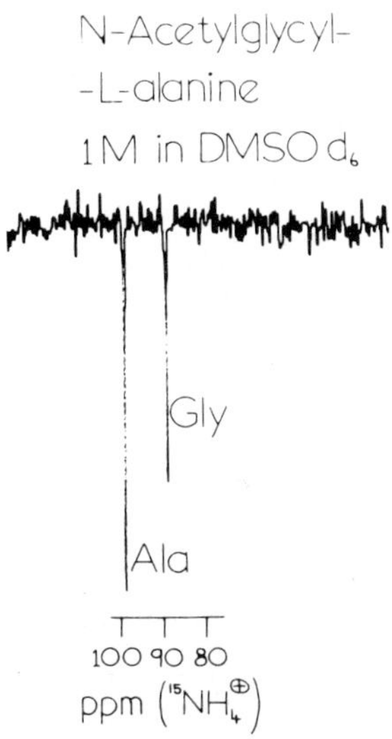

Fig. VII-4. Natural abundance, ^{1}H-decoupled FT ^{15}N-NMR spectrum at 9.1 MHz of N-acetylglycyl-L-alanine in d_6-DMSO. The spectrum is correctly phased and demonstrates the negative NOE (from Hawkes and Randall[10]).

using sample tubes of 13 to 15 mm diameter. Or, from Fig. VII-7, at a resonance frequency of ca. 20 MHz and using 25 mm sample tubes, ^{15}N spectra of less than millimolar solutions of enriched peptides should become accessible with accumulation times of several hours. The potential applications to quite large ^{15}N enriched peptides and proteins are all the more attractive, as the spectra of Figs. VII-6 and VII-7 indicate that the amide ^{15}N

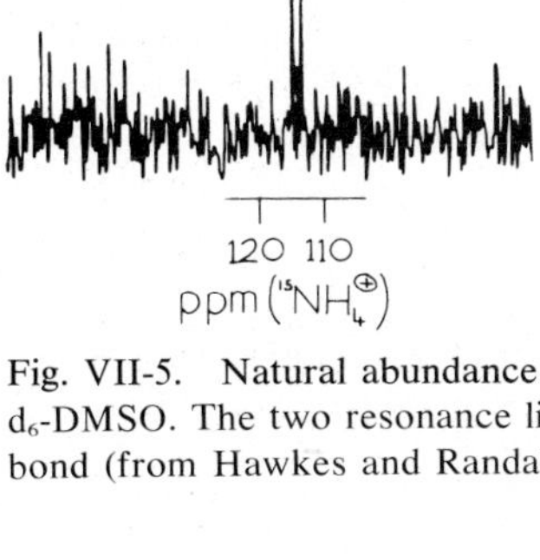

Fig. VII-5. Natural abundance ^{1}H-decoupled ^{15}N FT power spectrum of N-formylproline in d_6-DMSO. The two resonance lines were attributed to the cis and trans forms of the peptide bond (from Hawkes and Randall[10]).

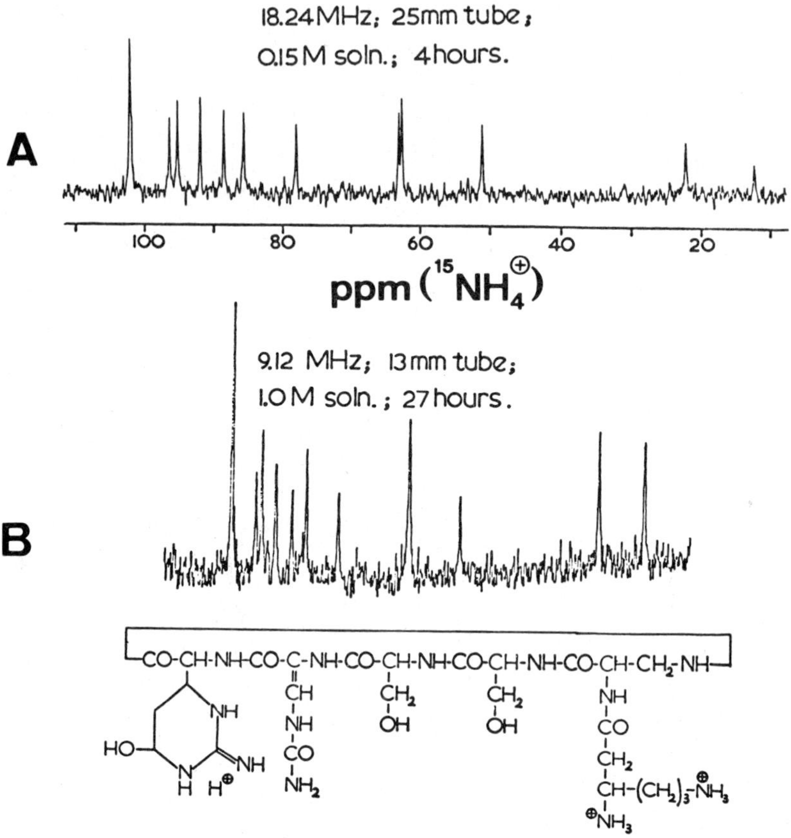

Fig. VII-6. Natural abundance ^{1}H-decoupled ^{15}N FT power spectra* of aqueous solution of viomycin sulphate. The structure of viomycin is indicated at the bottom of the figure. This figure provides a comparison of the signal : noise ratios and the spectral resolution attainable with two modern commercial spectrometers. Spectrum A was recorded with a wide bore superconducting magnet which accepts sample tubes of 25 mm outer diameter, and produces a magnetic field $H_0 = 42.2$ KG, corresponding to $\nu_{^{15}N} = 18.24$ MHz. Spectrum B was obtained with a conventional iron core magnet, $H_0 = 21.1$ KG, sample size 13 mm outer diameter (the same spectrometer was used to record the spectra in Figs. VII-3 to VII-5). Note that the spectrum A has a much better signal : noise ratio even though the solute concentration was ca. 6 fold lower, and the accumulation time ca. 7 fold shorter than for spectrum B. The total amount of substance used was ca. 5×10^{-3} moles for experiment A, where the sample volume was ca. 30 ml, and ca. 3×10^{-3} moles for experiment B, where the sample volume was ca. 3 ml (from Randall[7]).

resonances should be better resolved than e.g. those of the amide protons in the ^{1}H-NMR spectra.

To conclude this section, we consider some resonance assignments in the ^{15}N spectra presented above. In the spectrum of arginine methyl ester

*See footnote on p. 297.

dihydrochloride (Fig. VII-3), the highest field resonance at 21 ppm comes from the α-$NH_3^{\oplus}$ group, and the two resonances at 53.3 and 64.8 ppm from the nitrogens of the guanidine moiety, where the two $-NH_2$ groups are equivalent at 53.3 ppm. Compared with the resonance of the α-amino group in Fig. VII-3, the amide ^{15}N resonances in N-acetylglycyl-L-alanine (Fig. VII-4) are shifted approximately 80 ppm downfield, which was to be anticipated from the data in Table VII-1. The resonance of the amide nitrogen of glycine at approximately 89 ppm is 10 ppm to higher field than that of alanine. N-formyl-proline (Fig. VII-5) illustrates another aspect of ^{15}N-NMR studies of peptide groups. It is seen that this spectrum consists of two lines with a spacing of ca. 2 ppm. Earlier ^{13}C-NMR studies had shown that a solution of N-formylproline in DMSO contains comparable concentrations of the cis- and trans- forms of the peptide group (Fig. III-8). Fig. VII-5 thus implies that the cis–trans isomerism of the peptide group including the nitrogen atom of proline is also manifested in the ^{15}N chemical shifts[10]. In the spectrum of gramicidin S (Fig. VII-7), the high field resonance corresponds to the δ amino groups of the two ornithyl residues, and in the region from 90 to 115 ppm, five well separated resonances of the five pairs of amide nitrogens can be observed. Thus the C_2-symmetry of the molecular conformation of gramicidin S, which had previously been evidenced by X-ray, ^1H and ^{13}C-NMR, is also readily apparent in the ^{15}N-NMR spectrum. An interesting detail is the increased intensity of the ornithine side chain resonance, as compared to those of the backbone amide nitrogens. A likely

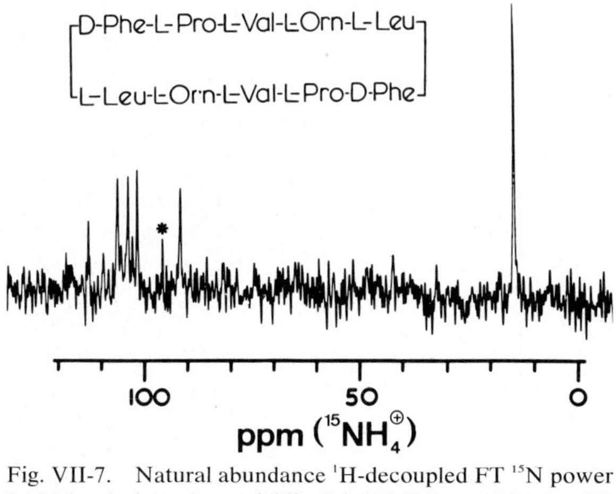

Fig. VII-7. Natural abundance ^1H-decoupled FT ^{15}N power NMR spectrum at 18.24 MHz of a 0.18-M solution of gramicidin S in DMSO, sample size 25 mm outer diameter, $T = 47°$. This spectrum was recorded in 40 minutes. The spectrometer was the same as that used for spectrum A of Fig. VII-6. From comparison with other ^{15}N-NMR spectra recorded for gramicidin S, the line marked * appears not to come from the peptide[7]. The structure of gramicidin S is also indicated in the figure (from Hawkes and Randall[10]).

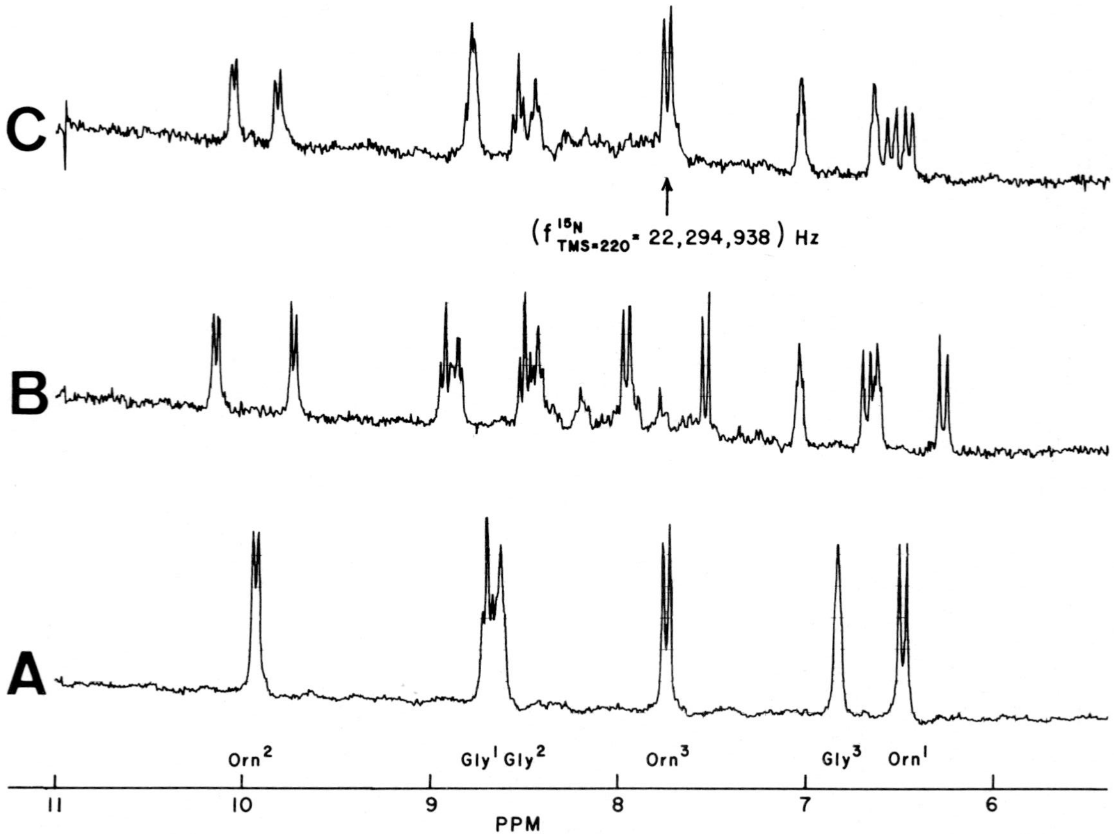

Fig. VII-8. FT ^{1}H-NMR spectra at 220 MHz of d_6-DMSO solutions of alumichrome, $T = 70°$. A. Natural abundance of the nitrogen isotopes. The resonance assignments (see Fig. IV-13 for the notation used, which does not correspond to the generally adopted rules for the numeration of amino acid residues in peptides) are also indicated (Refs. 77 to 79 of Chapter III). B. ^{15}N enriched to 99.2%. Each resonance is split into a doublet by the $^1J_{^{15}NH}$ coupling. C. Same as B except that double resonance irradiation was applied at 22.294938 MHz ($f^{15N}_{TMS=220}$). This frequency corresponds to the ^{15}N resonance frequency for Orn-3 at a field H_0 where the ^{1}H-NMR frequency of TMS is at 220 MHz. Note that the amount of substance needed for these experiments corresponds to that usually required for ^{1}H-NMR studies, i.e. 0.5 ml of a ca. 0.1-M solution were used in a 5 mm sample tube (from Llinás et al.[8]).

explanation seems to be that due to the intramolecular segmental motions, the effective rotational correlation time τ_c for the ornithine side chain nitrogens is sufficiently short for the NOE to be near the maximum value (Fig. VII-2), whereas for the backbone nitrogens, where the rotational motions are more nearly restricted to the overall rotations of the whole molecule, a reduced NOE would result from the relatively long correlation time τ_c.[10]

Measurements of the ^{15}N Parameters by ^{1}H—^{15}N Double Resonance Techniques

As an alternative way to the direct observation of the ^{15}N-NMR spectra, one can in certain cases obtain the ^{15}N-NMR parameters from investigations of the changes arising in the ^{1}H-NMR spectra as a result of double resonance irradiation of the ^{15}N signals. The application of these techniques is usually restricted to nitrogen atoms which are directly bonded to one or several hydrogen atoms. They are then generally applicable whenever the ^{1}H-NMR spectrum contains features which can be attributed to ^{15}N—^{1}H coupling. Since in these experiments one observes exclusively the proton resonances, the difficulties inherent in the low NMR sensitivity of ^{15}N can be evaded. In the following the method is illustrated with some experiments on ferrichrome peptides[8].

Fig. VII-8 shows some ^{1}H-NMR spectra of ferrichrome (see Fig. IV-13 for the structure and the somewhat unorthodox numerotation of the amino acid residues in this molecule), which were recorded to study the ^{15}N resonances. As in the previously discussed experiments with ferrichrome peptides (Section IV.5), the iron of ferrichrome had for these studies been replaced by the diamagnetic aluminium(III) ion, and we will therefore in the following adopt the term "alumichrome"[8]. Spectrum A of Fig. VII-8 represents the amide proton resonances of alumichrome with the natural isotope distribution. The observed multiplet structures arise from spin–spin coupling with the C^{α}-protons, while, as is usually the case in peptides, the coupling with ^{14}N is not manifested. The assignments of the amide proton resonances are also indicated (Refs. 77–79 of Chapter III). Spectrum B shows the corresponding resonances for alumichrome enriched to 99.2% in ^{15}N. The ^{15}N enriched peptide had been obtained from the growth medium of *Ustilago sphaerogena* supplied with ^{15}N-ammonium acetate as the sole nitrogen source. As a result of the ^{15}N—^{1}H spin–spin coupling, each resonance is now split into two components. In accordance with Table VII-2, the $^{1}J_{^{15}NH}$ are of the order of 90–95 Hz, and the geminal and vicinal ^{15}N—^{1}H couplings are so small that the regions from 0 to 6 ppm were essentially identical in the spectra A and B (not shown in Fig. VII-8)[8]. By selective double resonance irradiation of the ^{15}N resonances, the individual

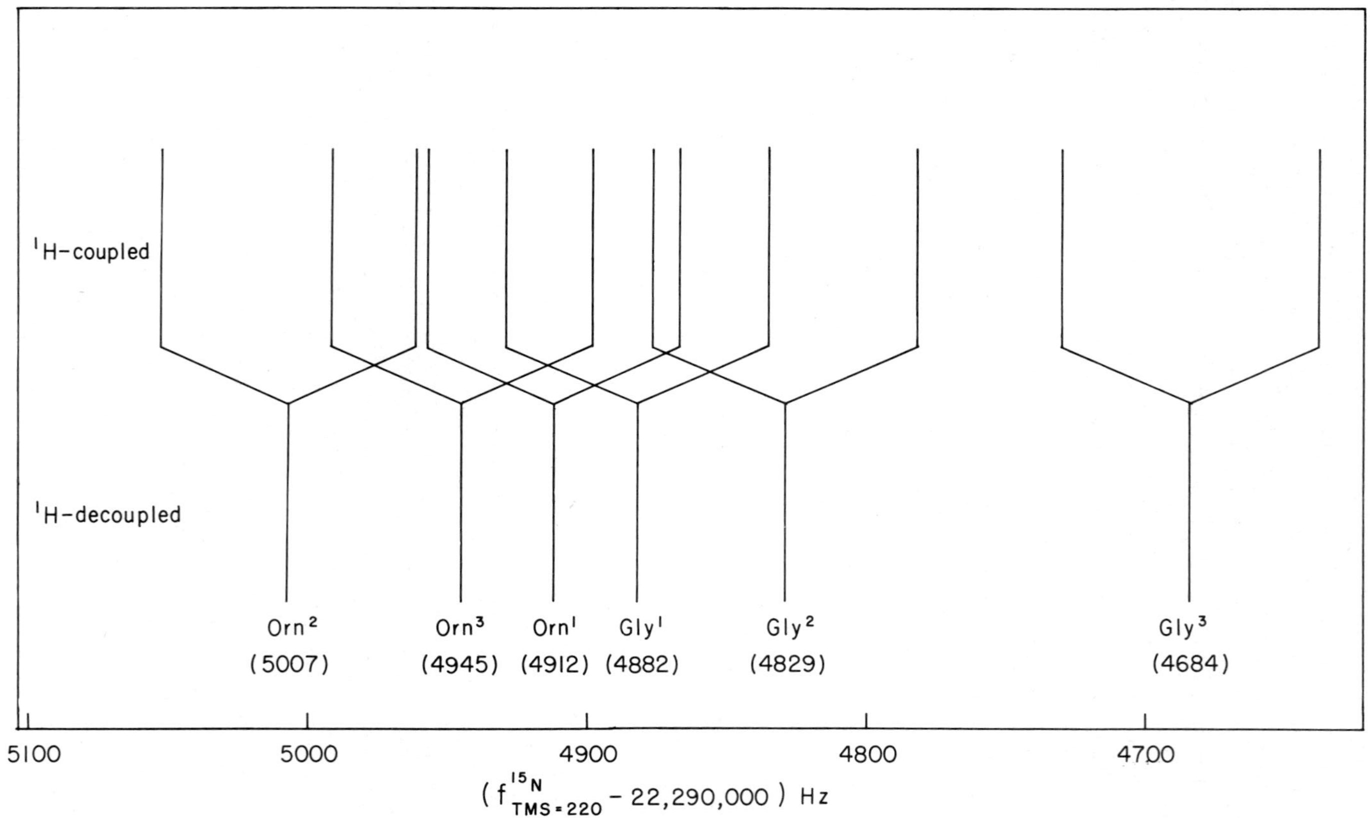

Fig. VII-9. (See caption on following page.)

${}^1J_{^{15}NH}$ couplings may now be eliminated, and the ^{15}N chemical shifts evaluated from the radio frequency at which the doublets collapse. In spectrum C of Fig. VII-8, the ^{15}N resonant frequency for Orn-3 corresponding to the magnetic field H_0 where the ^1H resonance frequency for TMS is 220 MHz, $f^{15N}_{TMS=220}$, was thus found to be 22.294938 MHz. Note that the spin–spin couplings of the other resonances have also been reduced in spectrum VII-8C. Analogous to the situation for off-resonance proton irradiation in ^{13}C-NMR (Section V.2, see also Fig. VI-8), the residual ^{15}N—^1H couplings depend on the spacings between the ^{15}N resonance frequencies, and the applied frequency.

The complete spectrum of the amide ^{15}N resonances in alumichrome obtained from these ^{15}N—^1H double resonance experiments is schematically represented in Fig. VII-9, and the ^{15}N-NMR parameters are indicated in the caption to Fig. VII-9. The figure shows that the six resonances are spread over a chemical shift range of approximately 15 ppm. Since the ^1H-NMR lines had previously been identified (Refs. 77 to 79 of Chapter III), the ^{15}N resonance assignments followed readily from the double resonance experiments of Fig. VII-8. It was thus found that in contrast to the ^1H-NMR (Fig. VII-8), the ^{15}N resonances of all the three ornithines are at lower field than any of the three glycines. The chemical shift differences within the two groups of identical residues are a readily apparent manifestation of the molecular conformation, as will be further discussed below.

3. ^{15}N-NMR PARAMETERS OF AMINO ACIDS AND PEPTIDES

Most of the amino acid ^{15}N-NMR spectra have by now become available, and a selection of model peptides have also been studied. It thus appears that the general traits of the ^{15}N-NMR parameters for this class of com-

---

Fig. VII-9. Schematic representation of the ^{15}N-NMR spectrum of the backbone amide groups in alumichrome (Fig. IV-13) obtained with the double resonance method illustrated in Fig. VII-8. Solvent = DMSO, $T = 19°$. In the figure, the resonance positions are indicated in terms of the resonance frequencies $f^{15N}_{TMS=220}$. These correspond to the following values for the relative chemical shifts of the individual resonances (in ppm), and the ^{15}N—^1H spin–spin coupling constants:

	δ (ppm relative to Gly-3)	${}^1J_{^{15}NH}$ (Hz)
Gly-3	0.0	92.9
Gly-2	6.5	95.1
Gly-1	8.9	94.4
Orn-1	10.2	90.2
Orn-3	11.7	93.4
Orn-2	14.5	91.5

(from Llinás et al.[8]).

pounds are quite well established. Nonetheless, biological ^{15}N-NMR work is still at a very early stage, and it is to be expected that the data presented here will in the near future be completed with more comprehensive studies.

^{15}N Chemical Shifts in Amino Acids and Peptides

The ^{15}N chemical shifts for the common amino acids are listed in Table VII-3. Measurements have so far been reported for several free amino acids in acidic, neutral, or basic aqueous solution[9a, 12, 12a], and for some amino acid methyl esters in slightly acidic H_2O.[6] It is seen that compared to the overall range of 900 ppm for ^{15}N shifts (Table VII-1), the chemical shifts of the α-$NH_3^{\oplus}$ groups in the different amino acids vary relatively little, i.e. over a range of approximately 30 ppm if one includes glycine and proline, and over a range of less than 10 ppm if glycine and proline are excluded. Protonation–deprotonation, or esterification of the α-carboxylic acid group produces only minor changes of the ^{15}N chemical shifts. Upfield shifts of 4.6 ppm[12] to 12.5 ppm[9a, 12a] have been reported to arise from the deprotonation of the α-$NH_3^{\oplus}$ group in glycine.

It has been suggested that, in analogy to ^{13}C (Equation V-1), the ^{15}N chemical shifts of the different amino acids relative to glycine could be accounted for by the following set of empirical parameters for substitution of protons by carbon atoms in the β-, γ-, and δ-positions of the amino acid side chains[6, 10, 12].

$$\begin{aligned} A_{\beta}^{CH_3}(^{15}N) &= 12.9 \text{ ppm} \\ A_{\gamma}^{CH_3}(^{15}N) &= -3.5 \text{ ppm} \\ A_{\delta}^{CH_3}(^{15}N) &= 1.2 \text{ ppm} \end{aligned} \qquad \text{(VII-2)}$$

A comparison of the ^{15}N parameters in Equation VII-2 with the corresponding parameters for ^{13}C (Equation V-2, Table V-5) reveals a close similarity in signs and relative magnitudes. A similar correlation between the empirical substituent parameters for ^{15}N and ^{13}C has also been observed for various other amines[13].

The ^{15}N chemical shifts for peptide groups in flexible linear peptides occupy the spectral region from approximately 83 to 116 ppm if Gly and Pro are included, or from 95 to 110 ppm for the common amino acids without Gly and Pro (Table VII-4). The substituent effects on the peptide nitrogen chemical shifts can be rationalized with a set of empirical parameters which are nearly identical to those in Equation VII-2 for the free amino acids[10].

The amino acid side chain ^{15}N chemical shifts included in Table VII-3 indicate that these resonances will be well separated from those of the backbone peptide nitrogens, and should therefore be well-resolved "natural

TABLE VII-3

^{15}N Chemical shifts of the common amino acids*.

Amino acid	Solvent	δ (ppm ($^{15}NH_4^{\oplus}$))	Ref.
Glycine[a]	H_2O	10.0	12
	6-N HCl	9.7	12
	6-N NaOH	5.4	12
Glycyl methyl ester	~0.1-N HCl	10.0	6
Alanine	H_2O	22.8	12
	6-N HCl	23.0	12
Alanyl methyl ester	~0.1-N HCl	23.2	6
Valine	H_2O	16.0	12
Isoleucyl methyl ester	~0.1-N HCl	18.9	6
Leucine	6-N HCl	21.4	12
Seryl methyl ester	~0.1-N HCl	15.8	10
Threonine			
Prolyl methyl ester	~0.1-N HCl	34.6	10
Aspartic acid	6-N HCl	19.8	12
Glutamic acid	6-N HCl	21.2	12
Lysyl methyl ester	~0.1-N HCl	21.5	6
ϵ-$NH_3^{\oplus}$		5.8	
Arginyl methyl ester	~0.1-N HCl	21.0	6
guanidine NH_2		53.3	
NH		64.8	
Asparagine			
Glutamine			
Methionyl methyl ester	~0.1-N HCl	21.2	10
Cysteinyl methyl ester	~0.1-N HCl	19.4	10
Tryptophan			
Phenylalanine	H_2O	19.1	12
Tyrosine	6-N HCl	19.6	12
Histidyl methyl ester	0.1-N HCl	21.7	6
imidazole ring		{144.2, 146.3}	

*The shifts are given in ppm relative to external $^{15}NH_4^{\oplus}$ (5-M $^{15}NH_4NO_3$ in 2-M HNO_3)[6], where positive numbers indicate shifts to lower field.

[a]Somewhat different ^{15}N chemical shifts for glycine have been reported by different authors[9a, 12a]:

$^{\oplus}H_3N—CH_2—COOH$	5.6 ppm ($^{15}NH_4^{\oplus}$)
$^{\oplus}H_3N—CH_2—COO^{\ominus}$	7.8 ppm
$H_2N—CH_2—COO^{\ominus}$	−4.5 ppm

probes" in the ^{15}N-NMR spectra of peptides and proteins. It should thus e.g. be possible in future experiments, to determine the pK_a of histidyl or lysyl residues in complex molecules from the pH-dependence of the ^{15}N resonances.

TABLE VII-4

^{15}N Chemical shifts of the peptide nitrogen atoms in some N-formyl- and N-acetyl-amino acids, and some dipeptides*.

Compound		Solvent	δ (ppm ($^{15}NH_4^{\oplus}$))		Ref.
N-formyl-glycine		DMSO	93.7		10
N-acetyl-glycine		DMSO	89.4		10
N-formyl-alanine		DMSO	108.1		10
N-acetyl-alanine		DMSO	104.4		10
N-formyl-valine		DMSO	101.5		10
N-acetyl-valine		DMSO	98.0		10
N-formyl-proline	cis	DMSO	115.3		10
	trans		113.7		
Ac-Gly-Gly-OH		DMSO	89.1	84.0	10
Ac-Gly-L-Leu-OH		DMSO	89.1	96.4	10
Ac-L-Leu-Gly-OH		DMSO	102.2	84.6	10
Ac-Gly-L-Ala-OH		DMSO	89.1	98.8	10
Ac-L-Ala-Gly-OH		DMSO	104.7	83.5	10
H-L-Val-Gly(^{15}N)-OH		H_2O, pH 2		83.0	12

*The shifts are given in ppm relative to external $^{15}NH_4^{\oplus}$,[6] where positive numbers indicate shifts to lower field.

Spin–Spin Couplings with ^{15}N

Three types of $^{15}N—^1H$ spin–spin coupling constants have been reported for a number of ^{15}N enriched linear and cyclic dipeptides[12, 14]. Direct coupling constants $^1J_{^{15}NH}$ between the amide nitrogen and hydrogen atoms (Table VII-5) were found to vary from 89 to 95 Hz. The direct couplings in

TABLE VII-5

$^{15}N—^1H$ spin–spin couplings $^1J_{^{15}NH}$ in some linear and cyclic dipeptides enriched to 95% in ^{15}N (from Ref. 12).

Compound	Solvent	$^1J_{NH}$ (Hz)
N-Acetyl-DL-Ala(^{15}N)-OH	H_2O, pH 6	93.3
H-L-Ala-L-Phe(^{15}N)-OH	H_2O, pH 2	92.1
H-L-Ala-L-Ala(^{15}N)-OH	H_2O, pH 2	93.2
H-L-Phe-Gly(^{15}N)-OH	H_2O, pH 2	93.1
H-L-Val-Gly(^{15}N)-OH	H_2O, pH 2	94.4
c-[-L-Ala-L-Ala(^{15}N)-]	H_2O, pH 2	91.1
c-[-L-Val-Gly(^{15}N)-]	H_2O, pH 2	91.1
c-[-L-Val-Gly(^{15}N)-]	DMSO	90.0
c-[-L-Phe-Gly(^{15}N)-]	DMSO	90.5
c-[-L-Ala-L-Phe(^{15}N)-]	DMSO	89.3

alumichrome are also within this range (Fig. VII-9). The vicinal coupling constants $^3J_{^{15}NH^\alpha}$ between the C_i^α-proton and the $(i+1)$th amide nitrogen (Fig. III-9) were found to be between $\leqslant 1.0$ and 1.8 Hz (Table VII-6), and the vicinal coupling constants $^3J_{^{15}NH^\beta}$ between the ith amide nitrogen and the protons attached to the β_i-carbon atom (Fig. III-10) vary between 1.9 and 3.7 Hz (Table VII-7). These values for the ^{15}N—^1H spin–spin couplings in peptides agree quite closely with the corresponding spin couplings in the amides listed in Table VII-2.

TABLE VII-6

Vicinal ^{15}N—^1H spin–spin couplings $^3J_{^{15}NH^\alpha}$ between the C_i^α-proton and the $(i+1)$th amide nitrogen (see Fig. III-9) in some linear and cyclic dipeptides enriched to 95% in ^{15}N (from Ref. 12).

Compound	Solvent	$^3J_{^{15}NH^\alpha}$
H-L-Ala-L-Ala(^{15}N)-OH	D_2O, pD 2	1.2
H-L-Val-Gly(^{15}N)-OH	D_2O, pD 2	<1
H-L-Ala-L-Phe(^{15}N)-OH	D_2O, pD 2	<1
c-[-L-Val-Gly(^{15}N)-]	D_2O, pD 2	1.0
c-[-L-Ala-L-Phe(^{15}N)-]	DMSO	1.8
c-[-L-Val-Gly(^{15}N)-]	DMSO	<1
c-[-L-Phe-Gly(^{15}N)-]	DMSO	<1

TABLE VII-7

Vicinal ^{15}N—^1H spin–spin couplings $^3J_{^{15}NH^\beta}$ between the α-amino nitrogen atom and the C^β-protons of some amino acids, and between the amide nitrogen i and the C_i^β-protons in some linear and cyclic dipeptides*.

Compound	Solvent	$^3J_{^{15}NH^\beta}$
Tyrosine	D_2O, pD 10	2.7, 2.9
Aspartic Acid	D_2O, pD 10	2.8, 3.7
Phenylalanine	D_2O, pD 1	2.2, 2.3
Alanine	D_2O, pD 1	3.1,
N-Acetyl-DL-Ala(^{15}N)-OH	H_2O, pH 6	3.0,
H-L-Ala-L-Ala(^{15}N)-OH	H_2O, pH 2	3.0,
H-L-Ala-L-Phe(^{15}N)-OH	H_2O, pH 2	2.3, 3.4
c-[-L-Ala-L-Phe(^{15}N)-]	DMSO	1.9, 3.2

*The compounds studied were enriched to 95% in ^{15}N (from Ref. 12).

4. MANIFESTATIONS OF DIFFERENT PEPTIDE CONFORMATIONS IN THE ^{15}N-NMR PARAMETERS

Conformation Dependent Spin–Spin Couplings with ^{15}N

The three types of $^{15}N—^{1}H$ couplings listed in the Tables VII-5 to VII-7 represent, at least theoretically, conformationally sensitive data, which contain information on individual ones of the dihedral angles determining the polypeptide conformation (Figs. III-9 and III-10). From the data of Table VII-5, and similar observations in various amides, it has been suggested that $^{1}J_{^{15}NH}$ could be employed as an indicator of cis–trans isomerism about the peptide bond[12]. For the trans peptide groups in the linear dipeptides of Table VII-5, $^{1}J_{^{15}NH}$ is between 92.1 and 94.5 Hz, whereas the cis peptide groups in the cyclo-dipeptides exhibit $^{1}J_{^{15}NH}$ couplings in the range 89.3 to 91.1 Hz*.

As was already discussed in Section III.2, the vicinal coupling $^{3}J_{^{15}NH^{\alpha}}$ is conceptionally of particular relevance, since it would provide a direct measure of the dihedral angles ψ_i in polypeptide chains (Figs. III-9 and III-10). Theoretical and experimental investigations[16–18] have shown that the angular dependence of $^{3}J_{^{15}NH^{\alpha}}$ can be described by a Karplus type relation (Fig. II-4), and the following numerical expression has been proposed[18] on the basis of the available experimental observations, which include the data of Table VII-6, N-Acetyl-glycine(^{15}N),[16] and valinomycin(L-Val(^{15}N)[12]

$$^{3}J_{^{15}NH^{\alpha}} = 5.4\cos^2\theta + 3.0\cos\theta + 0.8\sin^2\theta \qquad \text{(VII-3)}$$

where θ is the dihedral angle between the two planes defined by $H—C_i^{\alpha}—C_i'$ and $C_i^{\alpha}—C_i'—{}^{15}N_{(i+1)}$ (Fig. III-9), and $\theta = |\psi - 120°|$ for L-amino acid residues. From Equation VII-3 it follows that in the range of the torsion angle ψ from 0° to 180°, and -120 to $-180°$, $|^{3}J_{^{15}NH^{\alpha}}|$ is less than 1.5 Hz, which will be difficult to measure. Moreover, for $|^{3}J_{^{15}NH^{\alpha}}| \leqslant 1.5$ Hz, the sign of the coupling constants must be determined before a conclusion on the corresponding ψ-angle may be drawn. On the other hand, for ψ-angles in the range $-120°$ to 0°, Equation VII-3 predicts values of 1.5 to 7.0 Hz for $|^{3}J_{^{15}NH^{\alpha}}|$, which should in principle be accessible for experimental verification. Overall, from the

---

*A more general "cyclic" relation between $^{1}J_{^{15}NH}$ and ω was proposed by Llinás et al.[8], according to which $^{1}J_{^{15}NH}$ decreases from ≈ 95 Hz for planar trans peptide groups ($\omega = 0°$) to ≈ 67 Hz for $\omega = 90°$, and increases again to ≈ 90 Hz for planar cis peptide groups ($\omega = 180°$). In alumichrome (Fig. VII-8), the variations of $^{1}J_{^{15}NH}$ would thus reflect small transoid distortions of the peptide bonds, with $\omega \approx 0°$ for Gly-2, and $\omega \approx 10°$ for Orn-1 and Orn-2. This ω-dependence of $^{1}J_{^{15}NH}$ would be explained by the change in the hybridization of the atomic orbitals when going from a strictly planar peptide group, with partial sp^2-hybridization of the nitrogen orbitals, to a non-planar form, where there is predominantly sp^3-hybridization of the nitrogen valence electrons[8, 15].

small magnitudes of $|^3J_{^{15}NH^\alpha}|$ obtained so far in the experimental measurements (Tables VII-2 and VII-6) and predicted by Equation VII-3 for most angles ψ, it appears that the experimental determination of the ψ angles in polypeptide chains will only become practicable if more precise values for $^3J_{^{15}NH^\alpha}$ can be obtained by future improvements of the experimental techniques.

The vicinal ^{15}N—^1H couplings $^3J_{^{15}NH^\beta}$ (Table VII-7) can be used for studies of the dihedral angles χ_i^1 (Fig. III-10). Combining the data of Table VII-7, and similar data obtained in different experiments[14], with an analysis of the amino acid side chain rotamer populations from the homonuclear couplings $^3J_{H^\alpha H^\beta}$, gave $^3J_{^{15}NH^\beta})_{trans} = 4.8$ Hz, and $^3J_{^{15}NH^\beta})_{gauche} = 1.8$ Hz.[12] Note that because of the different character of the intervening covalent bonds, these coupling constants are quite different from $^3J_{^{15}NH^\alpha}$, and it appears quite likely that measurements of $^3J_{^{15}NH^\beta}$ could become a valuable complement to $^3J_{H^\alpha H^\beta}$, and $^3J_{^{13}C'H^\beta}$ (Section V.5) for studies of the amino acid side chain conformations.

In addition to $^3J_{^{15}NH^\alpha}$, the application of $^3J_{^{15}N^{13}C^\beta}$ between the amide nitrogen i and the $C^\beta_{(i-1)}$ carbon atom for studies of the dihedral angle ψ_i (Fig. III-9) was also investigated[19]. As one might have anticipated, these couplings are also rather small, generally $\leqslant 1$ Hz, and hence difficult to be determined.

Conformation Dependent ^{15}N Chemical Shifts

We have already noted the occurrence of conformation dependent ^{15}N chemical shifts in N-formylproline (Fig. VII-5) and in alumichrome (Fig. VII-9), and it is to be expected that the ^{15}N shifts will quite generally be useful for conformational studies in much the same way as the ^1H and ^{13}C chemical shifts (Chapters III and V). This is in the following illustrated with a particular phenomenon previously also used in ^1H and ^{13}C-NMR, i.e. the solvent dependence of the chemical shifts. In Fig. VII-10, the ^{15}N-NMR spectra of alumichrome in two solvents, trifluoroethanol and dimethylsulfoxide, are compared. It is seen that four resonances are shifted downfield when going from DMSO to TFE, one line is independent of the solvent, and one resonance is shifted upfield. These solvent dependences were rationalized in terms of the relative stability in solvents with different acid-base properties (Table II-4), of the different resonant structures of the peptide groups

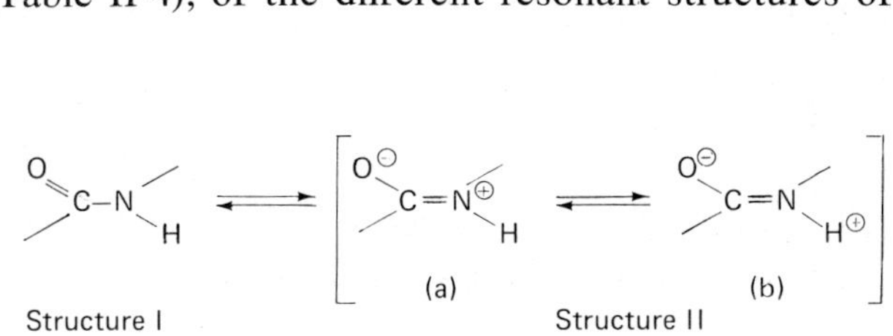

(VII-4)

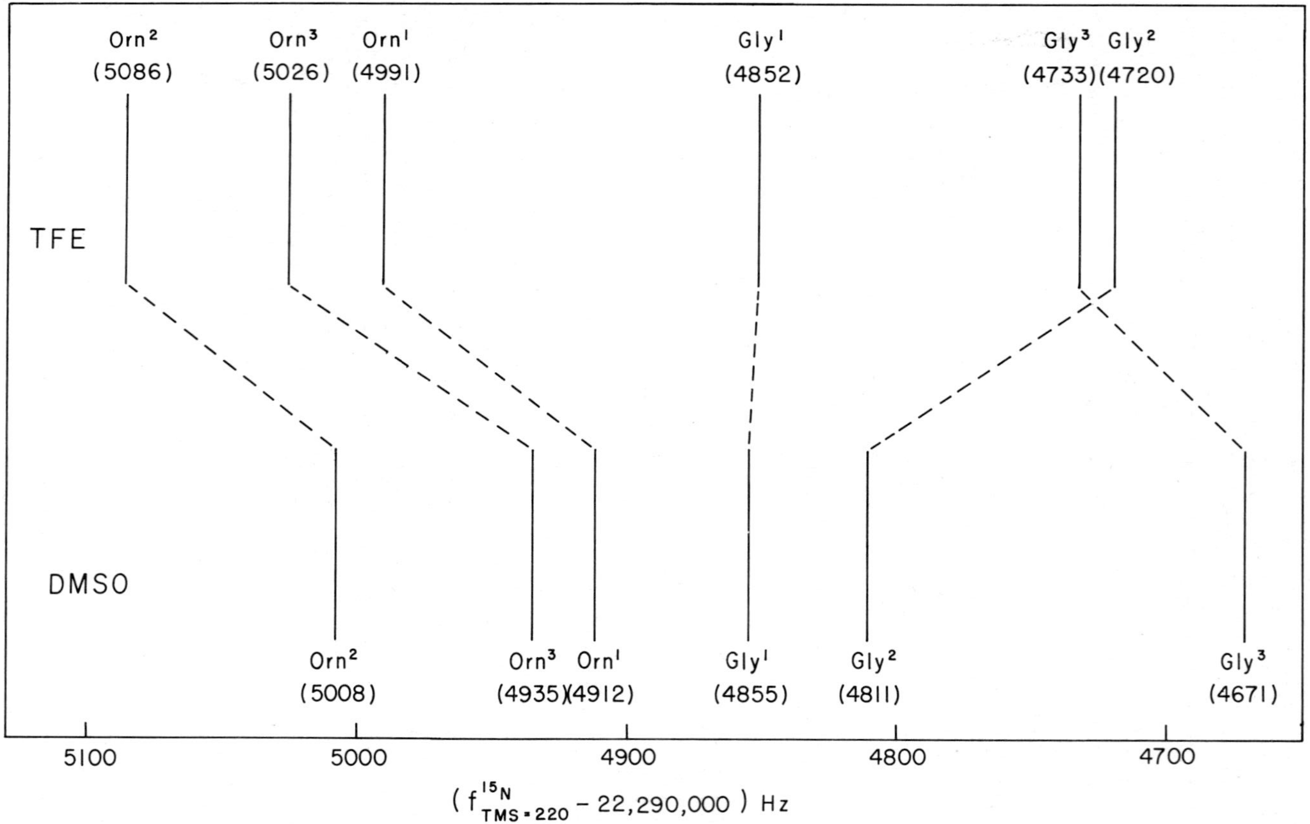

Fig. VII-10. Schematic representation of the ^{15}N-NMR chemical shifts of the backbone amide groups of alumichrome in two different solvents, trifluoroethanol (TFE) and dimethyl sulfoxide (DMSO), at 70°. The spectra were obtained with the double resonance technique of Fig. VII-8. The resonance assignments (see Fig. IV-13 for the notation used) are also indicated. The frequency scale is referred to the ^{1}H resonance frequency of TMS as explained in the caption to Fig. VII-8 (from Llinás et al.[8]).

Hydrogen bond formation of the carbonyl oxygen with protic solvents tends to favour the structures II, and as a result one would expect deshielding of the nitrogen and the hydrogen nuclei, i.e. both the amide ^{1}H and ^{15}N resonances should be shifted downfield. On the other hand, since protic acids are poor bases, they have a lesser tendency to form hydrogen bonds involving the amide proton, which would also stabilize the structures II, and thus cause downfield shifts of the amide ^{1}H and ^{15}N resonances. In alumichrome these two types of solvent interaction can be observed separately for individual ones of the amide groups. The amide protons of Gly-3 and the three ornithyl residues are protected from the solvent by their location in the interior of the molecule, and at the same time the carbonyl oxygens of these four peptide groups are exposed to the solvent. On going from DMSO ($pK_{a1} \approx 0$) to the protic acid TFE ($pK_{a1} \approx -8.2$), these four ^{15}N resonances are, as a consequence of the hydrogen bond formation between TFE and the carbonyl oxygen, shifted of lower field. On the average of the four peptide groups, the extent to the shift is 82 Hz, or 3.7 ppm. Gly-2 is unique among the amino acid residues in alumichrome in that the amide proton is solvent exposed, and the carbonyl oxygen protected from the solvent by the formation of an intramolecular hydrogen bond. Because of the loss of deshielding through amide proton-solvent hydrogen bonding, this ^{15}N resonance is shifted upfield by 91 Hz (4.1 ppm) when going from DMSO to TFE. It is seen that between DMSO and TFE, the ^{15}N deshielding arising from the stronger C=O protonation has almost exactly the same magnitude as the shielding gained because of the weaker solvent interactions with the amide proton. This is further corroborated by the extremely small solvent dependence of the ^{15}N resonance of Gly-1 in alumichrome, where both the amide proton and the carbonyl oxygen of the peptide group are exposed to the solvent. The ferrichrome peptides were particularly suitable for these pioneering studies[8], since earlier investigations had shown that the solution conformation of these metallo-peptides is very rigid, and essentially identical to the conformation in single crystals (Ref. 77–79 of Chapter III).

5. ^{15}N ISOTOPE ENRICHMENT AS A SUPPORTING TECHNIQUE FOR ^{1}H AND ^{13}C-NMR STUDIES OF PEPTIDES

In Section III.3 we have seen that in cases where certain amino acid residues occur repeatedly in a peptide molecule, the ^{1}H-NMR lines cannot in general be assigned solely on the basis of the spectra obtained with the fully protonated compounds. It is readily apparent from Fig. VII-8 that selective ^{15}N labelling provides an unambiguous assignment of the amide proton resonances, from which the other ^{1}H-NMR lines may then be identified through homonuclear double resonance experiments[20]. This approach would be essentially analogous to the selective deuterium labelling technique illustrated in Figs. III-13 and III-17.

In Chapter V we have encountered another difficult resonance assignment of crucial importance for studies of peptides, i.e. the identification of the carbonyl ^{13}C resonance lines (Figs. V-8, V-9, V-17). ^{15}N enrichment opens a novel avenue to this end. One-bond couplings $^1J_{^{15}N^{13}C}$ are of the order 10 to 20 Hz (Table VII-2), and therefore the carbonyl carbon lines in the proton noise-decoupled ^{13}C-NMR spectra of ^{15}N enriched peptides appear as well discernible doublets. The carbonyl ^{13}C lines may thus be identified by selective ^{15}N labelling. An alternative technique may be applied when, as was the case for alumichrome, the ^{1}H and ^{15}N-NMR lines could independently be identified (Figs. VII-8 and VII-9). The carbonyl ^{13}C resonances can then be assigned in the non-selectively ^{15}N enriched peptide by selective ^{15}N—^{13}C spin decoupling. This is illustrated in Fig. VII-11 with the carbonyl region of the ^{13}C-NMR spectrum of alumichrome. The top trace shows the proton noise-decoupled spectrum recorded without ^{15}N double resonance irradiation. The doublets arising from the 6 carbonyl carbon atoms are also indicated, together with the resonance assignments obtained from the spectra B to G. The lowest ^{15}N double resonance irradiation frequency used (spectrum B) corresponds to the low frequency end of the ^{15}N resonance frequency range in alumichrome (Fig. VII-9)*. All the doublet splittings are reduced, and, as is generally the case in off-resonance double resonance irradiation experiments, the residual ^{15}N—^{13}C couplings depend on the spacing between the resonance frequencies of the individual ^{15}N lines (Fig. VII-9)* and the indicated ^{15}N irradiation frequency. Therefore, we expect that the residual spin coupling in spectrum B is smallest for the carbonyl carbon atom directly bound to the nitrogen atom of Gly-3 (Fig. VII-9), i.e. the carbonyl carbon of Orn-1 (Fig. IV-13). Actually, the doublet splitting of this resonance line is completely collapsed in the spectra B and C of Fig. VII-11. When the irradiation frequency is further increased, one expects from Figs. VII-9 and IV-13 that the remaining five doublets collapse successively in the order Gly-3 (spectrum D of Fig. VII-11), Gly-2 (spectrum E), Orn-2, Gly-1 (spectrum F), Orn-3 (spectrum G). Thus the resonance assignments indicated in the top trace of Fig. VII-11 were obtained.

---

*Since the spectra in Figs. VII-9 and VII-11 were recorded at fields $H_0 = 51.7$ and 23.5 KG, respectively, the ^{15}N resonance frequencies in Fig. VII-11 are smaller by a factor 2.2. Additional small differences arise because the ^{15}N chemical shifts vary with temperature[8], and the two sets of data were recorded at 19°, and 50°, respectively.

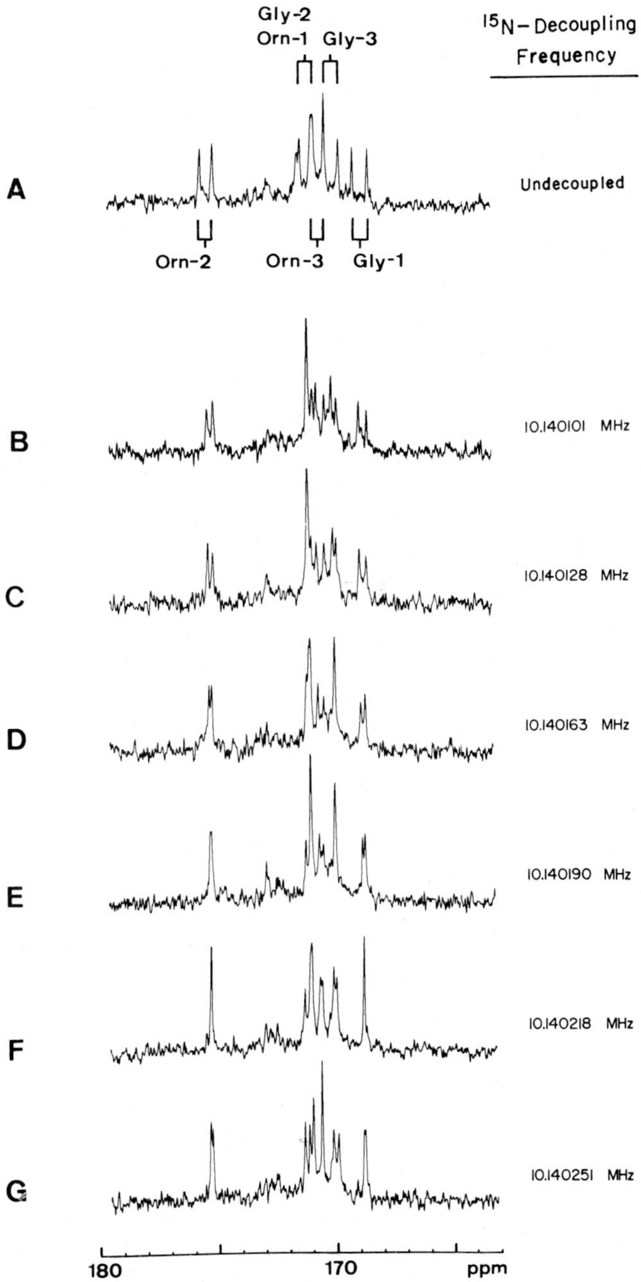

Fig. VII-11. Carbonyl carbon region from 165 to 180 ppm (from TMS) in the proton noise-decoupled FT ^{13}C-NMR spectra at 25.16 MHz of 99.2% ^{15}N enriched alumichrome,

6. BIBLIOGRAPHY

1 Richards R. E. and Thomas N. A., J. Chem. Soc., Perkin II, 1974, 368.
2 Axenrod T. and Webb G. A. (eds.), Nuclear Magnetic Resonance Spectroscopy of Nuclei other than Protons, J. Wiley, New York, 1974
3 Randall E. W. and Gillies D. G., Progress in NMR Spectroscopy *6*, 119 (1971)
4 Lichter R. L., in Determination of Organic Structures by Physical Methods, F. Nachod and J. J. Zuckerman (eds.), Academic Press, New York 1971, Vol. 4, p. 195
5 Witanowski M. and Webb G. A. (eds.), Nitrogen NMR, Plenum Press, New York, 1973
6 Pregosin P. S., Randall E. W. and White A. I., Chem. Comm. *1971*, 1602
7 Randall E. W., private communication
8 Llinás M., Horsley W. and Klein M. P., Proc. Nat. Acad. Sci. 1975 (submitted)
9 Lippmaa E., Saluvere T. and Laissaar S., Chem. Phys. Letters *11*, 120 (1971)
9a Leipert T. K. and Noggle J. H., J. Amer. Chem. Soc. *97*, 269 (1975)
10 Hawkes G. E. and Randall E. W. 1975 (to be submitted)
11 Lichter R. L. and Roberts J. D., J. Amer. Chem. Soc. *93*, 3200 (1971)
12 Sogn J. A., Gibbons W. A. and Randall E. W., Biochemistry *12*, 2100 (1973)
12a Cooper R. A., Lichter R. L. and Roberts J. D., J. Amer. Chem. Soc. *95*, 3724 (1973)
13 Lichter R. L. and Roberts J. D., J. Amer. Chem. Soc. *94*, 2495 (1972)
14 Lichter R. L. and Roberts J. D., J. Org. Chem. *35*, 2806 (1970)
15 Binsch G., Lambert J. B., Roberts B. W. and Roberts J. D., J. Amer. Chem. Soc. *86*, 5564 (1964)
16 Karplus S. and Karplus M., Proc. Nat. Acad. Sci. *69*, 3204 (1972)
17 Barfield M. and Gearhart H. L., Mol. Phys. *27*, 899 (1974)
18 Bystrov V. F., Gavrilov Y. D. and Solkan V. N., J. Magn. Res. *18*, (1975) (in press)
19 Di Blasi R. and Kopple K. D., Chem. Commun. *1975*, 33
20 Bradbury A. F., Burgen A. S. V., Feeney J., Roberts G. C. K. and Smyth D. G., FEBS Lett. *42*, 179 (1974)

---

solvent DMSO, $T = 50°$. A. No ^{15}N double resonance irradiation. The resonance assignments obtained from the spectra B to G are also indicated (see Fig. IV-13 for the notation used). B–G. Spectra obtained with selective CW^{15}N double resonance irradiation at the indicated radio frequency. Note that the amount of substance needed for these experiments corresponds to what is usually required for ^{13}C-NMR studies, i.e. 2 ml of a ca. 0.1-M solution were used in a 12 mm sample tube (M. Llinás, D. M. Wilson and M. P. Klein, private communiiation).

CHAPTER VIII

BIOLOGICAL NMR EXPERIMENTS WITH OTHER NUCLEI

As was briefly discussed in Section I.1 (Table I-1), nature provides us with a considerable variety of "NMR nuclei" in addition to ^{1}H, ^{13}C, and ^{15}N. Many of these nuclei share certain characteristics which make the NMR observation difficult, i.e. low natural isotope abundance, low intrinsic NMR sensitivity, and, for quadrupole nuclei, generally rather large natural line widths. Today, modern NMR spectrometers are not only equipped with Fourier transform accessories and large-diameter sample tubes to overcome the sensitivity problem, but may also include facilities for experiments with different nuclei. NMR observation of a number of the more exotic isotopes has thus become almost routine, and is now a real possibility for structural studies[1].

The construction of this chapter is quite different from the preceding ones in that it presents a selection of special techniques and particular experiments which do not necessarily always relate to a common theme. The material covered has mainly been selected so as to present the reader with some typical new situations and potentialities of the NMR method, which have not been encountered in the foregoing chapters on ^{1}H, ^{13}C, and ^{15}N high resolution NMR. Therefore, experiments with quadrupole nuclei are quite prominently represented, whereas a rather scant account is given of the use of fluorine-19, which is a typical high resolution nucleus (Table I-1). This does on the other hand by no means reflect the importance that the different nuclei had thus far in NMR work with proteins, where ^{19}F-NMR probes have played a very significant role. Furthermore, the material covered includes, in addition to studies with proteins, some experiments with membranes and intact tissues, illustrating techniques which may be of potential interest also for protein research.

1. DEUTERIUM

Our discussions of deuterium have so far been restricted to the use of deuterated solvents, and as an isotope label for resonance assignments in the NMR spectra of other nuclei. The use of D as a NMR nucleus in its own right has been limited by its low sensitivity for NMR detection, the low natural abundance (Table I-1), and the line broadening believed to arise from quadrupole relaxation. In addition, there was the question whether anything could be learned from D-NMR that cannot be obtained from the routinely available ^{1}H-NMR spectra[2]. The development of D-NMR until 1972 was reviewed by Diehl[2], who found that of a total of 90 papers on work with this nucleus, 10 were "predominantly of a biological nature". In the meantime, additional work was presented, and it is by now well established both that D-NMR spectra may be recorded even at natural isotope abundance[3], and that the information content of the D-NMR spectra can be of much interest for certain biological investigations.

Special Traits of D-NMR as Compared to ^{1}H-NMR

Two specific advantages of D-NMR over ^{1}H-NMR, which are particularly relevant for biological applications, arise from the presence of the nuclear quadrupole moment in D (Table I-1). First, the relaxation of deuterons in diamagnetic systems is overwhelmingly dominated by quadrupole coupling (Equation IV-38)[2], and the relaxation times are usually more readily interpretable than the corresponding ^{1}H-NMR data (Section IV.5). Second, D-NMR spectra of oriented molecules allow measurements of the degree of orientation if the quadrupole coupling constant is known[4]. To this it might be added that the line broadening from quadrupole coupling is sufficiently small so that "high resolution D-NMR spectra" are often quite well resolved[3], and advantages of high resolution D-NMR over ^{1}H-NMR studies may in certain situations arise from the small magnetic moment of D (Table I-1). Thus, deuterons are less sensitive to relaxation by paramagnetic ions than protons, and one may achieve an improved resolution in the D-NMR spectra of paramagnetic compounds as compared to the corresponding ^{1}H-NMR spectra[2]. Furthermore, as a consequence of the small spin–spin couplings with D, and since the chemical shifts in ppm are identical for ^{1}H and D, the D-NMR spectra of fully deuterated molecules are quite generally less complex than the corresponding ^{1}H-NMR spectra[2].

D Spin Relaxation Studies of the Molecular Dynamics

The advantage of D spin relaxation measurements for studies of the molecular dynamics is based on similar grounds as the preference given to ^{13}C relaxation studies over ^{1}H relaxation experiments (Sections IV.5 and V.5). For ^{13}C nuclei we had that the relaxation was essentially entirely due to dipolar coupling with the covalently bonded protons, and therefore different relaxation times of a given molecular segment in different peptides and proteins could be attributed to the different effective rotational correlation times τ_c. In analogy, we have for D that the field gradient q and the asymmetry parameter η (Equation IV-38) for a given type of carbon-deuterium single bond are remarkably insensitive to the overall molecular structure[5], and hence different relaxation times for individual deuterons can be attributed to the different effective correlation times τ_c. This has been qualitatively confirmed by measurements of the longitudinal relaxation times of D in a selection of small organic molecules, where T_1 values in the range from 0.1 to 5.0 sec were observed[6].

In peptides and proteins, τ_c is usually determined by the combination of the overall rotational motions of the molecule with the intramolecular segmental motions (Section IV.5, Appendix V), and thus contains information on the dynamics of the molecular conformation. As an illustration, let us consider an investigation of the neurohypophyseal hormone oxytocin[5]. In these studies, selectively deuterated analogs of oxytocin, which is a nonapeptide with the structure

```
 ┌─────────S-S─────────┐
H-Cys-Tyr-Ile-Gln-Asn-Cys-Pro-Leu-Gly-NH2
```

were prepared by chemical synthesis, and their interactions with neurophysin II were investigated by comparison of the D spin relaxation times in the presence and absence of this hormone binding protein. Fig. VIII-1 shows the D-NMR spectra of 2-L-Tyr(3,5-d_2)-oxytocin in the presence and absence of neurophysin II. It is seen that the D-NMR spectrum of this molecule consists of a single line corresponding to the 3,5-deuterons of the aromatic ring of Tyr. This resonance was quite markedly broadened after the addition of neurophysin II. From this the authors[5] concluded that on the one hand, in the absence of the hormone binding protein, the tyrosine ring undergoes intramolecular rotation relative to the oxytocin skeleton, and is on the other hand essentially immobilized when oxytocin interacts with neurophysin II. Similarly, deuterons in different regions of the oxytocin molecule, e.g. in 1-hemi-L-Cys(α,β,β-d_3)-oxytocin and 9-Gly(α,α-d_2)-oxytocin, were observed, and thus an overall microdynamical picture of motion in the hormone and the hormone-protein complex was developed[5]. From the NMR technical standpoint, it should be noted that the specific deuterations were so chosen that in each experiment the D-NMR spectrum consisted of one, or at most two lines, and values for the relaxation times T_2 were then obtained from the resonance line widths (Fig. I-1).

D-NMR Studies of the Molecular Orientations in Liquid Crystals

For studies of oriented molecules, one uses a variant of the conventional high resolution NMR techniques, i.e. the molecules of interest are examined

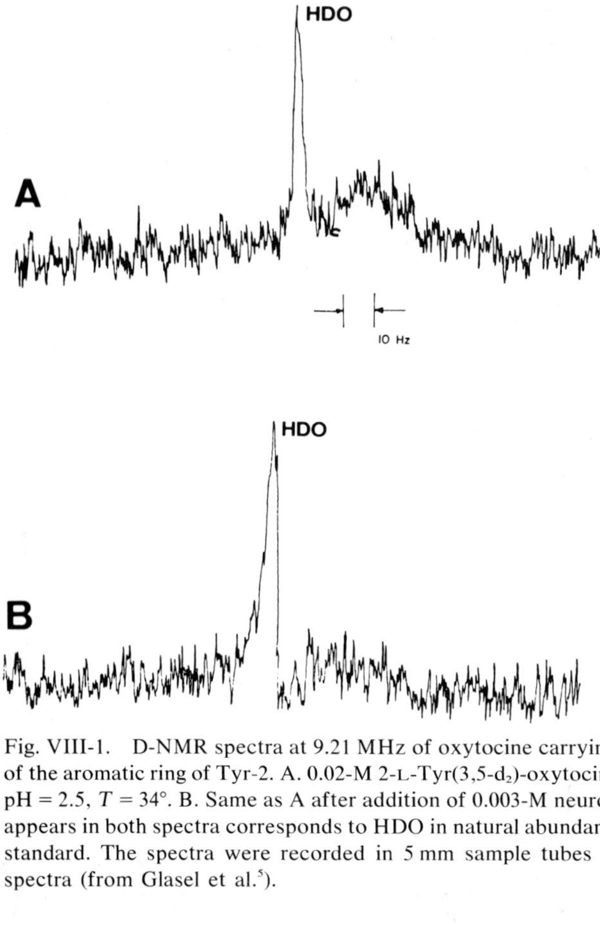

Fig. VIII-1. D-NMR spectra at 9.21 MHz of oxytocine carrying deuterons in the 3,5-positions of the aromatic ring of Tyr-2. A. 0.02-M 2-L-Tyr(3,5-d_2)-oxytocin in H_2O containing 0.45-M KCl, pH = 2.5, $T = 34°$. B. Same as A after addition of 0.003-M neurophysin II. The sharp line which appears in both spectra corresponds to HDO in natural abundance, and was used as a line width standard. The spectra were recorded in 5 mm sample tubes by computer averaging of CW spectra (from Glasel et al.[5]).

in a nematic liquid crystalline solvent. Excellent reviews of this technique have recently appeared[7,8]. For practical reasons, liquid crystal studies have so far been restricted to relatively simple spin systems. In principle, molecules of any size can thus qualify, provided one observes the D-NMR signals* of isotope labels introduced into a limited number of positions, preferably only one. To illustrate this technique, we shall consider a system from outside the protein field, i.e. a model study of lipid bilayer structures.

*Studies of the degree of orientation of oriented molecules may in principle in a similar way be based on the observation of any of the quadrupole nuclei[4,7]. However, D is quite generally preferred because of the relatively small line widths. For the experiments to be discussed in the following, it has the additional essential advantage that it can be introduced in the place of protons essentially without perturbation of the systems examined.

In these experiments[9], selectively deuterated lipids were used as building stones for liquid crystalline bilayers, and D-NMR was then employed to investigate the orientation of the hydrocarbon chains at various positions in the bilayer. Details of the experimental set-up are described in Fig. VIII-2. The observed spectral feature was the quadrupole coupling fine structure of the D-NMR lines. For rapid isotropic rotational tumbling of the molecules, and in the absence of spin–spin coupling, the D resonances consist of single lines. On the other hand, if the rotational motions are anisotropic, the D-NMR signal will be a doublet due to the quadrupole moment of the deuterium nucleus. The frequency difference, $\Delta\nu$, between the two doublet components depends on the anisotropy of motion, and on the angle θ

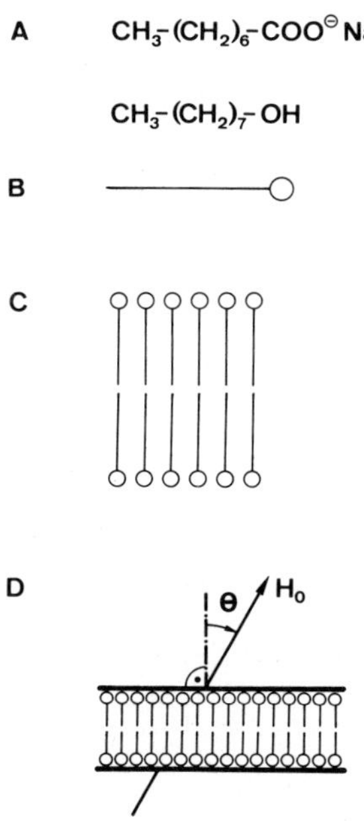

Fig. VIII-2. Some details of the experimental set-up for the lipid bilayer studies of Seelig and Niederberger[9]. A. Structure of amphiphilic components used as integral constituents of the lipid region, *n*-octanoic acid sodium salt and *n*-octanol. In successive experiments, the protons of individual methylene groups along the paraffin chains were replaced by deuterons. B. Schematic representation of the amphiphilic components A, where the circle represents the polar head group. C. Schematic representation of a lipid bilayer structure. In a certain concentration range of the three components *n*-octanoic acid, *n*-octanol, and water, the lipid molecules are arranged in such bilayers, which are separated from each other by layers of water. D. Sample preparation for the D-NMR experiments. The double layer structured material was sandwiched between glass plates. In the actual experiment, a stack of 20 glass plates with the dimensions $20 \times 6.6 \times 0.15$ mm was placed in a 10 mm NMR sample tube. The normal to the glass plates, which is also the normal to the oriented bilayers, was then adjusted at different angles θ relative to the external magnetic field H_0.

between the axis of motional averaging and the magnetic field H_0. For a homogeneously oriented sample of lipid bilayers (Fig. VIII-2D), $\Delta\nu(\theta)$ is given by

$$\Delta\nu(\theta) = \frac{3}{4}\left(\frac{e^2qQ}{h}\right) U(3\cos^2\theta - 1) \qquad \text{(VIII-1)}$$

where the quadrupole splitting constant (e^2qQ/h) is to a good approximation a constant for C–D bonds in paraffins, i.e. 170 KHz,[10] θ is preselected for a given experiment (Fig. VIII-2), and the order parameter U can thus be evaluated from the measured doublet splitting $\Delta\nu(\theta)$ (Fig. VIII-3). The order parameters U obtained from these experiments may be interpreted on the

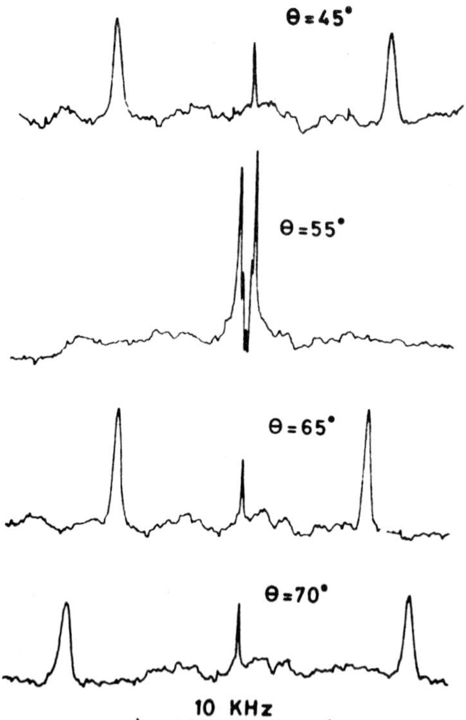

Fig. VIII-3. FT D-NMR spectra at 13.8 MHz of oriented bilayers in a phase consisting of 30 $wt\%$ 1,1-dideuterooctanol, 34 $wt\%$ sodium octanate, and 36 $wt\%$ H_2O for different angles θ. A total of ca. 20 mg of the material was used in the sample preparation described in Fig. VIII-2D. Each spectrum resulted from signal accumulation for 16 minutes. The sharp line in the center of the spectra arises from deuterated molecules in an isotropic phase, which formed during the sample preparation. Note that the line width of the resonances depends also on θ. Particularly sharp lines are observed near the "magic angle" of 54.7°, where the dipolar couplings to the surrounding protons are drastically reduced[11] (from Seelig and Niederberger[9]).

following basis: In the limit of rapid isotropic rotational motions, U will vanish, and as a result the doublet structure of the D signal will collapse to a single line. On the other hand, the more the rotational motions are restricted to the axis perpendicular to the bilayer plane (Fig. VIII-2), the order parameter U takes on larger values. By successive measurements of U for deuterons attached to methylene carbons at different spacings from the polar head group (Fig. VIII-2), the orientation of the paraffin chains at various positions in the bilayer structure could thus be established[9].

2. PHOSPHOROUS-31

^{31}P is the only naturally occurring isotope of phosphorous, has a spin $I = \frac{1}{2}$, and a relative sensitivity of 0.06 of that for ^{1}H (Tables I-1 and I-2). From the foregoing discussions of other nuclei, we conclude that the observation of ^{31}P high resolution NMR spectra presents no major technical problems. Furthermore, the chemical shift range for ^{31}P extends over more than 700 ppm, and the natural line widths are smaller than for protons.

^{31}P is, therefore, well suited for NMR experiments, and a large number of studies have been carried out[12–14].

The applications of ^{31}P-NMR for studies of proteins are rather limited, since phosphorous is a constituent of only relatively few conjugated proteins. On the other hand, it is one of the most challenging prospects, which we shall discuss some more at the end of this section, to use ^{31}P-NMR for investigations of metabolites in intact tissue specimen, and ^{31}P-NMR studies have also been described for nucleic acids[15,16], and phospholipids in membranes[17–19].

^{31}P-NMR Spectra of Lipoproteins

^{31}P-NMR spectra of lipoproteins have been reported by different groups[20–22]. These studies indicate the potential of ^{31}P-NMR to contribute substantially to the investigations of the distribution of different groups of phospholipids in lipoproteins, and the arrangement and spatial organization of the constituent molecular species in these particles. As an illustration, Fig. VIII-4 shows the ^{31}P-NMR spectra of native low density and high density human serum lipoproteins[22]. It is seen that both spectra contain two dominant resonances at 112.9 and 113.4 ppm. From comparison with the spectra of phosphatidylcholine and sphingomyelin (Fig. VIII-5), these resonances were attributed to those two lipids, which appear to be present in quite different relative concentrations in the preparations of low density and high density lipoproteins examined in Fig. VIII-4.

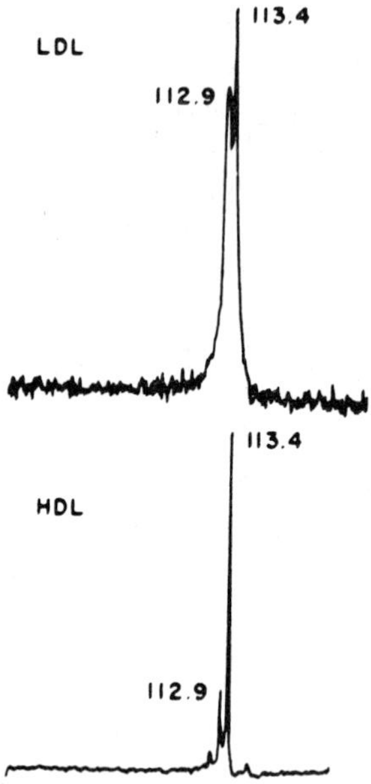

Fig. VIII-4. FT ^{31}P-NMR spectra at 40.48 MHz of low density (LDL) and high density (HDL) human serum lipoproteins in D_2O solution, pD = 8.6. The two numbers in the spectra indicate the chemical shifts in ppm from external P_4O_6 (from Assmann et al.[22]).

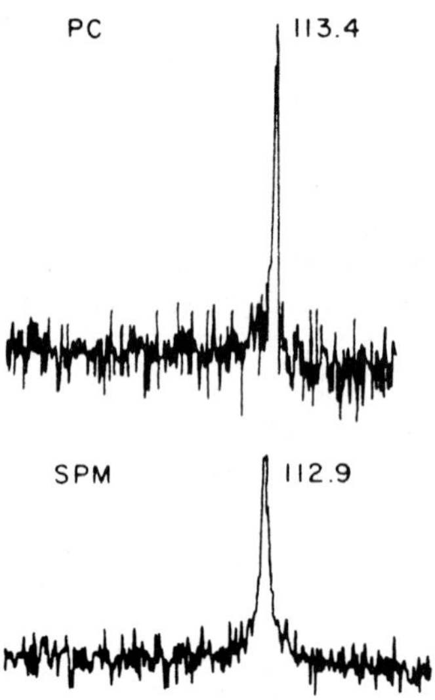

Fig. VIII-5. FT ^{31}P-NMR spectra at 40.48 MHz of phosphatidylcholine (PC) and sphingomyelin (SPM), each sonicated in D_2O for 5 minutes at 4°, pD = 8.6. The numbers in the spectra indicate the chemical shifts in ppm from external P_4O_6 (from Assmann et al.[22]).

^{31}P Compounds as Natural Extrinsic NMR Probes for Studies of Proteins

Since low molecular weight phosphorous containing compounds are omnipresent in biological systems, they occur often also associated with proteins, e.g. as cofactors, substrates or inhibitors of enzymes, or more generally as effectors controlling protein functions. Observation of the ^{31}P-NMR spectra can then provide information on certain features of the protein structure and the protein-effector interactions, and in this sense the ^{31}P containing molecules may be used as natural extrinsic NMR probes. In the following this is illustrated with some studies of hemoglobin.

One of the physiological controls of the oxygen affinity of human hemoglobin is through its interactions with 2,3-diphosphoglycerate (DPG). DPG lowers the ligand affinity by binding preferentially to the unliganded form of hemoglobin.

$$Hb^{II}, O_2 \underset{DPG}{\rightleftharpoons} Hb^{II}(O_2)_4 \qquad \text{(VIII-2)}$$

The ^{31}P-NMR study discussed here was undertaken to determine the stage of ligand binding at which the Hb-DPG complex dissociates, i.e. whether this occurs after one, two, three, or all four ligand binding sites of Hb^{II} are occupied (Equation VIII-2)[23]. In these experiments, CO was actually used as a ligand in place of O_2. The fractional ligand saturation was determined by measurement of the optical spectrum, and the fraction of DPG bound was measured by ^{31}P-NMR spectroscopy. The measurements were repeated for different degrees of CO saturation of the hemoglobin. Fig. VIII-6 shows the ^{31}P-NMR spectra of DPG in solution with fully liganded $Hb^{II}(CO)_4$ (A), and bound to deoxyhemoglobin (C). It is seen that binding to deoxyhemoglobin produces chemical shift changes of 0.42 and 0.75 ppm for the two ^{31}P resonance lines of DPG. DPG solutions with partially liganded hemoglobin gave intermediate chemical shifts (Fig. VIII-6B), indicating that there is rapid interconversion between the different complexes of Hb^{II}, O_2, and DPG present in the solution. Both the binding of CO measured in the optical spectrum, and the release of DPG monitored with ^{31}P-NMR are plotted vs. the partial pressure of CO in Fig. VIII-7. It is seen that the release of DPG lags behind the binding of CO by 5–10% throughout most of the range of p_{CO} over which ligand binding occurs. From this it was then concluded that DPG binds not exclusively to deoxyhemoglobin, but also, with intermediate affinity, to the partially liganded species[23].

^{31}P-NMR Studies of Metabolites in Intact Tissue Specimen

Low molecular weight phosphorous compounds give generally quite well resolved ^{31}P-NMR spectra[12–14]. As an illustration, Fig. VIII-8 shows the

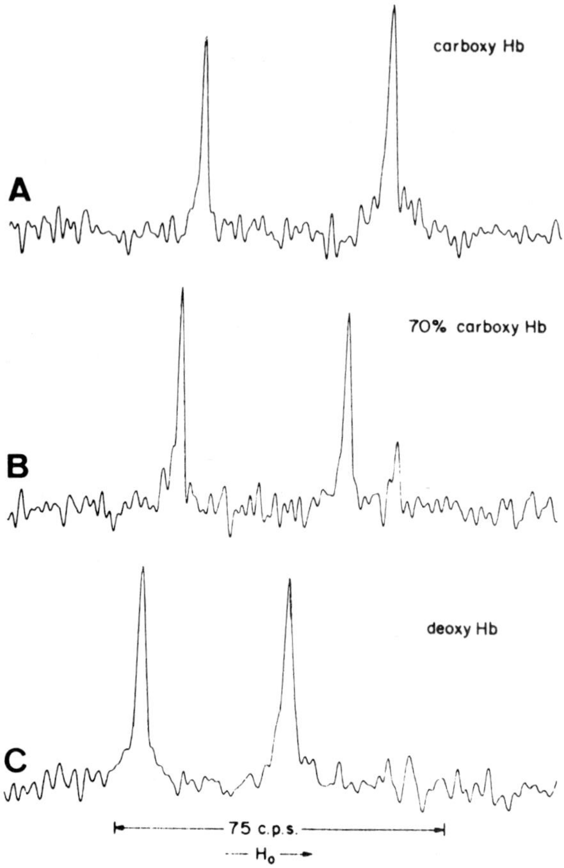

Fig. VIII-6. FT ^{31}P-NMR spectra at 40.48 MHz of $3 \cdot 10^{-3}$-M aqueous solutions of 2,3-diphosphoglycerate (DPG) containing $2 \cdot 10^{-3}$-M hemoglobin A, pH = 7.4, $T = 32°$. A. Deoxy-hemoglobin, B. 70% CO-hemoglobin; C. 100% CO-hemoglobin (from Huestis and Raftery[23]).

spectrum of adenosine triphosphate, ATP. In physiological systems, ATP and related small phosphorous containing molecules play an important role as intermediary metabolites, and in the regulation of enzyme activities. Therefore, knowledge of the metabolite levels, metabolite turnover, interactions and compartmentation in intact tissue and in model systems is of great interest for the delineation of the metabolic pathways. Recent experiments using a high field Fourier transform spectrometer have shown that ^{31}P-NMR is a very promising technique for these investigations. The ^{31}P resonances of a large number of different biologically important phosphate-containing compounds could be resolved, and it was demonstrated that their resonances can be detected in intact biological tissues[24].

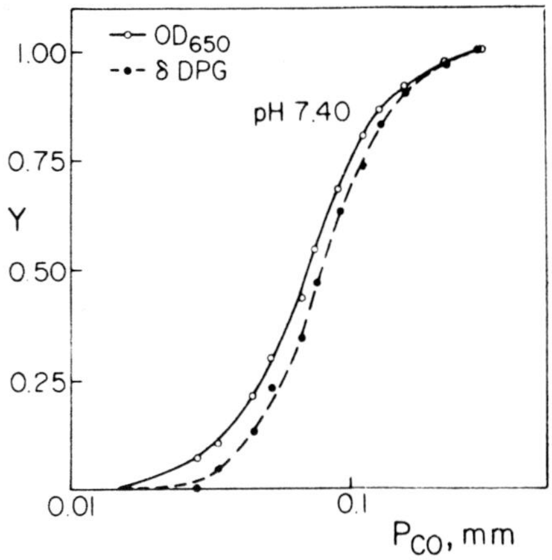

Fig. VIII-7. CO-binding curves of hemoglobin A obtained concurrently from fractional change in the visible spectrum (–○–), and fractional change in the ^{31}P-NMR chemical shifts (–●–). Y is the fraction of liganded hemoglobin (for Y = 1, all the hemoglobin molecules are in the form $Hb^{II}(CO)_4$), and p_{CO} is the partial pressure of CO in mm mercury (from Huestis and Raftery[23]).

No.	Height	Freq. (HZ)
1	1408	244·14
2	1510	264·16
3	1757	448·73
4	1929	468·26
5	720	881·83
6	1147	901·85
7	512	921·87

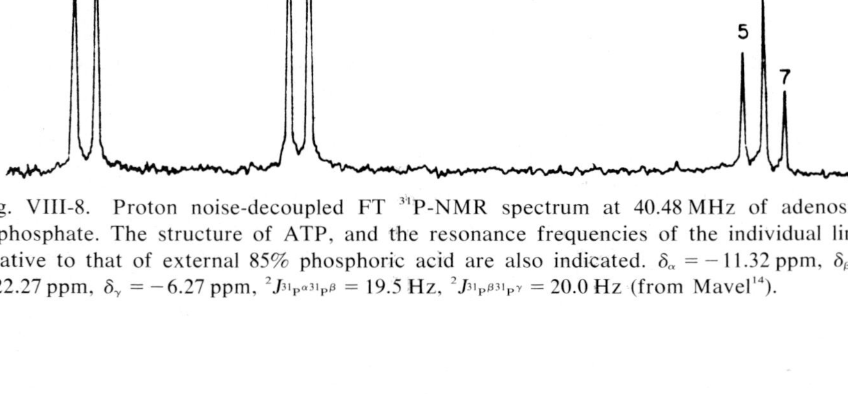

Fig. VIII-8. Proton noise-decoupled FT ^{31}P-NMR spectrum at 40.48 MHz of adenosine triphosphate. The structure of ATP, and the resonance frequencies of the individual lines relative to that of external 85% phosphoric acid are also indicated. $\delta_\alpha = -11.32$ ppm, $\delta_\beta = -22.27$ ppm, $\delta_\gamma = -6.27$ ppm, $^2J_{^{31}P\alpha ^{31}P\beta} = 19.5$ Hz, $^2J_{^{31}P\beta ^{31}P\gamma} = 20.0$ Hz (from Mavel[14]).

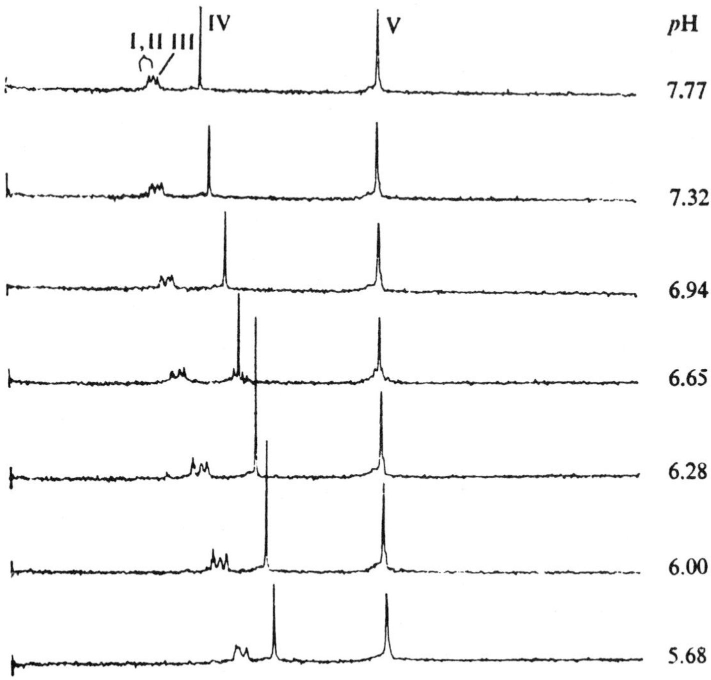

Fig. VIII-9. FT ^{31}P-NMR spectrum at 129 MHz of an aqueous solution containing a mixture of fructose-1,6-diphosphate (I,II), inosine monophosphate (III), inorganic phosphate (IV), and creatine phosphate (V) at various pH values (from Hoult et al.[24]).

Two illustrative experiments are reproduced in the Figs. VIII-9 and VIII-10. The technical details are indicated in the figure captions. Fig. VIII-9 shows the ^{31}P-NMR spectra of an aqueous solution containing fructose-1,6-diphosphate, inosine monophosphate, inorganic phosphate, and creatine phosphate at various pH values. It is seen that the resonances of the different compounds, which had been identified from comparison with solutions of the individual components, are quite well resolved, and some resonances are markedly dependent on pH. The pH dependence can be useful for resonance assignments in unknown systems. Furthermore it has been suggested that measurements of ^{31}P chemical shifts present a nondestructive technique for determining intracellular pH-values[25], and that extracellular and intracellular materials might be distinguished on the basis of the pH dependence of the ^{31}P chemical shifts. The latter aspect appears to indicate an avenue for studies of certain intercellular transport processes[24]. Fig. VIII-10 shows ^{31}P-NMR spectra of an intact muscle from a rat leg at different time intervals after the excision of the muscle. Again, the resonances of four different kinds of phosphate containing compounds are well

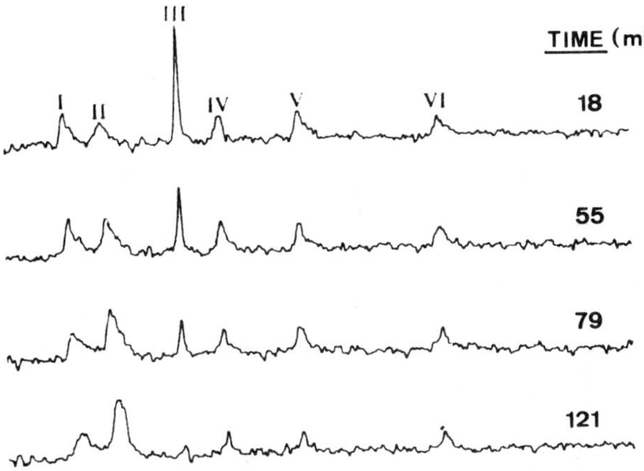

Fig. VIII-10. FT ^{31}P-NMR spectrum at 129 MHz of an intact muscle from the hind leg of a rat. The different spectra were recorded at the indicated time intervals after the excision of the muscle. The muscle was kept in a minimum volume of calcium-free Locke ringer. The following resonance assignments were made from comparison with model systems at different pH values (Fig. VIII-9): I. sugar phosphate and phospholipid; II. inorganic phosphate; III. creatine phosphate; IV to VI. γ-, α-, and β-phosphate of ATP (see Fig. VIII-8) (from Hoult et al.[24]).

resolved. From the chemical shifts of the lines IV to VI, it was concluded that the ATP in the muscle was predominatly complexed with Mg^{2+}, and from the chemical shift of the inorganic phosphate(II), the intracellular pH was estimated at 7.1 ppm.[24] With time elapsing after the excision of the muscle, the creatine phosphate concentration is seen to decrease, and the concentration of inorganic phosphate increases. In the trace recorded after 121 minutes, one notices also a decrease of the intensity of the ATP lines. In addition to the in vivo measurements of the metabolite levels, such experiments should also enable one to draw inferences on the activities of enzymes which are regulated by phosphate-containing compounds.

3. QUADRUPOLE NUCLEI AS NATURAL EXTRINSIC NMR PROBES OF BIOPOLYMERS

Ionic species with nuclear spins $I \geqslant 1$ are ubiquitous components in body fluids, and are often associated with protein molecules. Therefore it was suggestive that species like $^{23}Na^+$, $^{35}Cl^-$, $^{37}Cl^-$, $^{39}K^+$, $^{79}Br^-$, $^{81}Br^-$, and others could be used as extrinsic NMR probes without causing major perturbations of the proteins examined. The physical basis for the use of these nuclei as NMR probes is the spin relaxation by electric nuclear quadrupole interactions (Equation IV-38). For the present semiquantitative discussion of the principles of the technique, we may rewrite Equation IV-38 in the more compact form of Equation VIII-3

$$\frac{1}{T_1^Q} = \frac{1}{T_2^Q} = K\left(\frac{eQq}{\hbar}\right)^2 \tau_c \qquad \text{(VIII-3)}$$

K is taken to be a constant for a given nucleus, and we then have the two variable quantities q, the electric field gradient, and τ_c, the effective correlation time. In the "free" hydrated ions, the electron distribution has nearly cubic symmetry, i.e. q is small, and $\tau_c = \tau_R$ (Equation IV-31) is usually very short. Relatively narrow lines may thus result, as illustrated in Fig. VIII-11A, where the full width at half height of the chloride NMR signal

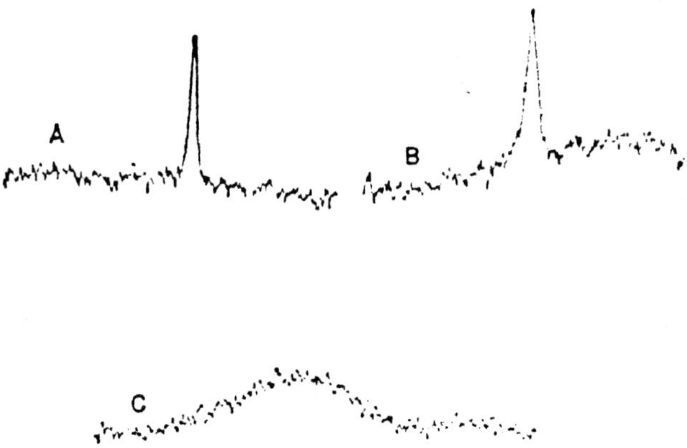

Fig. VIII-11. Chloride ion NMR lines in A. 0.5-M aqueous NaCl, B. 0.5-M aqueous NaCl plus $7 \cdot 10^{-5}$-M hemoglobin, C. 0.5-M aqueous NaCl plus $7 \cdot 10^{-5}$-M hemoglobin plus $14 \cdot 10^{-5}$-M $HgCl_2$ (from Stengle and Baldeschwieler[26]).

in an aqueous solution of NaCl is seen to be approximately 16 Hz.[26] The binding to macromolecules can greatly increase the effective correlation time for rotational motions of the ions, say from 10^{-12} sec for the free ion to 10^{-9} sec for the ion which is tightly bound to a medium-sized protein molecule, and the field gradient q may also be increased as a result of the interactions with the macromolecule. Direct NMR observation of the bound ions is usually not practicable, both because of the low NMR sensitivity of the nuclei in question (Table I-1), and the often extremely broad lines arising from quadrupole relaxation. Therefore, analogous to the technique used with paramagnetic relaxation probes (Equations VI-27 to VI-34), one usually observes the resonances of the bulk of the ions in a fast chemical exchange situation, e.g. in the system described by Equation VIII-4.

$$\text{Protein}(\text{Cl}^{\ominus}) \underset{}{\overset{\tau_M}{\rightleftharpoons}} \text{Cl}^{\ominus}(\text{Bulk}) \qquad \text{(VIII-4)}$$
$$T^Q_{2\text{bound}} \qquad\qquad T^Q_{2\text{free}}$$

$T^Q_{2\text{bound}}$ and $T^Q_{2\text{free}}$ are the transverse relaxation times for bound and free chloride ions, and τ_M is the life time in the binding sites of the protein with respect to chemical exchange. The relaxation time for the bulk of the chloride ions in the presence of the protein, $T_{2\,\text{obs}}$, is then given by

$$\frac{1}{T_{2\,\text{obs}}} = \frac{1}{T^Q_{2\text{free}}} + \frac{P_M \cdot n}{T^Q_{2\text{bound}} + \tau_M} \qquad \text{(VIII-5)}$$

$P_M = [\text{Protein}]/[\text{Cl}^-]$, and n is the number of ion binding sites per protein molecule. The effective correlation time τ_c for the protein-bound nuclei (Equation VIII-3) in the presence of chemical exchange is given by

$$\frac{1}{\tau_c} = \frac{1}{\tau_r} + \frac{1}{\tau_M} \qquad \text{(VIII-6)}$$

where $1/\tau_r = 1/\tau_R + 1/\tau_g$, and τ_R and τ_g are, respectively, the correlation times for the overall rotational motions of the protein-ion adduct, and the intramolecular motions of the bound ions.

In Section VIII.1 we have seen that the field gradient q for covalently attached deuterium labels in polypeptide chains is to a good approximation determined by the covalent structure, and different relaxation times may thus often be directly related to variations of τ_c. In the presently discussed experiments, a quantitative interpretation of the measured values of $T_{2\,\text{obs}}$ will generally be more ambiguous. Once it has been established that there is a fast exchange situation, the observed relaxation time will still depend on the combination of the three quantities q, τ_c, and n. The main use of quadrupole-ion probes will therefore in most cases be to detect the presence of specific binding sites, which have a dominant influence on the relaxation rate as compared to that of non-specific binding to the protein surface. As will be illustrated in the following, the method is in particular well suited for studies of certain metalloproteins.

In the examples described below, chloride ion was used as the NMR probe. Other nuclei employed in related experiments include ^{81}Br,[27] ^{23}Na,[28–30] and ^{39}K,[31] and ^{17}O in isotope enriched water[32] might also be mentioned in this context.

Chloride Ion Relaxation in Protein Solutions

The binding of chloride ions to the surface of protein molecules may mainly be through interactions with positively charged groups, e.g. lysine or histidine side chains. As we have seen previously (Section IV.5, Appendix

V), the rotational motions of these amino acid side chains are not usually restricted to the overall rotations of the protein molecule, and hence the effective correlation times in these binding sites may be quite short. Since the field gradient arising from predominantly electrostatic interactions is usually also rather small, the resulting effect on the relaxation time $T_{2\,obs}$ is not necessarily very large. As an illustration, Fig. VIII-11B shows that the line width of the chloride ion resonance in 0.5-M aqueous NaCl solution increases from 16 Hz to 22 Hz, when $7 \cdot 10^{-5}$-M hemoglobin is added[26]. Much care has to be taken in the interpretation of such a result, since line broadening of this magnitude might also arise from changes of the solvent structure. Yet, the technique has e.g. successfully been applied in a detailed chloride ion binding study of human oxy- and deoxyhemoglobin, which revealed preferential chloride binding to the deoxygenated form of the protein[33].

Chloride Ion Probe Studies of Mercury Labelled Proteins

Large relaxation rates usually arise when the quadrupole nuclei are in covalent bonds. This was used by Stengle and Baldeschwieler[26] to introduce artificial specific chloride binding sites into protein molecules. The technique was based on the rapid exchange of chloride ions in and out of the first coordination sphere of $Hg^{2\oplus}$

$$HgCl_4^{2\ominus} \rightleftharpoons Cl^{\ominus}\ (\text{Bulk}) \qquad \text{(VIII-7)}$$

$$T_{2\,\text{bound}}^{Q} \approx 4 \cdot 10^{-5}\ \text{sec} \qquad T_{2\,\text{free}}^{Q} \approx 2 \cdot 10^{-2}\ \text{sec}$$

Even though $HgCl_4^{2\ominus}$ is a small molecule with a rather short rotational correlation time τ_c, the large field gradient q arising from the covalent mercury-chloride bonds causes a marked line broadening. As indicated by $T_{2\,\text{bound}}^{Q}$ in Equation VIII-7, the full width of the chloride NMR line of $HgCl_4^{2\ominus}$ is thus approximately 8000 Hz. From Equation VIII-5 one estimates that in the fast exchange case, the addition of $3 \cdot 10^{-5}$-M $HgCl_4^{2\ominus}$ to a 0.5-M NaCl solution causes an increase of the observed chloride NMR line width of 1 Hz. This relaxation effect can be greatly enhanced if the mercury is attached to a macromolecule, since the effective correlation time may thus be increased by several orders of magnitude. This is illustrated in Fig. VIII-11C, which shows that in the presence of two equivalents of $HgCl_2$, the addition of $7 \cdot 10^{-5}$-M hemoglobin causes the line width of a 0.5-M NaCl solution to increase from 16 Hz to 210 Hz.[26]

Practical uses of the mercury label technique include the evaluation of the rotational correlation time for the labelled macromolecule. For example, if one assumes in the system of Fig. VIII-11 that the mercuric ions react to completion with the two accessible –SH groups of hemoglobin, and that q is about the same for the hemoglobin-mercuric chloride complexes and $HgCl_4^{2\ominus}$, then τ_c for the $Hg^{2\oplus}$ site in hemoglobin must be approximately 200 times as long as that for $HgCl_4^{2\ominus}$. Other potential applications include the determination of the number of $Hg^{2\oplus}$-accessible –SH groups in a protein[34], and studies of the interactions between macromolecules and mercury-labelled small ligands[35].

Chloride Ion Probe Studies of Metalloproteins

Large q-values occur also for chloride ions bound to other metal ions than Hg^{2+}, e.g. Zn^{2+} or Cd^{2+}. Therefore the chloride ion probe technique may also be applied for investigations of natural metalloproteins. This is illustrated in Fig. VIII-12 with a study of a zinc enzyme, alkaline phosphatase[36]. These experiments had been undertaken to determine the number of Zn^{2+}-ions specifically bound by the protein, and to investigate the state of the protein-bound metal ions. The figure shows that 4 Zn^{2+}-ions are specifically bound to the apo-alkaline phosphatase. These four Zn^{2+}-ions cannot be detected by the NMR of the bulk of the chloride, whereas the addition of an

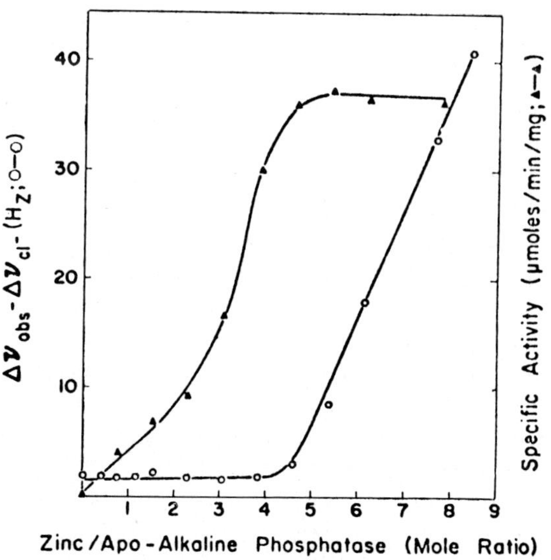

Fig. VIII-12. Titration of apo-alkaline phosphatase with zinc ion. The reaction mixtures contained 3.1 mg/ml of apo-alkaline phosphatase, 0.57-M NaCl, 0.01-M Tris-HCl, pH 8.0, and the indicated ratio of zinc : protein. The line broadening of the chloride ion NMR signal arising from the addition of Zn^{2+} in the presence of protein (–○–○–) and the specific enzymatic activity (–▲–▲–) are plotted vs. the zinc : apo-protein ratio (from Cottam and Ward[36]).

excess of Zn^{2+} causes line broadening of the resonance. Comparison with the simultaneously recorded dependence of the enzymatic activity on the Zn^{2+}-content shows that all the four specifically bound metal ions are required to obtain the full activity. The absence of a relaxation effect from the protein-bound Zn^{2+}-ions indicates that these are either not accessible for coordination of chloride ions, or that the exchange rate of chloride ions between the coordination sites of protein-bound Zn^{2+} and the bulk of the solution is too slow to cause a detectable relaxation effect $T_{2\,obs}$ (Equation VIII-5).

4. FLUORINE-19 NMR PROBES IN PEPTIDES AND PROTEINS

Fluorine-19 has several attributes which make it a suitable NMR label for studies of biological systems: ^{19}F has a spin $I = \frac{1}{2}$, and the sensitivity for NMR detection of ^{19}F is only slightly smaller than that for ^{1}H (Table I-2). ^{19}F chemical shifts span a much larger range than those of ^{1}H, i.e. approximately 300 ppm,[37] and hence readily discernible relative shifts may be expected to arise from perturbations which might otherwise hardly be detectable in the ^{1}H-NMR shifts. ^{19}F—^{19}F couplings and ^{1}H—^{19}F couplings are also sizeably larger than the corresponding ^{1}H—^{1}H couplings[1]. In addition to these favorable NMR properties, fluorine is also a good label from the viewpoint of its chemical structural properties. The fluorine atoms are rather small, about the size of an OH-group, and carbon-fluorine bonds are very unreactive. It is therefore not surprising that in many systems the substitution of hydrogen atoms or hydroxyl groups with fluorine atoms appears to cause only minor perturbations of the molecular conformations, and does not cause a complete loss of the biological activity in the labelled molecules.

A great number of biological applications of ^{19}F-NMR labels have been described. These may be classified into two groups. The first group includes fluorine labels which are covalently attached to the macromolecule examined. Since the natural systems do not usually contain fluorine atoms, the NMR lines of the ^{19}F-labels are in most cases readily identified. They may then be monitored e.g. during interactions of the labelled protein with other molecules, during protein denaturation, etc. Two illustrative applications of covalent ^{19}F-NMR labels are described below. The second group includes ^{19}F-labelled small molecules which are studied during their interactions with macromolecular species. They may include substrates and inhibitors of enzymes, other effectors which control protein functions in diverse ways, etc. Some specific examples will be presented in Chapter IX.

Covalent ¹⁹F-NMR Labels for Product Control in Peptide Synthesis

This technique was based on the observation that certain N-trifluoroacetylated amino acids can be identified from the ^{19}F chemical shifts (Table VIII-1)[38]. Since on the other hand the penultimate amino acid influences the ^{19}F chemical shift of a N-trifluoroacetylated peptide only slightly, the N-terminal amino acid residue may be identified from the ^{19}F-NMR spectrum even in relatively long polypeptide chains. Fig. VIII-13 illustrates tne use of this technique for product control in a solid-phase synthesis of a tetrapeptide. Details are described in the figure caption. The figure shows that the different intermediary reaction products in the step-wise synthesis of H-Ile-Ala-Val-Gly-OH could readily be distinguished in the ^{19}F-NMR spectra. The purification of the final product could thus be monitored by ^{19}F-NMR[38].

TABLE VIII-1

^{19}F chemical shifts δ of N-trifluoro-acetylated amino acids. The chemical shifts are in ppm relative to internal methyl trifluoro-acetate, CF_3COOCH_3 (from Ref. 38).

Compound	δ (ppm)
N-TFA-Gly-OH	−0.61
N-TFA-Ala-OH	−0.49
N-TFA-Val-OH	−0.08
N-TFA-Leu-OH	−0.31
N-TFA-Ile-OH	−0.10
N-TFA-Asp-OH	−0.56
N-TFA-Glu-OH	−0.47
N-TFA-Asn-OH	−0.72
N-TFA-Gln-OH	−0.76
N-TFA-Phe-OH	−0.49
N-TFA-Tyr-OH	−0.46
N-TFA-Try-OH	−0.44
N-TFA-Ser(OBzl)-OH	−0.29
N-TFA-Thr(OBzl)-OH	−0.22
N-TFA-Cys(SBzl)-OH	−0.33
N-TFA-cystine	−0.36
N-TFA-Met-OH	−0.37
N-TFA-Arg-OH	−0.58

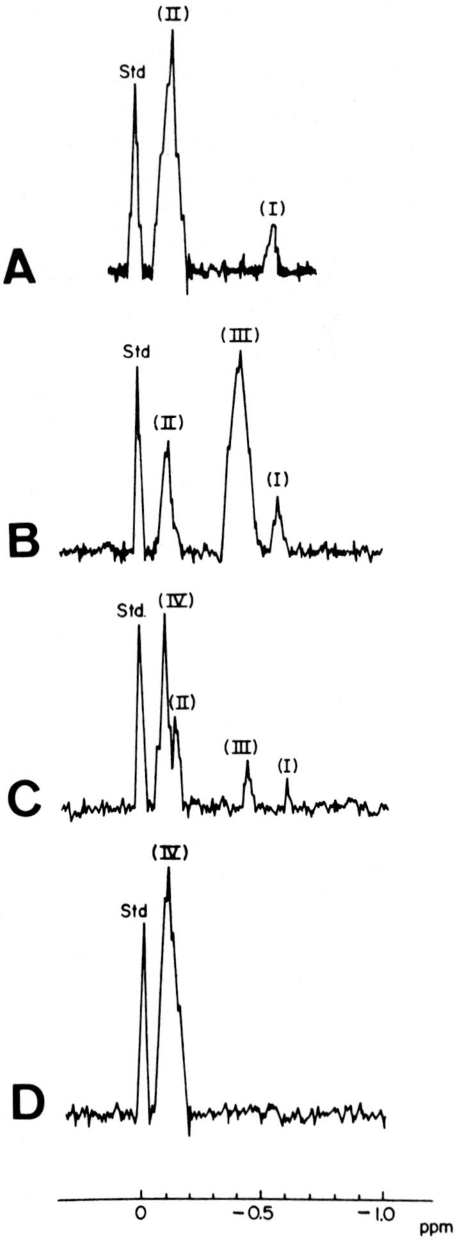

Fig. VIII-13. ^{19}F-NMR control of a solid state synthesis of the linear tetrapeptide H-Ile-Ala-Val-Gly-OH. A. ^{19}F-NMR spectrum at 56.4 MHz of the N-trifluoroacetylated reaction products after the first coupling step to form Val-Gly. Resonance assignments: (I) TFA-Gly-OH,

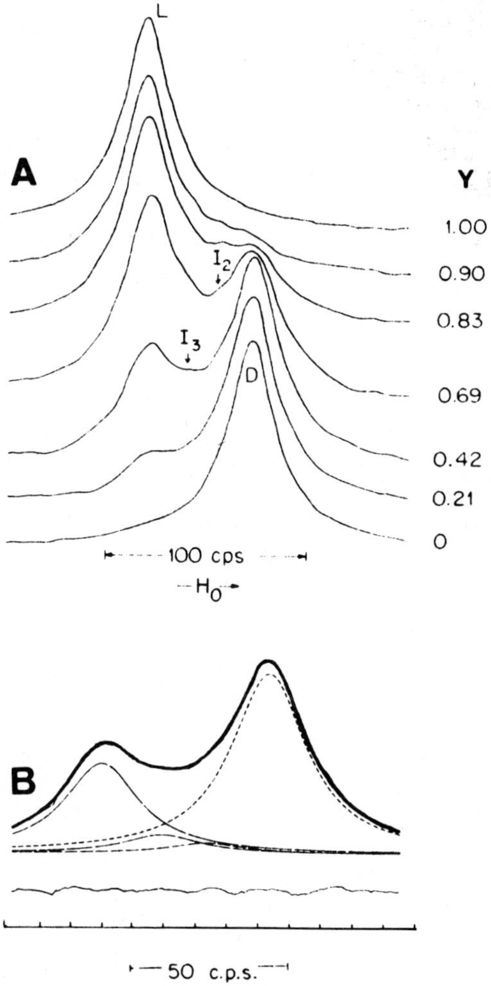

Fig. VIII-14. A. FT ^{19}F-NMR spectra at 94 MHz of trifluoroacetylated hemoglobin A at the various stages of fractional ligandation Y indicated by the numbers on the right (see Fig. VIII-7 for the definition of Y). B. ^{19}F-NMR spectrum of TFA-Hb at the ligandation state $Y = 0.42$ (solid line), and the computer fitted Lorentzians for the spectra of the constituent species D, I_2, I_3, and L (broken lines) (from Huestis and Raftery[39]).

(II)TFA-Val-Gly-OH. B. ^{19}F-NMR spectrum after the second coupling step. (III) = TFA-Ala-Val-Gly-OH. C. ^{19}F-NMR spectrum after the third coupling step. (IV) = TFA-Ile-Ala-Val-Gly-OH. D. ^{19}F-NMR spectrum of the purified N-trifluoroacetylated tetrapeptide. "Std" indicates the chemical shift of internal CF_3COOCH_3 (from Bayer et al.[38]).

Covalent ^{19}F-NMR Labels for Studies of Protein Function

One of the proteins studied by this technique is human hemoglobin[39]. Previously we had discussed the relations between ligand binding to hemoglobin (Equation VIII-2) and the interactions with DPG (Figs. VIII-6 and VIII-7). The present experiment was undertaken to investigate if there was preferential oxygen binding to the α- or β-chains of hemoglobin, and if the occurrence of partially liganded molecules could be detected. The hemoglobin used for these studies had been specifically labelled at cysteinyl-β93 by reaction with 1-bromo-3-trifluoroacetone. The ^{19}F-NMR spectra of the fluorine-labelled hemoglobin at various stages of ligandation are shown in Fig. VIII-14A. The ^{19}F chemical shift is seen to be sensitive to the ligand binding, and the spectra of deoxyhemoglobin (D) and oxyhemoglobin (L) can readily be distinguished. The ligand induced ^{19}F chemical shift changes were found to be different depending on whether an additional oxygen molecule was bound to the α- or β-chains. From a computer fit of the spectra of the partially liganded species to Lorentzian lines (Fig. VIII-14B), the authors concluded that there was preferential initial ligand binding to the α-chains, and that two partially liganded species (I_2 and I_3) could be identified in the spectra at intermediate states of ligandation[39].

5. BIBLIOGRAPHY

1 Axenrod T. and Webb G. A. (Eds.), Nuclear Magnetic Resonance Spectroscopy of Nuclei other than Protons, J. Wiley, New York, 1974
2 Diehl P., in Ref. 1, p. 275
3 Briggs J. M., Farnell L. F. and Randall E. W., Chem. Comm. *1973*, 70
4 Diehl P. in Ref. 1, p. 377
5 Glasel J. A., Hruby V. J., McKelvy J. F. and Spatola A. F., J. Mol. Biol. *79*, 555 (1973)
6 Mantsch H. H., Saito H., Leitch L. C. and Smith I. C. P., J. Amer. Chem. Soc. *96*, 256 (1974)
7 Diehl P. and Khetrapal C. L., in NMR-Basic Principles and Progress, P. Diehl, E. Fluck and R. Kosfeld (Eds.), Springer Verlag, New York, 1969, Vol. 1
8 Meiboom S. and Snyder L. C., Accts Chem. Res. *4*, 81 (1971)
9 Seelig J. and Niederberger W., J. Amer. Chem. Soc. *96*, 2069 (1974)
10 Burnett L. J. and Muller B. H., J. Chem. Phys. *55*, 5829 (1971)
11 Andrew E. R., Progr. Nuclear Magnetic Resonance Spectroscopy *8*, Part I (1971)
12 Crutchfield M. M., Dungan C. H., Letcher J. H., Mark V. and Van Wazer J. R., ^{31}P Nuclear Magnetic Resonance, J. Wiley, New York, 1967
13 Van Wazer J. R., in Determination of Organic Structures by Physical Methods, F. Nachod and J. J. Zuckerman (Eds.), Academic Press, New York, 1971, Vol. 4, p. 323
14 Mavel G., Annual Reports on NMR Spectroscopy, Vol. 5B (1973)
15 Guéron M., FEBS Lett. *19*, 264 (1971)
16 Shulman R. G., Sternlicht H. and Wyluda B. J., J. Chem. Phys. *43*, 3116 (1965)
17 Henderson T. O., Glonek T. and Myers T. C., Biochemistry *13*, 623 (1974)

18 Bystrov V. F., Shapiro Y. E., Viktorov A. V., Barsukov L. I. and Bergelson L. D., FEBS Lett. *25*, 337 (1972)
19 Berden J. A., Cullis P. R., Hoult D. I., McLaughlin A. C., Radda G. K. and Richards R. E., FEBS Lett. *46*, 55 (1974)
20 Ho C. and Kurland R. J., J. Biol. Chem. *241*, 3002 (1966)
21 Glonek T., Henderson T. O., Kruski A. W. and Scanu A. M., Biochim. Biophys. Acta *348*, 155 (1974)
22 Assmann G., Sokoloski E. A. and Brewer H. B., Proc. Nat. Acad. Sci. *71*, 549 (1974)
23 Huestis W. H. and Raftery M. A., Biochem. Biophys. Res. Comm. *49*, 428 (1972)
24 Hoult D. I., Busby S. J. W., Gadian D. G., Radda G. K., Richards R. E. and Seely P. J., Nature *252*, 285 (1974)
25 Moon R. B. and Richards J. H., J. Biol. Chem. *248*, 7276 (1973)
26 Stengle T. R. and Baldeschwieler J. D., Proc. Nat. Acad. Sci. *55*, 1020 (1966)
27 Ward R. L. and Whitney P. L., Biochem. Biophys. Res. Comm. *51*, 343 (1973)
28 Haynes D. H., Pressman B. C. and Kowalsky A., Biochemistry *10*, 852 (1971)
29 Shporer M., Zemel H. and Luz Z., FEBS Lett. *40*, 357 (1974)
30 Ostroy F., James T. L., Noggle J. H., Sarrif A. and Hokin L. E., Arch. Biochem. Biophys. *162*, 421 (1974)
31 Bryant R. G., Biochem. Biophys. Res. Comm. *40*, 1162 (1970)
32 Civan M. M. and Shporer M., Biophys. J. *12*, 404 (1972)
33 Chiancone E., Norne J. E., Forsén S., Antonini E. and Wyman J., J. Mol. Biol. *70*, 675 (1972)
34 Man M. and Bryant R. G., J. Biol. Chem. *249*, 1109 (1974)
35 Haughland R. P., Stryer L., Stengle T. R. and Baldeschwieler J. D., Biochemistry *6*, 498 (1967)
36 Cottam G. L. and Ward R. L., Arch. Biochem. Biophys. *141*, 768 (1970)
37 Emsley J. W. and Phillips L., Progress in Nuclear Magnetic Resonance Spectroscopy, Vol. 7 (1971)
38 Bayer E., Hunziker P., Mutter M., Sievers R. E. and Uhmann R., J. Amer. Chem. Soc. *94*, 265 (1972)
39 Huestis W. H. and Raftery M. A., Biochem. Biophys. Res. Comm. *49*, 1358 (1972)

CHAPTER IX

MULTIMOLECULAR SYSTEMS: INTERMOLECULAR INTERACTIONS OF PEPTIDES AND PROTEINS

Many of the recent advances in biological NMR applications clearly indicate that one trend for the future developments will be to use sophisticated NMR techniques for studies of ever more complex multimolecular biological systems. On the basis of the preceding Chapters I–VIII, we are now in a position to appreciate the potentialities of such uses of NMR. The intention of this last chapter is to provide the reader with selected examples of NMR studies in multicomponent systems, and possibly to stimulate him to tackle such ambitious tasks in his own research.

For the sake of presentation, the experiments described in this chapter are grouped into four sections, i.e. binding of small molecules to macromolecules, interactions of peptides and proteins with nucleic acids, proteins in membranes, and studies of intact cells and tissues. It is a common trait of most of these experiments that certain components in the systems are more readily accessible for NMR observation than others, which may be present only in very low abundance. Therefore the "probe concept", in the widest sense of the word, plays an important role. Quite often, the NMR spectral properties of the more readily observable components can be used to investigate certain structural and dynamic features of the less abundant molecules, thus providing naturally built-in NMR probes. In many cases the conditions for such experiments can be greatly improved by selective isotope labelling of individual components. Also, natural or external paramagnetic NMR probes may quite often be a crucial help to obtain the required spectral resolution, or amplification of certain NMR spectral phenomena.

1. BINDING OF SMALL MOLECULES

In this section, we consider systems of the general type

$$\text{Macromolecule(L)}_q \underset{\tau_M}{\overset{L}{\rightleftharpoons}} \text{Bulk of L (Free ligands)} \qquad \text{(IX-1)}$$

where L indicates small ligand molecules which are exchanged between the binding sites of the macromolecule and the bulk of the solution containing the free ligands L. q is the number of binding sites per macromolecule, and the ligand exchange rate is characterized by the life time with respect to chemical exchange, τ_M, of the ligand molecules in the binding sites of the macromolecule. The situation IX-1 is essentially analogous to those previously encountered in solutions of Co^{2+} or Mn^{2+} ions (Equations VI-23 and VI-27), except that the binding sites on the macromolecule may be either diamagnetic or paramagnetic. In general, both the chemical shifts and the spin relaxation times T_1 and T_2 will be different for the nuclei of the bound and free ligand molecules. As indicated previously (Sections VI.5 and VIII.3), information on the binding sites of the macromolecule may in principle be obtained from direct observation of the resonances of the bound ligand molecules. In the situation of slow ligand exchange this is actually the only relevant observable quantity. An example are the studies of ^{13}C enriched carbon monoxide bound to the heme groups of hemoglobin, where the CO molecules on the α- and β-subunits could be distinguished from their different chemical shifts[1–5]. In practice, however, systems with rapid exchange of the ligand molecules (Equation IX-1) are of more general interest. As was discussed in more detail in Section VI.5, the influence of the macromolecule on the chemical shifts and relaxation times of the bound ligands is averaged over all the molecules L in the system by the rapid exchange in and out of the binding sites. From the NMR parameters of the free ligand molecules in the bulk of the solution containing the macromolecules. One can then draw inferences on the kinetics of the ligand exchange reactions IX-1, or the nature of the ligand binding sites, or both (Equations VI-24, VI-25, VI-30 to VI-36).

In most studies of interactions between macromolecules and rapidly exchanging ligands, only the NMR spectra of the low molecular weight components are observed. In these experiments, the macromolecular components may actually be present in much lower molar concentrations than what would be required for the direct observation of their NMR spectra in high resolution NMR experiments. This enables one to apply such studies to quite large molecules with molecular weights up to several hundred thousand, or, more generally, to systems containing very low molar concentrations of the macromolecule of interest. Another attractive feature is that the NMR observation of the small ligand molecules depends to a lesser degree on optimal resolution and sensitivity of the apparatus than the direct observation of macromolecules, and therefore experiments of this type were of practical interest for certain areas of protein research since the early 1960's (Section I.3). The present account of these techniques is restricted to a brief qualitative discussion of some fundamental aspects, and the indication of some representative references to original papers. This seemed all the more warranted, since the recent monograph by Dwek[6] includes a comprehensive treatment of this area of biological NMR work, and numerous excellent review articles are also available[7–11].

NMR Relaxation Studies of the Hydration of Diamagnetic Macromolecules

In aqueous solutions of biopolymers, a certain number of water molecules is bound to one or several different types of hydration sites of the macromolecule. Typically, the protein concentration is of the order 10^{-3}-M, while the bulk water concentration is approximately 55-M. In diamagnetic systems, the dominant relaxation mechanism for the water protons is proton–proton dipole–dipole coupling, so that the Equations IV-29 and IV-30 apply. Because of the factor $1/r^6$, where r is the distance between interacting nuclei, one may actually to a good approximation assume that for both the free and the protein-bound water molecules, the proton relaxation comes essentially entirely from dipolar interactions between the two water protons modulated by the rotational motions of the relaxation vector $\boldsymbol{r}$ (Fig. IV-11). The longitudinal relaxation times for free and protein-bound water molecules may thus be given by the following compact form of Equation IV-29

$$\frac{1}{T_1} = \text{const} \cdot f_1(\tau_c) \qquad \text{(IX-2)}$$

where "const." is a constant factor typical for the protons in a water molecule, and $f_1(\tau_c)$ is a function of the effective correlation time τ_c, with the general features given in Fig. IV-12. Different relaxation times for free and bound water molecules can thus arise as a consequence of the different rotational correlation times for free water, i.e. approx. $3 \cdot 10^{-12}$ sec, and the hydrated macromolecule, i.e. approx. $1 \cdot 10^{-7}$ sec for a spherical particle with the size of hemoglobin. The relaxation rate of the bulk water in the presence of the macromolecules is then

$$\frac{1}{T_{1\,\text{obs}}} = \frac{1}{T_{1\text{L}}} + \sum_i \frac{P_{\text{M}} q_i}{T_{1\text{M}i} + \tau_{\text{M}i}} \qquad \text{(IX-3)}$$

where $T_{1\text{L}}$ is the relaxation time for free water, P_{M} = [Macromolecules]/[H_2O], q_i is the number of hydration sites of type i, $T_{1\text{M}i}$ and $\tau_{\text{M}i}$ are the relaxation time and the life time with respect to chemical exchange for the water molecules in the hydration sites i, and the summation is over the i different types of hydration sites of the macromolecule. In the fast exchange situation, one has that $\tau_{\text{M}i} \ll T_{1\text{M}i}$ in Equation IX-3, and from the explicit form of $f(\tau_c)$ (Equation IV-29, Fig. IV-12) one expects that $T_{1\,\text{obs}}$ is dependent on the resonance frequency ω_{I}. The expected frequency dependence is readily apparent in the data on apotransferrin shown in Fig. IX-1[12]. It is seen that there is a dominant dispersion centered at approximately $6 \cdot 10^5$ Hz, and there is also a small frequency dependence in the range 10 MHz to 100 MHz.

A set of data corresponding to that of Fig. IX-1 can in principle yield information on the rotational correlation time τ_R of the hydrated protein, the life times τ_{Mi} of the bound water molecules, and the number of bound water molecules, q. Since the effective correlation time τ_c in Equation IX-2 is given by a combination of τ_R and τ_M, and possibly a characteristic time for intramolecular rate processes (Equation IV-40), the dispersion of $1/T_1$ may be observed at different frequencies for water molecules in different hydration sites. Thus in the example of Fig. IX-1 it was concluded that the nuclear magnetic relaxation dispersion centered around 0.6 MHz comes from hydration water for which $\tau_R \ll \tau_M \ll T_{1M}$, whereas the dispersion at higher frequencies was assigned to loosely bound water molecules for which τ_M would make an important contribution to τ_c.[12] From the major dispersion in Fig. IX-1, τ_R for the hydrated apo-transferrin was then found to be $\approx 3 \cdot 10^{-7}$ sec. Once the correlation time and the extent of the relaxation enhancement in the bulk water are known, the hydration number q can in

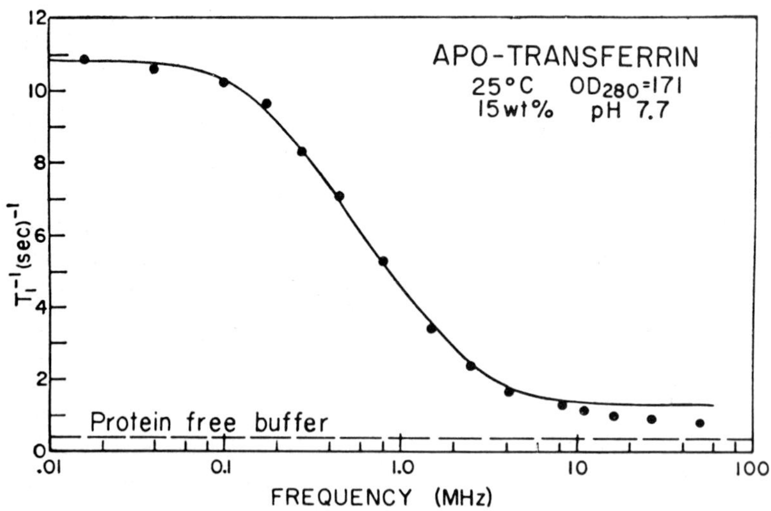

Fig. IX-1. Frequency dependence between 0.01 MHz and 50 MHz of the rate of the longitudinal proton relaxation of the bulk water in an aqueous solution of apo-transferrin. The experimental conditions are indicated in the figure, and the relaxation rate for the protein-free buffer solution is also given. For the technical details concerning the measurements at very low fields H_0 ($\nu_H = 0.1$ MHz corresponds to a field $H_0 = 23$ G), the reader is referred to the original literature[12, 15] (from Koenig and Schillinger[12]).

principle be computed with Equation IX-3. Assuming in the example of apotransferrin that the rotational motions of the more tightly bound water molecules are actually restricted to the overall rotational tumbling, $q_i \approx 15$ was thus obtained. From model considerations, this appeared to correspond only to a small fraction of the available hydration sites. If, on the other hand, the protein-bound water molecules were assumed to spin rapidly about an axis fixed with respect to the macromolecule, the resulting relaxation rates $1/T_{1M}$ would be greatly reduced[12, 13], which is in analogy to the previously discussed situation for rapidly spinning methyl groups (Section IV-5 and Appendix V), and hence a considerably larger hydration number q would be obtained from Equation IX-3.

In conclusion, the above discussion has shown that the relaxation enhancement in the NMR of the bulk water may in principle be employed to investigate the hydration of dissolved macromolecules. However, it also indicated that a detailed interpretation of the solvent relaxation data is by no means straightforward, not even in simple binary systems consisting of a diamagnetic macromolecular solute and the solvent. Despite these inherent limitations, studies of the solvent NMR properties are of considerable practical interest. First, more direct NMR studies of the hydration, as outlined e.g. in Section VI.5 (Equation VI-26) are not usually applicable, since the water exchange between the hydration sites of biopolymers and the bulk of the solution is generally rather fast. Second, the water resonance is often the most prominent feature in the NMR spectra of complex systems, e.g. also in intact tissues, and its use as a potential source of information on other constituent components is therefore quite attractive from the experimental point of view. One can readily anticipate that the interpretation of the data may become more ambiguous when additional potential ligand molecules besides H_2O are present. The following section will show that the presence of natural or extrinsic paramagnetic NMR probes can in many cases help to obtain certain quite definite information even in rather complex systems.

Experiments with Paramagnetic Relaxation Probes

The principles governing the nuclear spin relaxation in the bulk water of solutions of paramagnetic metal ions, which were discussed in Section VI.5 (Equations VI-27 to VI-36), apply also for paramagnetic centers in macromolecules. Thus, information may be obtained on both the kinetics of the ligand exchange reactions, and on the nature of the binding sites (Equations VI-30 and VI-31). From the NMR studies, the binding sites can in principle be characterized by the effective correlation times (Equations VI-28 and VI-29), the distances r_i between the paramagnetic center and individual nuclei of the ligand molecules, the contact coupling constants A_i (Equations VI-15 and VI-16), and the number of binding sites, q (Equations VI-30, VI-31, IX-3). Quite generally, the potentially most interesting situation arises when the NMR relaxation is dominated by electron-nucleus dipole–dipole coupling, where the relaxation rates can by the r^{-6}-dependence be related to the geometry of the protein-ligand complex. Again it should be added, however, that a detailed analysis of the experimental relaxation data is in general not straightforward. In particular, since the observed relaxation times depend at the same time on τ_c, r, and q, the separate evaluation of the individual parameters is often ambiguous, and additional data obtained with different techniques may be needed for a complete description of the system.

Because of the large magnetic moment of the unpaired electron (Table VI-1), the interactions of the ligand nuclei with the paramagnetic centers of macromolecules are in many cases the dominant factor governing the NMR relaxation in the ligands. On this basis, essential qualitative features of the systems studied can often be delineated. Some typical situations are in the following described with the aid of the schematic drawing in Fig. IX-2, where M is a typical paramagnetic relaxation probe, e.g. a paramagnetic metal ion with a long electronic relaxation time T_{1e}, or a nitroxide spin label. (i) In an aqueous solution of the protein, the water proton relaxation rate is greatly affected if there is rapid exchange of water molecules in and out of the first coordination sphere of the paramagnetic center. On the other hand, because of the r^{-6}-dependence of the dipolar interactions, a paramagnetic center located in the interior of the protein has a much lesser impact on the

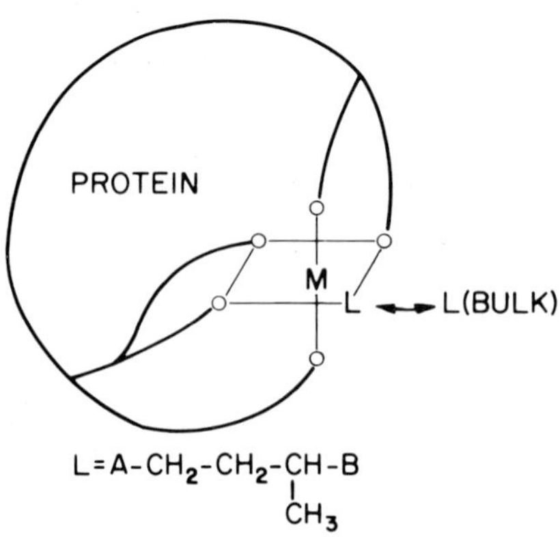

Fig. IX-2. Schematic representation of a paramagnetic protein interacting with low molecular weight ligands. The ligand L contains two potential binding sites A and B for attachment to the paramagnetic center M. In the situations discussed in the text, there is rapid exchange of ligand molecules between the paramagnetic center of the macromolecule and the bulk of the solution.

nuclear spin relaxation in the solvent molecules. Thus the accessibility of the paramagnetic center for binding of solvent molecules can be determined. (ii) One may be interested in the mode of binding of a ligand which contains several different potential binding sites, e.g. A and B in the ligand molecule shown in Fig. IX-2. It is readily apparent that the spin relaxation of the individual protons in this molecule will be affected differently by the paramagnetic center, depending on whether it is bound via the atoms A or B. (iii) In an aqueous solution of a metalloprotein and a ligand L, which may e.g. be a substrate or an inhibitor, studies of the water proton relaxation times can then also be used to investigate if the paramagnetic center in the enzyme-substrate or enzyme-inhibitor complex is accessible for binding of solvent molecules. Thus the formation of binary, ternary, or even quaternary complexes can in principle be detected. The method may also be used for studies of systems containing metal free enzymes, metal ions as a cofactor, and substrate molecules. NMR relaxation studies can then indicate whether any two of these components form a binary complex, or if on the contrary only the reactive ternary complex enzyme-metal ion-substrate, or possibly the quaternary complex enzyme-metal ion-substrate-water, is formed in the presence of all the different components.

The Relaxation Enhancement Factor

In an experiment using a paramagnetic metal ion as a relaxation probe, the system studied contains typically both free and protein-bound metal ions. One thus has to consider three different types of water molecules, i.e. those bound to the free metal ions, those bound to the metal ions associated with the protein, and the uncomplexed bulk water. A special notation has been introduced to present the results of water proton relaxation studies in such systems[14]. Since this notation is now customarily used[6,9], it is briefly described in the following.

The systems of interest contain always a large excess of bulk water, and therefore the chemical exchange of protons between the different paramagnetic sites can be neglected. The contribution of the paramagnetic ions to the observed bulk water proton relaxation rate, $1/T^*_{1p}$, is then the sum of the effects arising from the free and the protein-bound metal ions

$$\frac{1}{T^*_{1p}} = \frac{1}{T^*_{1\,\text{obs}}} - \frac{1}{T^*_{1(0)}} = \frac{1}{(T^*_{1p})_{\text{free}}} + \frac{1}{(T^*_{1p})_{\text{bound}}} \qquad \text{(IX-4)}$$

The asterix denotes the presence of the macromolecule, $T_{1(0)}$ is the relaxation time in the absence of the paramagnetic ions, and "free" and "bound" denote the paramagnetic relaxation effects from the free and the protein-bound metal ions. For a metal ion solution in the absence of the macromolecule, one has correspondingly (see Equation VI-30)

$$\frac{1}{T_{1p}} = \frac{1}{T_{1\,\mathrm{obs}}} - \frac{1}{T_{1(0)}} \tag{IX-5}$$

For reasons of convenience, the water proton relaxation data are now often described in terms of the "relaxation enhancement factor" ϵ^*, rather than by the different relaxation rates in Equations IX-4 and IX-5. ϵ^* refers the paramagnetic relaxation rate in the presence of the macromolecule to that measured when the latter is absent

$$\epsilon_1^* = \frac{\dfrac{1}{T_{1p}^*}}{\dfrac{1}{T_{1p}}} = \frac{\dfrac{1}{T_{1\,\mathrm{obs}}^*} - \dfrac{1}{T_{1(0)}^*}}{\dfrac{1}{T_{1\,\mathrm{obs}}} - \dfrac{1}{T_{1(0)}}} \tag{IX-6}$$

Values for ϵ_1^* in solutions containing a protein, Mn^{2+}-ions, and specific ligand molecules are typically in the range from 1 to approximately 20.[9] In most instances, the deviations of ϵ_1^* from 1 arise from the different effective correlation times for the electron-nucleus interactions in the complexes formed with the free and the protein-bound metal ions, respectively.

A corresponding notation is used for the transverse relaxation times, with the relaxation enhancement factor ϵ_2^*.

Selected Examples

Solvent water proton relaxation enhancement studies were extensively applied for investigations of phosphoryl transferring enzymes, e.g. creatine kinase and pyruvate kinase[7,9,10,16]. Almost all the enzymes of this class require divalent cations as added cofactors, in animal cells typically Mg^{2+}, or contain divalent cations as tightly bound components, e.g. Zn^{2+} in a variety of metallo-enzymes. For NMR studies, these diamagnetic metals may generally be replaced with Mn^{2+} or Co^{2+}, which provide comparable enzymatic activation, and are also suitable paramagnetic NMR probes (Section VI.5). Mn^{2+} has played a particularly prominent role as a NMR relaxation probe for studies of phosphoryl transferring enzymes. More recently, the water proton relaxation enhancement measurements with the manganous enzyme systems were complemented by studies with nitroxide spin labels as additional paramagnetic probes[9], ^{13}C and ^{31}P-NMR studies of ligand molecules[16–18], EPR studies of the paramagnetic centers[9,16], and with various other techniques. From the combined experimental data, quite detailed descriptions of the ternary and quaternary complexes formed during the enzymatic reactions were proposed[7,9,10,16].

Paramagnetic metalloproteins which were investigated by the solvent water proton relaxation enhancement technique include also Mn^{2+}-carboxypeptidase A,[19] transferrin[20], Co^{2+}-carbonic anhydrase[21], and various hemoproteins[22]. Other proteins were studied by relaxation enhancement measurements in various substrate or effector molecules. Examples include ^{19}F-NMR experiments in Mn^{2+}-carboxypeptidase A solutions containing an excess of fluoride ion[23], and ^{1}H, ^{13}C, and ^{19}F-NMR studies of the interactions of Co^{2+}-, Mn^{2+}-, and Zn^{2+}-carbonic anhydrase with various carboxylate anion inhibitors[24, 25], cyanide ion[26], and bicarbonate[27, 28].

Numerous NMR studies were reported of the binding of sugar molecules to proteins. Many of these investigations focused on the mechanism of action of lysozyme. Interactions of this protein with fully protonated and selectively fluorinated mono- and oligosacharides were studied with ^{1}H and ^{19}F-NMR, with and without application of external paramagnetic probes[29–34]. A ^{13}C-NMR relaxation study of ^{13}C-labelled α- and β-methyl-D-glucopyranoside in aqueous solutions of Zn^{2+}- and Mn^{2+}-concanavalin A resulted in a description of the protein-sacharide complexes formed in these solutions which provided also a valuable complementation of the single crystal X-ray data on this sytem[35].

Much work has been devoted to studies of interactions between proteins and peptides. In Section VIII.1 we already discussed a deuterium label approach for such investigations. Numerous experiments dealt with the mechanisms of action of chymotrypsin. These included NMR studies with fully protonated or selectively ^{19}F- or D-labelled inhibitor and substrate molecules, e.g. N-trifluoroacetyl-D-*p*-fluorophenylalanine and L-phenylalanine-tert-butyl ester[36–38], and the use of special stopped-flow Fourier transform NMR techniques for investigations of the reaction kinetics[39]. No paramagnetic probes were used with these systems, and the principal manifestations of the interactions with chymotrypsin were in the form of chemical shift differences and line broadening in the NMR spectra of the ligand molecules. A novel approach was employed for studies of the interactions of peptides with bovine neurophysin, where the measurements of the line broadening in the NMR spectrum of the peptide ligand were complemented by investigations of the nuclear Overhauser enhancement resulting for certain NMR lines from the interactions with the protein[40].

2. PEPTIDES AND PROTEINS INTERACTING WITH NUCLEIC ACIDS

From the NMR technical point of view, studies of the interactions between protein molecules and oligonucleotides (Fig. III-25), or between nucleic acids and small peptides[41] fall in most cases into the category of experiments discussed in Section IX.1. New aspects may arise when one investigates interactions between molecules of similar sizes, i.e. proteins and nucleic acids, or oligopeptides and oligonucleotides. Two phenomena have so far been of particular interest in practice. These are the manifestations in the NMR line shapes of the increased rigidity of the protein-nucleic acid complexes as compared to the individual components, and the chemical shifts arising from intermolecular interactions with the local ring current fields of the aromatic amino acid residues, and the pyrimidine and purine bases. Practical difficulties arise in such experiments because the bulk of the ^{1}H and ^{13}C resonance lines in proteins and nucleic acids fall into the same spectral regions, and the resonance lines of the individual components may thus not readily be resolved and identified. As will be seen in the examples described in the following, these experimental problems have been overcome in ^{1}H-NMR studies of several interesting systems. ^{31}P-NMR is of course particularly favorable in this respect, since the ^{31}P resonances of the nucleic acids[42–44] are not usually obscured by the presence of proteins. This might be an important aspect for future developments.

Interactions of Histones with DNA

Studies of the interactions between histones and DNA were based on the observation of line broadening effects in the ^{1}H-NMR spectra of the protein components. Under the conditions of these experiments, the NMR lines of DNA were broadened beyond detection, and thus did not interfere with the observation of the histone resonances. The NMR experiments made it possible to identify certain regions in the histone polypeptide chains which interact preferentially with DNA[46].

Two structural properties of the histones were of particular relevance for the NMR experiments[45,46]. First, histones differ from most other proteins in that the distribution of amino acid residues with different character, e.g. polar and non-polar, or acidic and basic, in the amino acid sequence is markedly non-uniform. Second, histones stand out among proteins in general in that they undergo large and reversible changes in conformation with relatively small changes of the solvent medium. These conformation changes are clearly manifested in the ^{1}H-NMR spectra. Making use of the non-uniform amino acid distribution, certain variations of the spatial structures could from the concomitant changes of the NMR spectra be shown to be confined to specific regions of the polypeptide chains. Histone F2B may serve as an illustration. F2B consists of one polypeptide chain with 125 amino acid residues. The non-uniform amino acid distribution includes for example that the amino terminal peptide 1–31 contains 4 of a total of 6 prolines, none of the 10 aromatic amino acids, one of the 30 apolar residues, and 11 of the 20 lysines. Correspondingly, non-uniform amino acid distributions characterize different segments of the polypeptide chain[45]. In spite of the limited spectral resolution of the ^{1}H-NMR spectrum, which is characteristic for proteins of this size (Chapter III), this allowed to assign the bulk of the resonance intensities in individual spectral regions to specific segments of the polypeptide chain. On this basis, evidence was obtained that the amino acid residues 60 to 102 were strongly affected by the salt-induced conformational changes of F2B. This was further corroborated by studies of the cleaved halves of the histone, i.e. the amino terminal half 1–58, and the carboxyl terminal half 63–125. ^{1}H-NMR spectra of the latter peptide at different NaCl concentrations are shown in Fig. IX-3. In the absence of salt, the spectrum is rather typical for a random coil polypeptide chain (see Chapter III). With increasing salt concentration, marked changes occurred in both overall and relative peak areas. The interpretation of the observations on the lysine and arginine resonances is quite representative

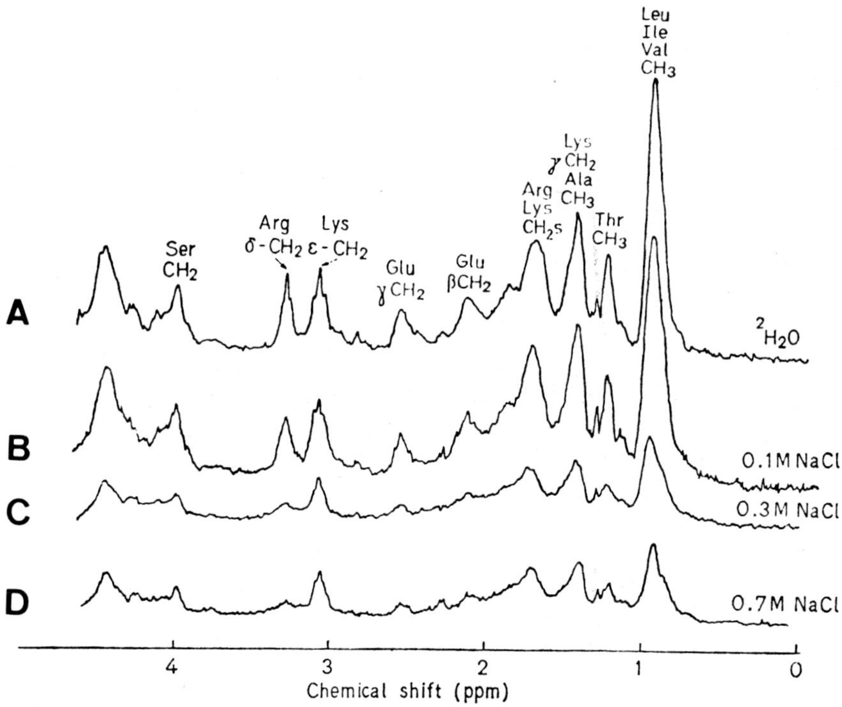

Fig. IX-3. ^{1}H-NMR spectra at 220 MHz of the carboxyl half of histone F2B in D_2O solutions containing variable concentrations of NaCl. The protein concentration was 10 mg/ml. The signal-to-noise ratio was improved by averaging 100 scans in a computer of average transients. Resonance assignments are indicated in the top trace (from Bradbury et al.[45]).

for these experiments[45]. The methylene resonances of lysine, in particular the ε-methylene resonance at 3.06 ppm, are only little affected by the salt induced conformational changes, whereas the δ-CH_2 line of arginine is broadened almost beyond detection at 0.7-M NaCl. In the amino acid sequence, the arginyl and lysyl residues are clustered in different regions. All the five arginines in the carboxyl half of F2B are in the segment from residues 72 to 99, while four of the five lysines are in the carboxyl terminal peptide 108 to 125. The NMR data on the lysine and arginine side chains thus led to the conclusion that high salt concentration greatly reduced the segmental mobility of the residues 70 to 100 of the carboxyl half of histone F2B, while the region from residues 108 to 125 retained the mobility typical for a random coil type extended conformation.

Fig. IX-4A shows the ^{1}H-NMR spectrum of histone F2B associated with DNA. It is readily apparent that this spectrum is quite markedly different from that of a random coil polypeptide chain. The traces B and C of Fig. IX-4 represent two different simulated spectra of the DNA-bound histone F2B. These hypothetical spectra were computed with a standard set of resonances for the individual amino acid residues (e.g. Table II-7). A good fit with the random coil type spectra observed in salt-free solutions of the histones (Fig. IX-3A) was thus obtained. To simulate the spectra of F2B at high salt concentration, or in the presence of DNA, respectively, the

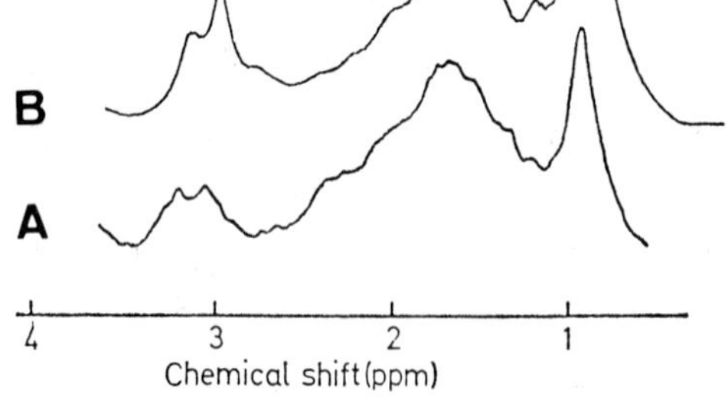

Fig. IX-4. ^{1}H-NMR spectra at 100 MHz of a solution in D_2O of histone F2B and DNA 1:1 ω/ω. A. Experimental spectrum recorded in a solution which contained also 1-M NaCl. B. Spectrum computed as the sum of the amino acid resonances in F2B with the assumption that the resonance lines of the peptide segments 1-30 and 103-125 are broadened by the interactions with DNA. C. Spectrum computed with the assumption that the resonance lines of the peptide segment 1-62 are broadened (from Bradbury and Rattle[46]).

resonances of certain segments of the polypeptide chain were assumed to be broadened beyond detection. Fig. IX-4 illustrates on the one hand a rather poor fit between the experimental spectrum A and the spectrum C, where the segment 1–62 was assumed to give rise to very broad lines, while a more satisfactory agreement was obtained with the assumption in spectrum B that the resonances of the segments 1–30 and 103–125 are broadened by the interactions with DNA.

Overall it was concluded from the spectra of Figs. IX-3 and IX-4, and from additional data obtained with a variety of different histone preparations and different solvent media, that the salt-induced conformational changes and interchain interactions of histone F2B affect mainly the region of the polypeptide chain from the amino acid residues 30 to 106. While the amino terminal segment 1–29 and the carboxy terminal segment 107–125 retain high mobility under high salt conditions, their resonances are broadened by the interactions with DNA, indicating that they are the principal contact regions with DNA. It is worth noting that in this particular study of a very complex system, essential clues for a quite detailed interpretation of the data were provided by the particular primary structural features of the histones. These served as useful "natural NMR probes" even at the very limited spectral resolution attained in these experiments.

Interactions of Proteins with t-RNA

In the experiments to be discussed here, the ^{1}H-NMR spectra of the *t*-RNA molecules were observed. As an introduction to the discussion of two specific examples, i.e. studies of the interactions of *t*-RNA with the elongation factor Tu and with a *t*-RNA synthetase, a brief survey of the ^{1}H-NMR features of *t*-RNA molecules will be presented.

Early high resolution ^{1}H-NMR studies of D_2O solutions of *t*-RNA showed that the resonances of the non-labile protons in these compounds are rather poorly resolved[47,48]. More recently it was found that the NMR spectra of solutions of *t*-RNA in H_2O contained in addition a number of much better resolved resonances in the spectral region from 11 to 15 ppm (Fig. IX-5A), which correspond to ring nitrogen protons of intramolecularly hydrogen-bonded base pairs[49]. The physical basis for the appearance of these resonance lines in extreme low field positions is in close analogy to that causing the well resolved spectrum of hydrogen bonded labile protons in the basic pancreatic trypsin inhibitor (Fig. III-26): The proton exchange between the hydrogen bonds of *t*-RNA and the bulk water is sufficiently slow so that separate resonances can be observed (Section II.5), and, as had been demonstrated with low molecular weight model systems[50], the hydrogen bond formation in the Watson-Crick base pairs causes sizeable downfield

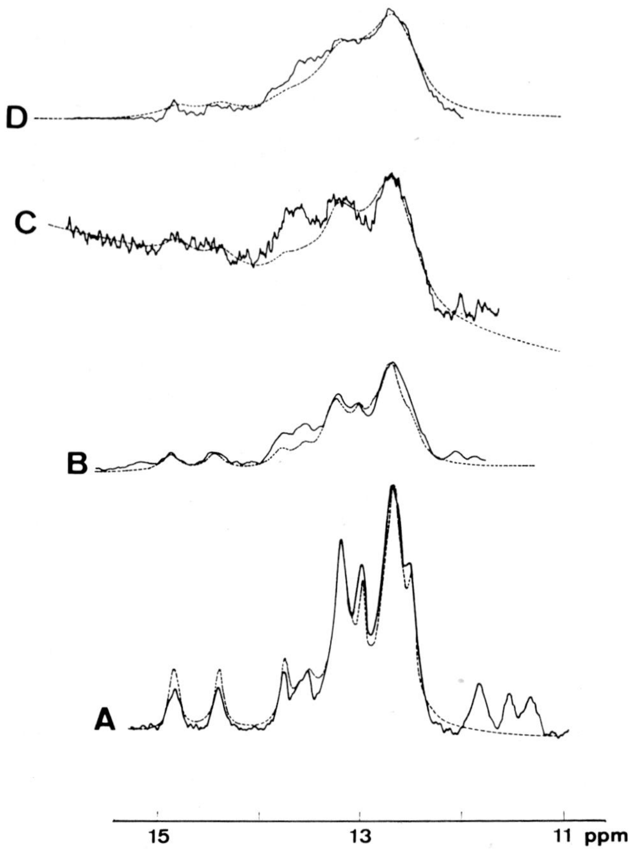

Fig. IX-5. Spectral region from 11 to 15 ppm of the ^{1}H-NMR spectra at 300 MHz of t-RNA^{Glu} with and without the presence of the elongation factor Tu (EFTu). The solid lines represent the experimental spectra. The dashed lines are spectra computed using the resonance positions for the hydrogen-bonded protons in free t-RNA^{Glu}, and the resonance line widths given below. The signal : noise ratio of the experimental spectra was improved by signal accumulation in a CAT. A. E. coli t-RNA^{Glu}. The linewidth in the simulated spectrum is 45 Hz. B. t-RNA^{Glu} + EFTu + GDP. Linewidth 60 Hz. C. t-RNA^{Glu} + EFTu + GTP. Linewidth 96 Hz. D. Aminoacyl-t-RNA^{Glu} + EFTu + GTP. Linewidth 120 Hz. (from Shulman et al.[55]).

shifts of the resonances. The low field ^{1}H-NMR lines provided convincing direct evidence for the occurrence of the "clover leaf" type secondary structure of t-RNA in solution, in that the intensity of these resonances was found to coincide within the accuracy of the experiment with the number of hydrogen bonded protons predicted by the clover leaf model[51]. From studies of partial nucleotide sequences of t-RNA, and from model calculations of the ring current shifts caused by the purine and pyrimidine rings, the individual resonance lines in the spectrum IX-5A were assigned to specific base pairs in the clover leaf structure[52,53]. Practical uses of these spectral features so far include studies of the thermal denaturation of *t*-RNA,[54] comparative studies of different *t*-RNA molecules[52,53], and investigations of the conformational changes caused by aminoacylation of *t*-RNA[55]. In the context of the present discussion, the low field resonances of *t*-RNA are also of interest as natural probes for studies of intermolecular interactions. In systems of *t*-RNA and proteins, these resonances are not usually obscured by the spectrum of the protein, and it should thus in principle be possible to differentiate between the effects of the binding to the protein on different sections of the *t*-RNA molecules.

Fig. IX-5 shows the changes of the resonances of labile protons of t-RNA^{Glu} which arise as a consequence of the interactions with the elongation factor Tu (EFTu) under different conditions[56]. In its biological role during protein synthesis in prokariotic organisms, the protein EFTu promotes the binding of animoacyl-*t*-RNA to ribosomes in the presence of *m*-RNA. There is evidence that the formation of a stable ternary complex with guanosine triphosphate, EFTu · GTP · aminoacyl-*t*-RNA, is an essential step in this process. In the present experiments, aqueous solutions of t-RNA^{Glu}, t-RNA^{Glu} + GDP + EFTu, t-RNA^{Glu} + GTP + EFTu, and aminoacyl-t-RNA^{Glu} + GTP + EFTu were compared. Fig. IX-5 shows that the principal spectral differences between the ^{1}H-NMR spectra of t-RNA^{Glu} in these four different systems could be explained with the assumptions that the number of one-proton lines and their chemical shifts are identical, while the resonance line width increased from 45 Hz in t-RNA^{Glu} to 60 and 96 Hz in the complexes of the uncharged *t*-RNA, and 120 Hz in the complex with the aminoacylated *t*-RNA. From the line width analysis it was concluded that stable complexes were formed between EFTu and both charged and uncharged t-RNA^{Glu} in the presence of GTP, while at most a loose association occurred in the presence of GDP. The increase of the line width when going from t-RNA^{Glu} ($M \approx 28000$) to t-RNA^{Glu} · GTP · EFTu ($M \approx 72000$) corresponds to what one would expect if the rotational motions of the bound *t*-RNA were essentially restricted to the overall rotational tumbling of the complex (Equation IV-31). It is of particular interest that the number of hydrogen bonded protons observed in the spectra IX-5, A–D, appears to be essentially identical, indicating that the clover leaf secondary structure of *t*-RNA is maintained in the complex.

Very similar results were obtained from studies of the interactions between t-RNA^{Glu} and glutamyl-*t*-RNA synthetase, which is a protein with molecular weight 56000 (Fig. IX-6)[57]. The observed line widths indicated that the rotational motions of the bound *t*-RNA were essentially restricted to the overall tumbling of the complex formed with the synthetase. From the

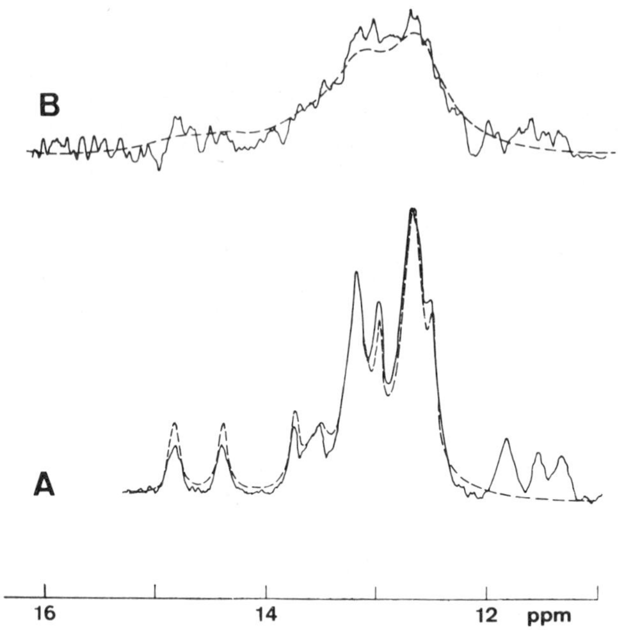

Fig. IX-6. Spectral region from 11 to 15 ppm of the ^{1}H-NMR spectra at 300 MHz of (A) E. coli t-RNA^{Glu}. The dashed line represents a spectrum computed with the chemical shifts obtained from the experiment, and line widths of 45 Hz. (B) E. coli t-RNA^{Glu} + E. coli glutamyl-t-RNA synthetase. The concentrations of t-RNA (M = 28000) and enzyme (M = 56000) were approximately 10^{-3}-M. 256 scans were accumulated in a CAT to improve the signal:noise ratio. The dashed line represents a spectrum computed with the same chemical shifts as in the free t-RNA, but with line widths of 150 Hz (from Shulman et al.[56]).

resonance intensities it was further concluded that the clover leaf structure of t-RNA was essentially maintained in the complex, and there was no other spectral evidence that additional hydrogen bonds with slowly exchanging protons were formed in the complex.

Peptide Antibiotic-Nucleic Acid Interactions

Actinomycin D is a peptide antibiotic with the structure given below. In its biological role it binds reversibly to double-helical DNA, and inhibits RNA synthesis. From studies with different techniques it had been suggested that the actinomycin D-DNA interactions involved intercalation of the phenoxazone ring into the DNA helix. In the following, evidence for the intercalation model obtained from ^{1}H, ^{13}C, and ^{31}P-NMR experiments will be briefly surveyed.

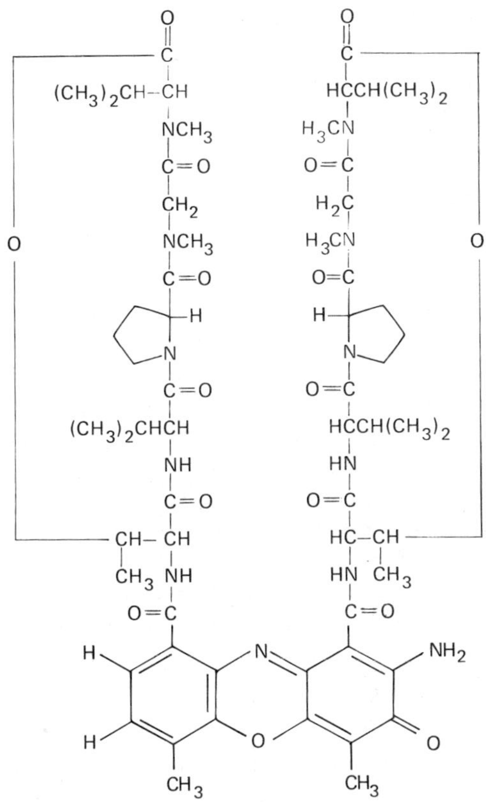

L–MeVal

Sar

L–Pro

(IX-7)

D–Val

L–Thr

Actinomycin D

In one of the ^{1}H-NMR studies[58], the chemical shifts of the non-labile protons of actinomycin D in a solution in D_2O were measured in the presence of variable concentrations of selected deoxydinucleotides, such as d-pG-pC, d-pC-pG, d-pT-pG, and d-pA-pG*. Stacking of the phenoxazone ring with the purine and pyrimidine bases should be manifested in chemical shifts arising from the local ring current fields of the heterocycles. From measurements of the phenoxazone chemical shifts during titrations of actinomycin D with the dinucleotides, it could be established that 1:2 complexes of actinomycin D with the deoxydinucleotides were formed. The sign and size of the chemical shifts of the actinomycin D protons arising from the complexation was found to correspond to what one would have expected from an intercalated ("sandwich-type") complex.

In a different approach, the NMR lines of deoxymononucleotides,

*p stands for a phosphate group, d stands for deoxy-, G = guanosine, C = cytidine, T = thymidine, A = adenosine.

deoxydinucleotides, and deoxyoligonucleotides were investigated in the complexes formed with actinomycin D.[44] From the molecular structure of actinomycin D, one expects that the ring current fields of the phenoxazone moiety would cause chemical shifts for certain resonances of the nucleotides in the complexes with the drug. From the analysis of the complexation dependent ^{1}H, ^{13}C, and ^{31}P chemical shifts of the nucleotide resonances, evidence leading to a rather detailed description of the solution structure of the intercalation complexes could be obtained. For the investigations of actinomycin D binding to the deoxyhexanucleotide d-Ap-Tp-Gp-Cp-Ap-T, it was of particular interest that the ^{1}H-NMR spectra of double-helical deoxyoligonucleotides in H_2O solution contain resonances of slowly exchanging labile protons involved in the Watson-Crick hydrogen bonds[59, 60], which is in analogy to the previously discussed *t*-RNA spectra. Mainly from studies of the resonances of these labile protons, it could be established that the intercalation with actinomycin D occurred at the site of the central GC base pairs[44].

It may be added at this point that high resolution ^{1}H-NMR studies of oligopeptides in the presence of DNA lead to the conclusion that intercalation complexes of the aromatic rings of polypeptide chains with double helical DNA might play a prominent role in the interactions between DNA and proteins[61].

3. PROTEINS IN MEMBRANES

It is by now generally accepted that combinations of lipid bilayers and proteins are integral structural elements of most biological membranes. Detailed knowledge of lipid-protein interactions, and of the molecular conformations of proteins associated with lipids is thus of much interest for a deeper insight into the mechanisms of membrane processes. While NMR techniques appear in principle to be well suited for studies in this domain, practical applications are still in an early stage. This section attempts primarily to outline some prospects for future development. Quite generally, the NMR lines of the lipid components in natural membranes or in membrane models may often be more readily observable than those of the proteins. By this reason the nuclei in the lipid molecules can play an essential role as natural NMR probes of protein-lipid interactions, and therefore a brief survey of some NMR features of lipids is also included.

NMR Spectra of Lipids

In Section VIII.1 a D-NMR study of the ordering of paraffin chains in model bilayer structures has already been described (Figs. VIII-2 and

VIII-3). Information on the lipid molecules, and in particular their mobilities in different multimolecular structures could also be derived from the ^{1}H, ^{13}C, and ^{31}P-NMR spectra. The ^{1}H-NMR spectrum of a sonicated dispersion of lipids in D_2O in Fig. IX-7 shows that under suitable experimental conditions a considerable number of resonance lines can be resolved at the presently available field strengths. In particular, the resonances of individual groups of protons both near the polar head group and in the non-polar paraffin chains

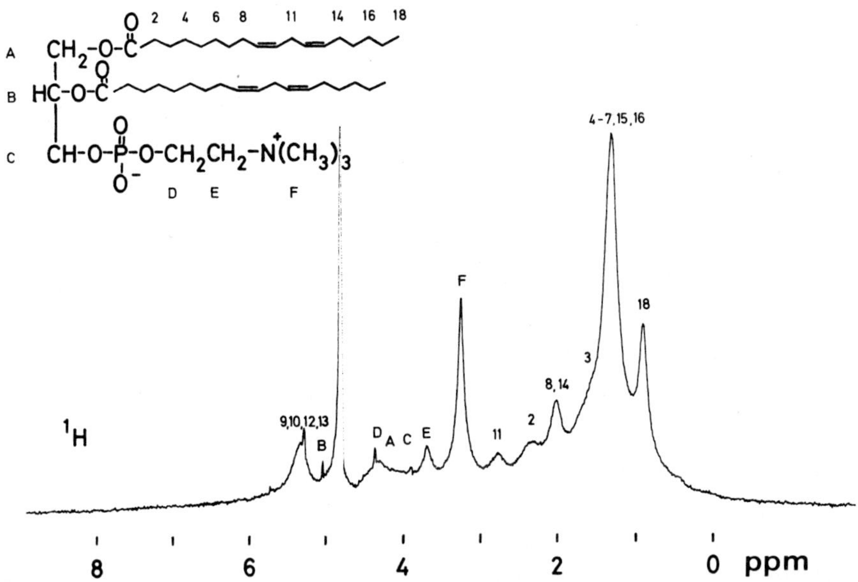

Fig. IX-7. ^{1}H-NMR spectrum at 220 MHz of mixed vesicles of egg yolk phosphatidyl choline and beef cardiolipin 4:1 in D_2O. The preparation had been sonicated for 5 minutes at 4°, and the sample was kept under a nitrogen atmosphere. The resonance assignments refer to the schematic structure of phosphatidyl choline shown in the figure, and correspond to those given in Ref. 62.

can readily be recognized. It is worth noting that the resonances of the lipid molecules cover only part of the spectral region occupied by resonance lines of proteins (Chapters III and VI). In particular the resonances of the aromatic amino acid residues, of aliphatic protons in conformation-dependent high field positions, and of hyperfine shifted lines in the spectra of proteins should in general not be obscured by the presence of lipids. Fig. IX-8 shows the ^{13}C-NMR spectrum of a sonicated lipid dispersion in D_2O. Compared to the ^{1}H-NMR spectrum of a corresponding sample (Fig. IX-7), an appreciably improved spectral resolution was obtained, and most of the carbon atoms along the paraffin chains and in the polar head group give rise to well separated lines. Some ^{31}P spectra of phospholipids were shown in the preceding chapter (see Fig. VIII-5).

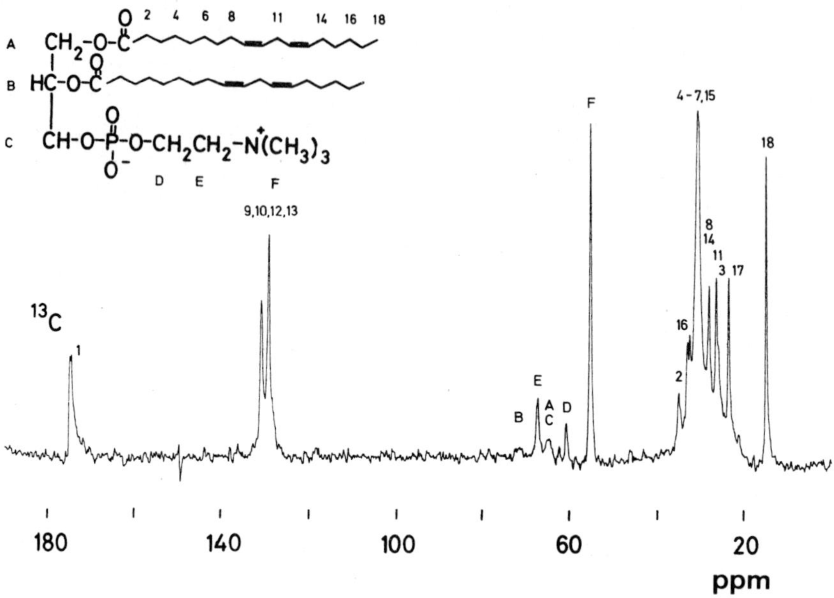

Fig. IX-8. ^{1}H noise-decoupled FT ^{13}C-NMR spectrum at 25.16 MHz of mixed vesicles of egg yolk phosphatidyl choline and beef cardiolipin 4:1 in D_2O. See Fig. IX-7 for details of the sample preparation. The resonance assignments refer to the schematic structure of phosphatidyl choline shown in the figure, and correspond to those given in Ref. 63.

The NMR spectra of lipids were used in a variety of different experiments to study the state of the lipid molecules in membrane models and intact biological membranes. Quite generally, the mobility of lipid molecules in multimolecular structures is accessible for investigation by measurements of the nuclear spin relaxation times (see Chapter IV and Appendix V). With the available spectral resolution (Figs. IX-7 and IX-8), both the rate of lateral diffusion of the entire molecules and the intramolecular segmental motions can be studied by NMR techniques[63,64]. Biosynthetic ^{13}C labelling of the lipids in natural membranes promises to provide systems which are particularly amenable to such approaches[65]. Paramagnetic NMR shift and relaxation probes have also been used in various ways for experiments with membranes, e.g. for studies of the fluidity of lipid layers[63,66], and for differentiation of the "inner" and "outer" surfaces of model membranes[67,68].

Membrane Proteins and Proteins Associated with Membranes

In the ^{1}H and ^{13}C-NMR spectra of biological membranes, resonance lines corresponding to part of the membrane proteins have in certain experiments been observed and not in others, depending on the membrane and on the mode of sample preparation[69–71]. Quite generally, the bulk of the membrane proteins seem to be highly immobilized so that much of the NMR resonance intensity contributes to extremely broad lines[70,71], which cannot usually be detected. To get more detailed information on membrane proteins than appears presently to be available solely from NMR studies of intact membranes, isolated membrane protein fractions were studied, and the binding properties for certain NMR markers compared in the membrane-bound and free proteins[72]. In different approaches, the NMR spectral properties of corresponding lipid molecules in intact membranes and in model lipid bilayer structures were compared to investigate the effects of the membrane proteins on the lipids[73]. NMR studies of model systems included also investigations of lipid bilayer interactions with simple model peptides e.g. with oligopeptides of alanine and glycine[74], and with poly-L-glutamic acid[75].

Numerous proteins with specific functions in transport processes, enzymatic reactions, or as receptor sites in cell regulation may also be associated with membranes. In many cases such proteins have been solubilized and studied in much detail in aqueous solution. Typical examples can be found among the cytochromes. The question may arise as to the biological relevance of the data obtained with the solubilized proteins. It thus remains to investigate the effects of lipid-protein interactions on the protein conformation, and possibly other structural features relating to the specific biological roles of such proteins. Since in the membranes the concentration of these functional proteins is usually rather small, experiments based on the direct NMR observation of these molecules in the natural systems will probably not be practicable in the immediate future. On the other hand, relevant data might possibly be obtained from experiments with model systems enriched in the protein of interest, and also from applications of suitable NMR probe techniques for studies of the natural systems.

4. STUDIES OF INTACT CELLS, TISSUES, AND LIVE ORGANISMS

In this section some ^{1}H and ^{13}C-NMR experiments are described in which one or several of the natural components in intact tissues were observed. Also included is a novel technique which makes it feasible to obtain two- and three-dimensional "NMR zeugmatograms" of macroscopic samples.

NMR studies of complex biological systems are of course in no way restricted to the observation of protons and carbon-13. On the contrary, work with other nuclei may in many cases appear more attractive because a smaller number of different components will contribute to the observed spectrum, which may thus be more straightforward to interpret than the ^{1}H and ^{13}C data. In Section VIII.2 we have already discussed the use of ^{31}P-NMR experiments with intact muscles for in vivo studies of phosphate containing metabolites. Another approach which was investigated in detail is the use of ^{23}Na-NMR for studies of sodium compartmentation in intact tissues[76, 77]. It may also be added that preliminary experiments with ^{15}N isotope enriched intact cells[78] indicate that ^{15}N-NMR might be of use for in vivo studies of nitrogen containing metabolites.

^{1}H-NMR Experiments

A low frequency "whole-body NMR spectrometer" operating at 40 KHz for protons and capable to accommodate live mammals has been developed by Jackson and Langham[79]. This spectrometer could admit sample volumes of up to 1.2 liters. Some representative data recorded with this instrument are shown in Fig. IX-9, which includes also a spectrum of a live anesthetized rat. The authors[79] pointed out that a broad and a somewhat narrower component could be distinguished in these spectra, where the narrower line was attributed mainly to the water present in the samples, and the broader line to the protein components.

As a second example, a spectrum obtained in one of the numerous studies of water ^{1}H-NMR signals in intact cells and tissues is presented in Fig. IX-10. The sharp line corresponds to intracellular water of a rabbit sciatic nerve prepared in the following way: The nerve fibers, which are approximately three centimeters long, were excised, rinsed, and cleaned from connective tissues. They were then tied at both ends to a capillary with nylon filaments. The capillary contained TMS as an external reference. The entire assembly was inserted with teflon spacers into a 5 mm NMR tube containing 99.5% deuterated modified Locke's physiolgical solution. The bathing solution contained no glucose, and Mn^{2+}-ions had been added to shift and broaden the resonance of the residual protons in the extracellular solvent. As a result, which is readily apparent in Fig. IX-10, the resonances of the comparatively small amount of intracellular water and the large volume of deuterated bathing solution can be observed as separate lines. Additional weaker resonance lines of the sciatic nerve between 0 and 4 ppm (not shown in Fig. IX-10) were assigned to the phospholipid components in the nerve, implying that the sciatic nerve contains hydrophobic regions which are sufficiently fluid to give a high resolution ^{1}H-NMR spectrum.

Measurements of the NMR relaxation times of the water resonance in muscle and other tissues were interpreted in terms of compartmentation of

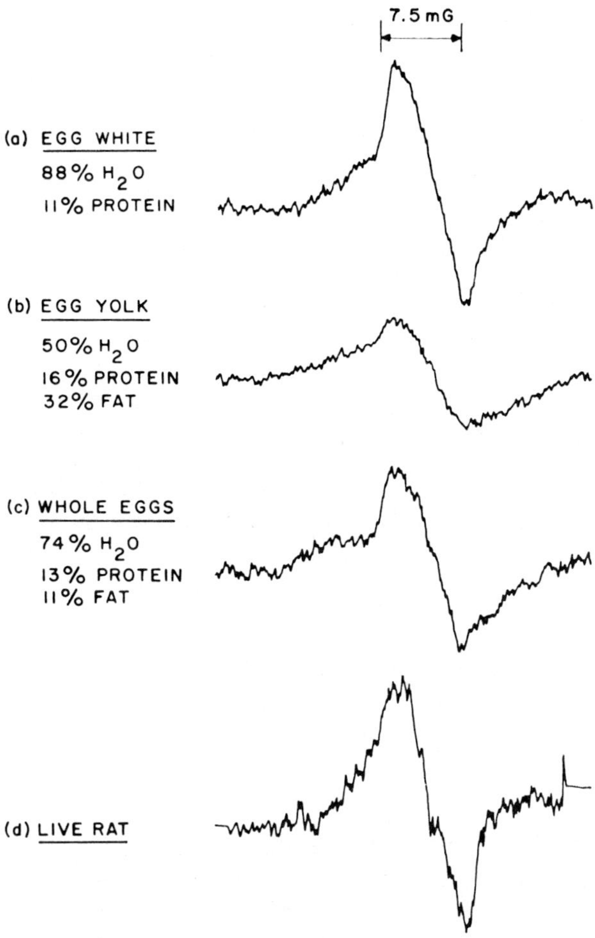

Fig. IX-9. ^{1}H-NMR spectra of several biological samples recorded with a "whole-body NMR spectrometer" operating at 40 KHz (from Jackson and Langham[79]).

water, hydration of certain components, and diffusion rates of water protons[81–85]. Much impetus for investigations of this type has recently come from the observation that the water proton relaxation times T_1 are significantly longer in cancerous tissues than in the corresponding normal tissues[86–90]. Some representative numbers for human organs are $T_1 = 1.0$ and 0.4 sec for cancerous and normal breast tissue, respectively, $T_1 = 0.8$ and 0.6 sec for liver tissue, and $T_1 = 1.4$ and 0.9 sec for uterus tissue[89]. Devices have been described for in vivo studies of water relaxation times[87], and

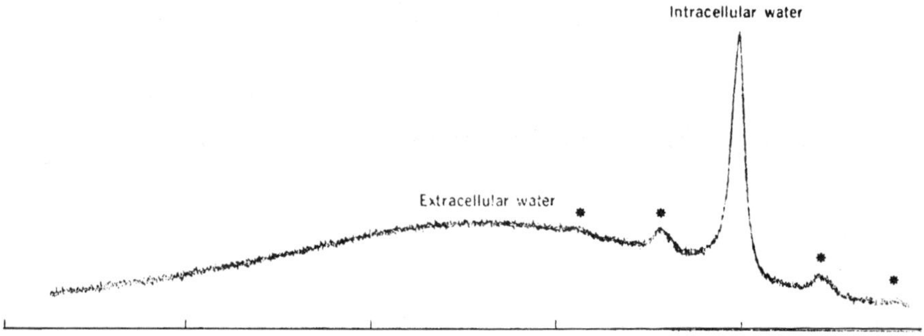

Fig. IX-10. ^1H-NMR spectrum at 220 MHz of intracellular and extracellular water in a sample containing a sciatic nerve trunk of rabbit immersed in a bathing solution doped with 0.002-M Mn^{2+}. The small peaks marked with an asterix are spinning side bands of the intracellular water resonance (from Dea et al.[80]).

NMR has also been suggested as a complementary technique for diagnosing malignancy[89].

^{13}C*-NMR Experiments—Studies of Biosynthetic Pathways*

^{13}C-NMR experiments with intact cells or tissue specimen labelled with ^{13}C present a promising approach e.g. for studies of the mobility of individual tissue components by measurements of the spin relaxation times, or, perhaps more important in practice, for in vivo and in vitro investigations of metabolic pathways. Fig. IX-11 shows some ^{13}C-NMR spectra obtained in experiments with intact *Candida utilis* cells[91]. Isotopically enriched cells were grown on a medium containing 20% ^{13}C-enriched acetic acid as the sole carbon source. A suspension of these cells in D_2O gave the spectrum of Fig. IX-11A. The cellular components remaining after exhaustive extraction with D_2O gave the spectrum B of Fig. IX-11, and spectrum C was obtained from a solution of the material released from the cells by an osmotic shock with D_2O. From comparison of the spectra of Fig. IX-11 with those of macromolecular and low molecular weight components in the cells, many of the resonances were associated with particular types of carbon atoms. From the line widths, inferences were then drawn on the mobilities of individual components in the cells[91].

The appearance of numerous rather sharp lines in the spectrum A of Fig. IX-11 implied that in vivo studies of the metabolism of a sizeable number of compounds should be feasable by metabolizing the specifically enriched substances in otherwise non-labelled cell suspensions. This was borne out by experiments in which the time course of the ^{13}C-NMR spectrum of

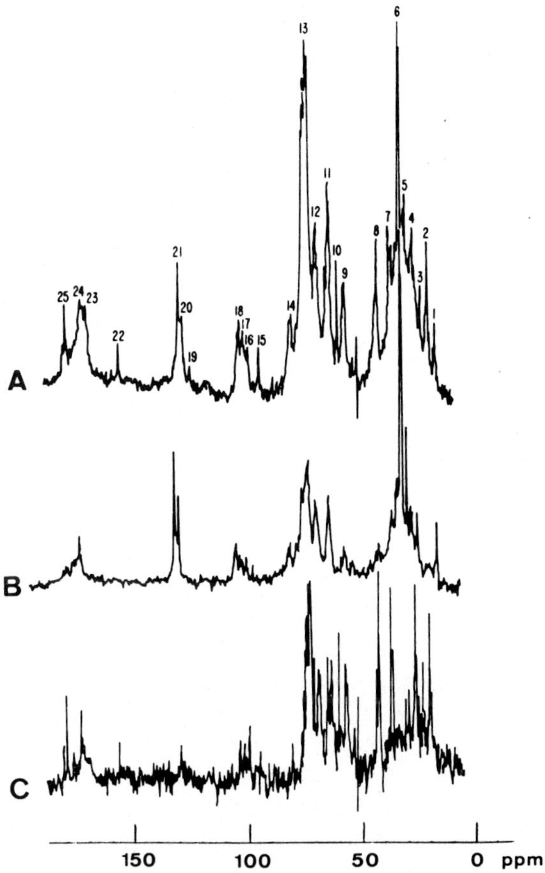

Fig. IX-11. ^{1}H noise-decoupled FT ^{13}C-NMR spectra at 25.16 MHz. (A) Suspension in D_2O of *Candida utilis* cells randomly enriched to 20% in ^{13}C. (B) Cellular material left after exhaustive extraction with D_2O of the cells studied in spectrum A. (C) Water soluble components released by osmotic shock with D_2O from the cells used in A (from Eakin et al.[91]).

unenriched *Candida utilis* cells was recorded after mixing the cell suspension with specifically labelled [1-^{13}C]-D-glucose[91].

One of the classical approaches for the elucidation of the biosynthetic pathways leading to secondary metabolites consists of the feeding of precursors labelled in specific positions with ^{14}C to the appropriate culture. The metabolite in question is then isolated, and the labelling pattern of the carbon skeleton established by standard tracer techniques, which usually include degradation of the metabolite. For some time now it has been recognized that with the use of ^{13}C instead of ^{14}C, the location of the labels in the isolated metabolites could in many cases be determined more directly by

^{13}C-NMR spectroscopy, i.e. in particular without chemical degradation of the substances[92–98]. Studies based on direct observation of the ^{13}C-NMR spectra of the metabolites, measurements of ^{13}C—^{13}C spin–spin couplings in double labelling experiments[95], and observation of the ^{13}C satellites in the ^{1}H-NMR spectra of the selectively ^{13}C labelled metabolites[98] have been described. Overall, the rapidly increasing number of papers on this subject seems to indicate that the introduction of ^{13}C-NMR has added substantially to the methodology for biosynthetic studies. It is to be expected that ^{13}C-NMR has not only attained a secure position in this field, but that its importance will further increase in the future with the availability of a larger selection of ^{13}C labelled biogenetic precursors[98].

NMR Zeugmatography

The term "NMR zeugmatography" was suggested for NMR studies based on experiments in which the sample is exposed to linear gradients of the polarizing magnetic field H_0.[99–101] If these field gradients are sufficiently large, structures above the molecular level can be investigated, and through combination with "conventional" NMR measurements the spatial distributions of NMR spectral properties in a macroscopic sample may be determined. NMR zeugmatography thus opens rather exciting prospects for studies of integral biological systems. In the following, some aspects of the technique will be discussed, and illustrated with studies of biological materials.

The experimental data needed to construct a NMR zeugmatogram may be recorded with a conventional NMR set-up (Fig. IX-12) which had been modified so that linear field gradients can be superimposed on the homogenous polarizing field H_0. With the application of a field gradient, nuclei in different locations in space will experience different magnetic fields H_0, and correspondingly the resonance frequencies will be different. If the line widths of the individual resonances are small compared to the applied field gradients, a plot of the signal intensity as a function of frequency yields the integrated intensities of the resonances in each surface of constant field, which represent the spatial distribution of the nuclear spins in the sample. Illustrative examples of such one-dimensional projections of the three-dimensional intensity distribution are shown in Figs. IX-13 and IX-14. When linear field gradients are applied to a homogeneous cylindrical sample giving rise to a single sharp line in a conventional NMR experiment (Fig. IX-13a), e.g. H_2O in a conventional NMR tube, the intensity profiles b to d in Fig. IX-13 may be generated. A sample in a z-gradient experiences vertical planes of constant magnetic field perpendicular to the z-axis (Fig. IX-12), and the signal shape b results if the coupling with the coil is the same for the

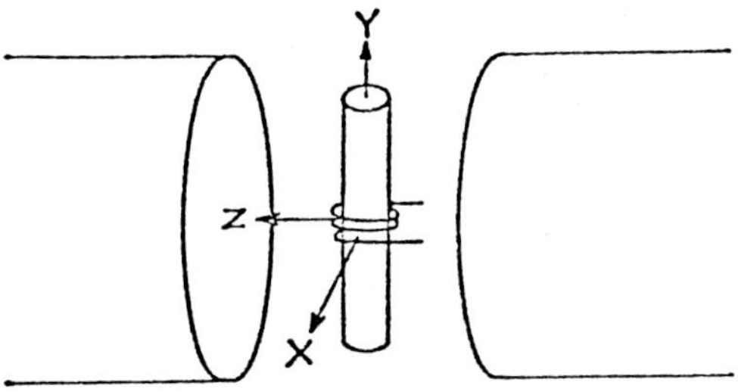

Fig. IX-12. NMR zeugmatography. Coordinate system and sample geometry (from Lauterbur[100]).

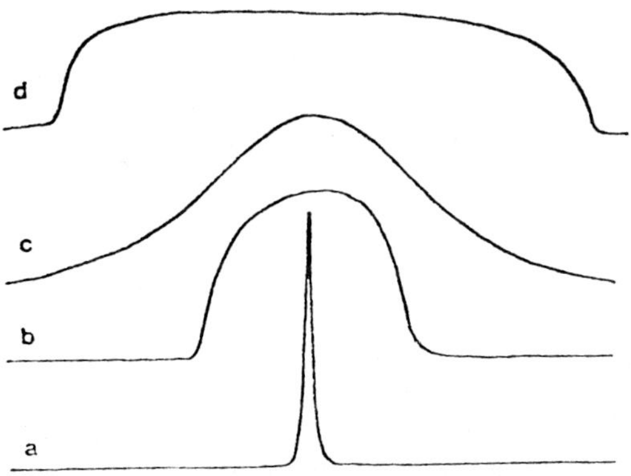

Fig. IX-13. NMR zeugmatography. Signal intensity distributions from a homogeneous cylindrical sample in linear magnetic field gradients. a) no gradient; b) z-gradient; c) y-gradient; d) x-gradient. See Fig. IX-12 for the sample geometry. Note that the traces b and d have the same general trends. The apparent differences reflect merely the application of different field gradients (from Lauterbur[100]).

nuclei within each vertical plane of constant field. The situation is the same with an x-gradient. On the other hand, for a y-gradient the signal falls off more rapidly (Fig. IX-13c). This reflects the lesser coupling to the receiver coil of the nuclei in the horizontal planes which are off center with respect to the coil. Fig. IX-14 shows four one-dimensional projections of a somewhat more complicated sample consisting of two capillaries containing H_2O, which had been fitted into a sample tube filled with D_2O. The inhomogeneous distribution of the density of nuclear spins in this sample is readily recognizeable in the four traces recorded with field gradients applied at 45° intervals in the xz-plane (Fig. IX-14).

Several one-dimensional projections of the three-dimensional intensity distribution may be combined to synthesize a two-dimensional picture, or in principle even a reconstructed three-dimensional representation of an

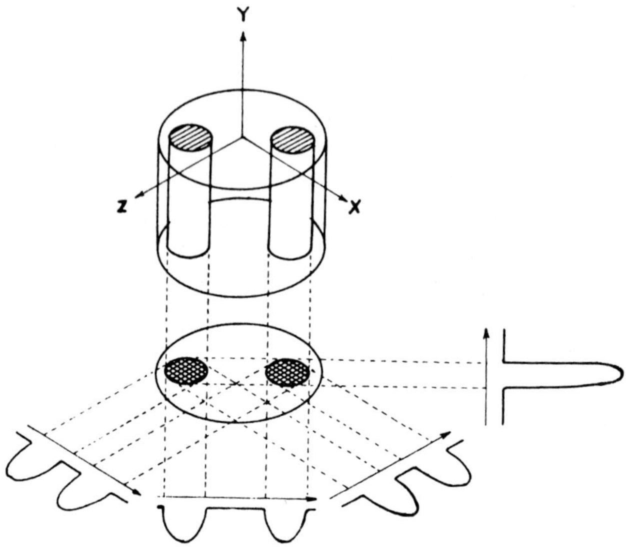

Fig. IX-14. NMR zeugmatography. Relationship between a three-dimensional object (here: a cylindrical sample tube containing two capillaries), its two-dimensional projection along the y-axis, and four one-dimensional projections at 45° intervals in the *xz*-plane. The arrows indicate the gradient directions (from Lauterbur[99]).

object, in terms of the density distribution of the observed NMR nuclei. The two-dimensional picture obtained by the application of image reconstruction techniques to the set of data in Fig. IX-14 is represented in Fig. IX-15. The areas of high proton density in the two capillaries can readily be recognized by the increased gray levels. Fig. IX-16 shows the distribution of oil in an intact pecan nut. Pecan nuts contain only little water, but give a readily detectable signal from the protons of the oils in the nut meat. It is seen that the nut shell did not affect the recording of the image of the inside. Fig. IX-17 shows a partial zeugmatogram of a live mouse. The receiver coil was centered on the thoraic region, and was sensitive only to that part of the mouse. The lungs may be clearly recognized, whereas no other details are visible (the white spot in the center is an artefact[101]).

The one-dimensional projections needed for the construction of a zeugmatogram (Fig. IX-14) can be obtained with CW or FT-NMR techniques. On the one hand, different image reconstruction techniques may be employed depending on the mode of data aquisition[101–104]. On the other hand, different NMR techniques may also be combined to complement results such as those in Figs. IX-15 and IX-17, which reflect the density distribution of NMR nuclei in the objects studied, with data responding to certain NMR spectral properties of these nuclei. For example, with the use

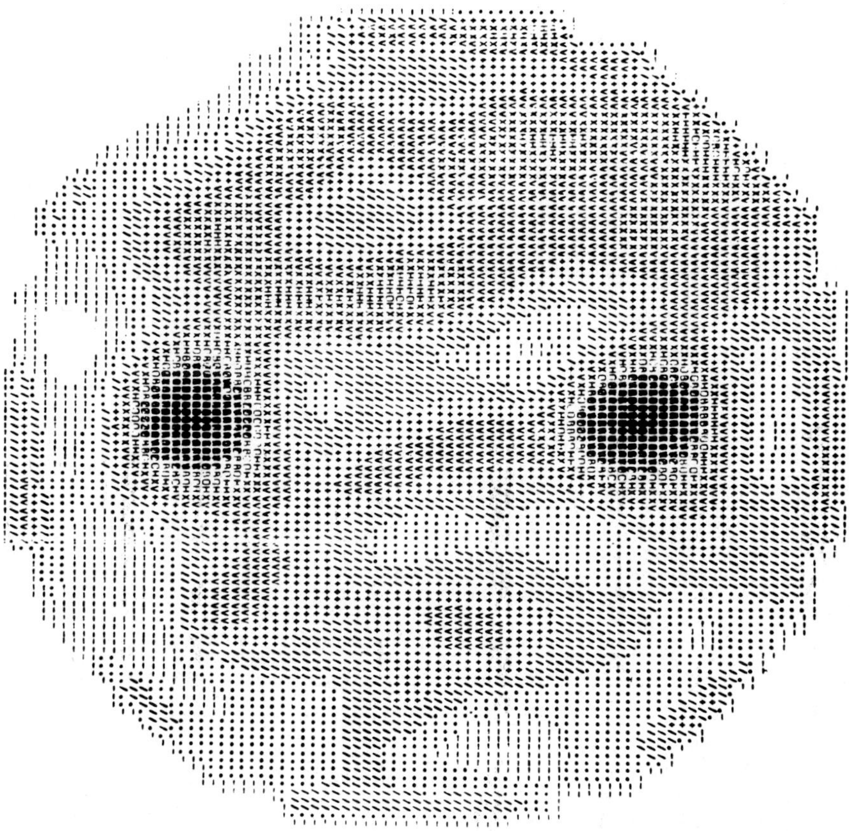

Fig. IX-15. ^{1}H-NMR zeugmatogram of two 1 mm capillaries of H_2O fitted as shown in Fig. IX-14 into a sample tube of 4.2 mm inner diameter containing a mixture of D_2O and H_2O (from Lauterbur[101]).

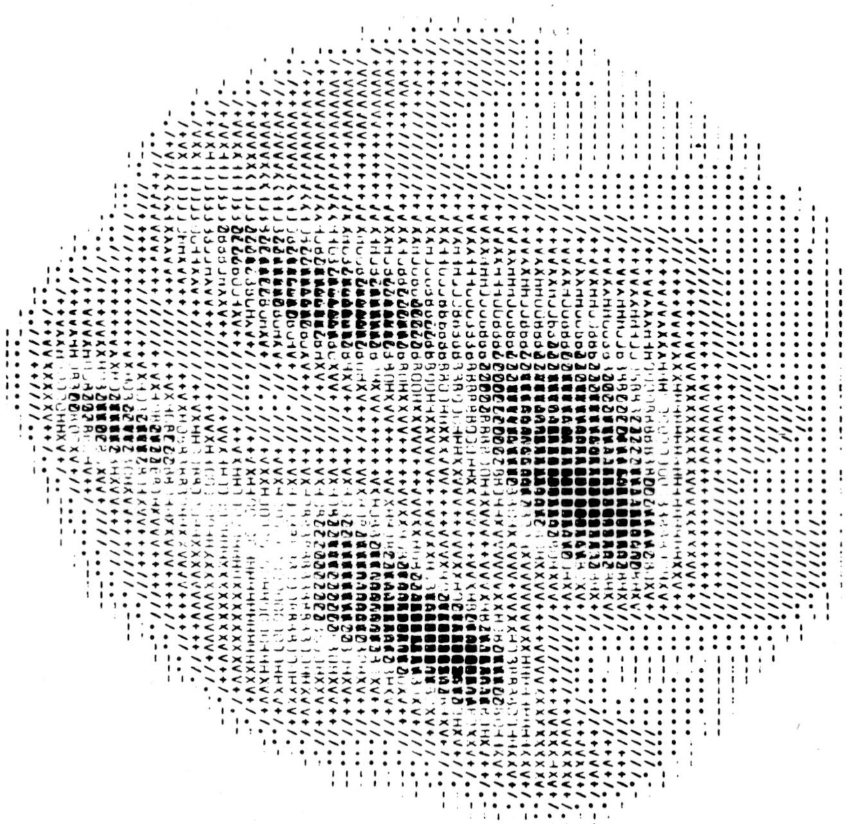

Fig. IX-16. ^{1}H-NMR zeugmatogram of the oil distribution within an intact pecan nut, *Carya illinoensis* (from Lauterbur[101]).

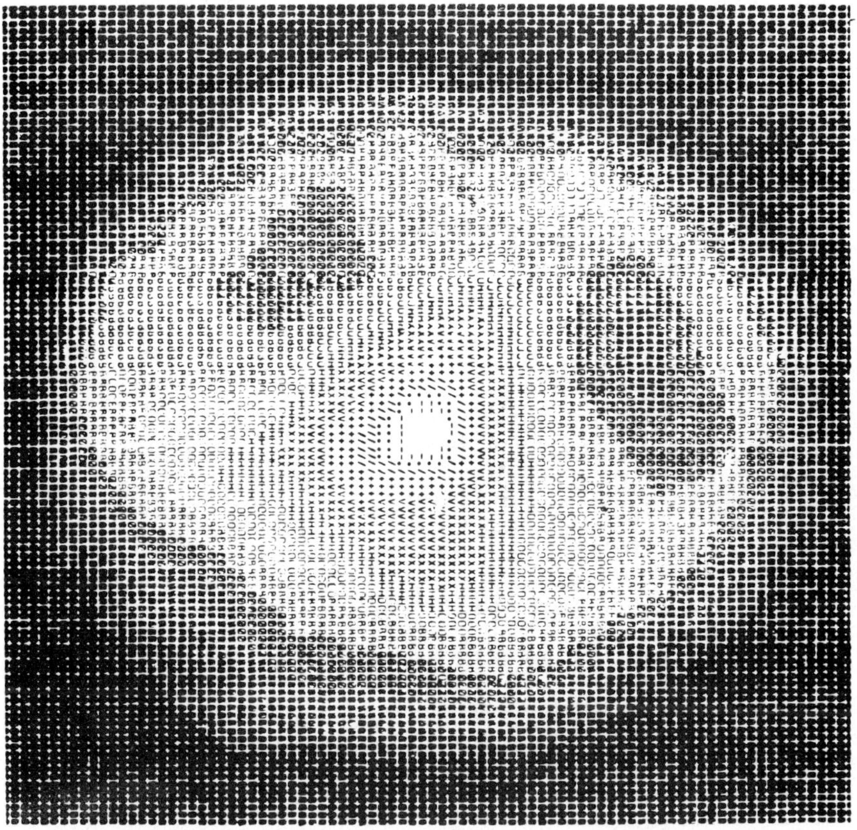

Fig. IX-17. ^{1}H-NMR zeugmatogram of the thoracic cavity of a live mouse. The plane of the receiver coil passed through the lungs at right angles to the long axis of the body, and the image is similar to that which would be obtained from a thick transverse section (from Lauterbur[101]).

of suitable pulse sequences (see Chapter IV) zeugmatograms reflecting different relaxation times in different locations of the object studied may be obtained. It has been suggested[101, 103] that such experiments might be used as a nondestructive method to locate cancerous growth in live tissues on the basis of the increased longitudinal water proton relaxation time as compared to normal tissue[86–90]. More generally, it would appear that NMR zeugmatography combined with studies of the "conventional" NMR properties of biological systems might provide information on a large variety of important phenomena, such as the mobilities of certain components in different locations of a biological object, compartmentation in space of certain metabolites, or two- and three-dimensional pathways of transport processes.

5. BIBLIOGRAPHY

1 Matwiyoff N. A., Vergamini P. J., Needham T. E., Gregg C. T., Volpe J. A. and Caughey W. S., J. Amer. Chem. Soc. *95*, 4429 (1973)
2 Antonini A., Brunori M., Conti F. and Geraci G., FEBS Lett. *34*, 69 (1973)
3 Vergamini P. J., Matwiyoff N. A., Wohl R. C. and Bradley T., Biochem. Biophys. Res. Comm., *55*, 453 (1973)
4 Moon R. B. and Richards J. H., Biochemistry *13*, 3437 (1974)
5 Wüthrich K., Pure Appl. Chem. *40*, 127 (1974)
6 Dwek R. A., Nuclear Magnetic Resonance in Biochemistry, Clarendon Press, Oxford, 1973 (see in particular the Chapters 6, and 8–12)
7 Mildvan A. S. and Cohn M., Adv. Enzymology *33*, 1 (1970)
8 Sheard B. and Bradbury E. M., Progr. in Biophysics and Molecular Biology *20*, 187 (1970)
9 Cohn M., Quart. Rev. of Biophysics *3*, 61 (1970)
10 Cohn M. and Reuben J., Accts. Chem. Res. *4*, 214 (1971)
11 Sykes B. D. and Scott M. D., Ann. Rev. Biophysics Bioeng. *1*, 25 (1972)
12 Koenig S. H. and Schillinger W. E., J. Biol. Chem. *244*, 3283 (1969)
13 Connick R. E. and Wüthrich K., J. Chem. Phys. *51*, 4506 (1969)
14 Eisinger J., Shulman R. G. and Blumberg W. E., Nature *192*, 963 (1961)
15 Anderson A. G. and Redfield A. G., Phys. Rev. *116*, 583 (1959)
16 Mildvan A. S. and Grisham C. M., Structure and Bonding *20*, 1 (1974)
17 Zukin R. S., Hollis D. P. and Gray G. A., Biochem. Biophys. Res. Comm. *53*, 238 (1973)
18 Fung C. H., Mildvan A. S., Allerhand A., Komoroski R. and Scrutton M. C., Biochemistry *12*, 620 (1973)
19 Shulman R. G., Navon G., Wyluda B. J., Douglass D. C. and Yamane T., Proc. Nat. Acad. Sci. *56*, 39 (1966)
20 Koenig S. H. and Schillinger W. E., J. Biol. Chem. *244*, 6520 (1969)
21 Fabry M. E., Koenig S. H. and Schillinger W. E., J. Biol. Chem., *245*, 4256 (1970)
22 A survey of relaxation enhancement studies with hemoproteins is given in Wüthrich K., Structure and Bonding *8*, 53 (1970)
23 Navon G., Shulman R. G., Wyluda B. J. and Yamane T., J. Mol. Biol. *51*, 15 (1970)
24 Taylor P. W., Feeney J. and Burgen A. S. V., Biochemistry *10*, 3866 (1971)
25 Lanir A. and Navon G., Biochim. Biophys. Acta *341*, 75 (1974)
26 Feeney J., Burgen A. S. V. and Grell E., Eur. J. Biochem. *34*, 107 (1973)
27 Koenig S. H. and Brown R. D., Proc. Nat. Acad. Sci. US *69*, 2422 (1972)
28 Yeagle P. L., Lochmüller C. H. and Henkens R. W., Proc. Nat. Acad. Sci. *72*, 454 (1975)
29 Dahlquist F. W. and Raftery M. A., Biochemistry *7*, 3269 (1968)
30 Dahlquist F. W. and Raftery M. A., Biochemistry *8*, 713 (1969)
31 Raftery M. A., Dahlquist F. W., Chan S. I. and Parsons S. M., J. Biol. Chem. *16*, 4175 (1968)
32 Raftery M. A., Dahlquist F. W., Parsons S. M. and Wolcott R. G., Proc. Nat. Acad. Sci. *62*, 44 (1969)
33 Millett F. and Raftery M. A., Biochemistry *11*, 1639 (1972)
34 Butchard C. G., Dwek R. A., Kent P. W., Williams R. J. P. and Xavier A. V., Eur. J. Biochem. *27*, 548 (1972)
35 Brewer C. F., Sternlicht H., Marcus D. M. and Grollman A. P., Biochemistry *12*, 4448 (1973)
36 Gammon K. L., Smallcombe S. H. and Richards J. H., J. Amer. Chem. Soc. *94*, 4573 (1972)
37 Gerig J. T. and Rimerman R. A., J. Amer. Chem. Soc. *94*, 7565 (1972)
38 Sykes B. D., J. Amer. Chem. Soc. *91*, 949 (1969)

39 Grimaldi J. J. and Sykes B. D., J. Amer. Chem. Soc. *97*, 273 (1975)
40 Balaram P., Bothner-By A. A. and Breslow E., Biochemistry *12*, 4695 (1973)
41 Fritzsche H., FEBS Letters *23*, 105 (1972)
42 Haar W., Thompson J. C., Maurer W. and Rüterjans H., Eur. J. Biochem. *40*, 259 (1973)
43 Guéron M., FEBS Letters *19*, 264 (1971)
44 Patel D. J., Biochemistry *13*, 1476, 2388, 2396 (1974)
45 Bradbury E. M., Cary P. D., Crane-Robinson C., Riches P. L. and Johns E. W., Eur. J. Biochem. *26*, 482 (1972)
46 Bradbury E. M. and Rattle H. W. E., Eur. J. Biochem. *27*, 270 (1972)
47 McDonald C. C., Phillips W. D. and Penswick J. R., Biopolymers *3*, 609 (1965)
48 Smith I. C. P., Yamane T. and Shulman R. G., Canad. J. Biochem. *47*, 480 (1969)
49 Kearns D. R., Patel D. J. and Shulman R. G., Nature *229*, 338 (1971)
50 Shoup R. R., Miles H. T. and Becker E. D., Biochem. Biophys. Res. Comm. *23*, 194 (1966)
51 Wong Y. P., Kearns D. R., Shulman R. G., Yamane T., Chang S., Chirikjian J. G. and Fresco J. R., J. Mol. Biol. *74*, 403 (1973)
52 Shulman R. G., Hilbers C. W., Wong Y. P., Wong K. L., Lightfoot D. R., Reid B. R. and Kearns D. R., Proc. Nat. Acad. Sci. *70*, 2042 (1973)
53 Shulman R. G., Hilbers C. W., Kearns D. R., Reid B. R. and Wong Y. P., J. Mol. Biol. *78*, 57 (1973)
54 Hilbers C. W., Shulman R. G. and Kim S. H., Biochem. Biophys. Res. Comm. *55*, 953 (1973)
55 Wong Y. P., Reid B. R. and Kearns D. R., Proc. Nat. Acad. Sci. *70*, 2193 (1973)
56 Shulman R. G., Hilbers C. W. and Miller D. L., J. Mol. Biol. *90*, 601 (1974)
57 Shulman R. G., Hilbers C. W., Söll D. and Yang S. K., J. Mol. Biol. *90*, 609 (1974)
58 Krugh T. R. and Neely J. W., Biochemistry *12*, 4418 (1973)
59 Crothers D. M., Hilbers C. W. and Shulman R. G., Proc. Nat. Acad. Sci. *70*, 2899 (1973)
60 Patel D. J. and Tonelli A. E., Proc. Nat. Acad. Sci. *71*, 1945 (1974)
61 Hélène C. and Dimicoli J. L., FEBS Letters *26*, 6 (1972)
62 Finer E. G., Flook A. G. and Hauser H., Biochim. Biophys. Acta *260*, 49 (1972)
63 Godici P. E. and Landsberger F. R., Biochemistry *13*, 362 (1974)
64 Lee A. G., Birdsall N. J. M. and Metcalfe J. C., Biochemistry *12*, 1650 (1973)
65 Metcalfe J. C., Birdsall N. J. M. and Lee A. G., FEBS Letters *21*, 335 (1972)
66 Kornberg R. D. and McConnell H. M., Proc. Nat. Acad. Sci. *68*, 2564 (1971)
67 Bystrov V. F., Shapiro Y. E., Viktorov A. V., Barsukov L. I. and Bergelson L. D., FEBS Letters *25*, 337 (1972)
68 Andrews S. B., Faller J. W., Gilliam J. M. and Barrnett R. J., Proc. Nat. Acad. Sci. US *70*, 1814 (1973)
69 Chapman D., Kamat V. B., De Gier J. and Penkett S. A., J. Mol. Biol. *31*, 101 (1968)
70 Sheetz M. P. and Chan S. I., Biochemistry *11*, 548 (1972)
71 Millett F., Hargrave P. A. and Raftery M. A., Biochemistry *12*, 3591 (1973)
72 Randall R. F., Stoddart R. W., Metcalfe S. M. and Metcalfe J. C., Biochim. Biophys. Acta *255*, 888 (1972)
73 Robinson J. D., Birdsall N. J. M., Lee A. G. and Metcalfe J. C., Biochemistry *11*, 2903 (1972)
74 Nicolau C., Dreeskamp H. and Schulte-Frohlinde D., FEBS Letters *43*, 148 (1974)
75 Chang C. A. and Chan S. I., Biochemistry *13*, 4381 (1974)
76 Shporer M. and Civan M. M., Biophys. J. *12*, 114 (1972)
77 Yeh H. J. C., Brinley F. J. and Becker E. D., Biophys. J. *13*, 56 (1973)
78 Llinás M., Schwotzer W., von Phillipsborn W. and Wüthrich K., Nature, in press
79 Jackson J. A. and Langham W. H., Rev. Sci. Instr. *39*, 510 (1968)
80 Dea P., Chan S. I. and Dea F. J., Science *175*, 206 (1972)

81 Outhred R. K. and George E. P., Biophys. J. *13*, 97 (1973)
82 Cooke R. and Wien R., Biophys. J. *11*, 1002 (1971)
83 Finch E. D., Arch. Biochem. Biophys. *147*, 299 (1971)
84 Chang D. C., Hazlewood C. F., Nichols B. L. and Rorschach H. E., Nature *235*, 170 (1972)
85 Belton P. S., Jackson R. R. and Packer K. J., Biochim. Biophys. Acta *286*, 16 (1972)
86 Damadian R., Science *171*, 1151 (1971)
87 Weisman I. D., Bennett L. H., Maxwell L. R., Woods M. W. and Burk D., Science *178*, 1288 (1972)
88 Block R. E., FEBS Letters *34*, 109 (1973)
89 Damadian R., Zaner K., Hor D. and Di Maio T., Proc. Nat. Acad. Sci. *71*, 1471 (1974)
90 Floyd R. A., Yoshida T. and Leigh J. S., Proc. Nat. Acad. Sci. *72*, 56 (1975)
91 Eakin R. T., Morgan L. O., Gregg C. T. and Matwiyoff N. A., FEBS Letters *28*, 259 (1972)
92 Battersby A. R., Hunt E. and McDonald E., Chem. Comm. *1973*, 442
93 Lacroix M., Mosora F., Pontus M., Lefèbre P., Luyckx A. and Lopez-Habib G., Science *181*, 445 (1973)
94 Brown C. E., Shemin D. and Katz J. J., J. Biol. Chem. *248*, 8015 (1973)
95 Seto H., Sato T. and Yonehara H., J. Amer. Chem. Soc. *95*, 8461 (1973)
96 Wasserman H. H., Sykes R. J., Peverada P., Shaw C. K., Cushley R. J. and Lipsky S. R., J. Amer. Chem. Soc. *95*, 6874 (1973)
97 Wilson D. M., Burlingame A. L., Cronholm T. and Sjövall J., Biochem. Biophys. Res. Comm. *56*, 828 (1974)
98 Séquin U. and Scott A. I., Science *186*, 101 (1974)
99 Lauterbur P. C., Nature *242*, 190 (1973)
100 Lauterbur P. C., Proceedings of the First International Conference on Stable Isotopes in Chemistry, Biology and Medicine, US Atomic Energy Commission CONF-730525 (1973) p. 255
101 Lauterbur P. C., Pure Appl. Chem. *40*, 149 (1974)
102 Hinshaw W. S., Physics Lett. *48A*, 87 (1974)
103 Kumar A., Welti D. and Ernst R. R., Naturwissenschaften *62*, 34 (1975)
104 Kumar A., Welti D. and Ernst R. R., J. Magn. Res. *18*, (1975)

SUBJECT INDEX